Lecture Notes in Mechanical Engineering

Lecture Notes in Mechanical Engineering (LNME) publishes the latest developments in Mechanical Engineering - quickly, informally and with high quality. Original research reported in proceedings and post-proceedings represents the core of LNME. Volumes published in LNME embrace all aspects, subfields and new challenges of mechanical engineering. Topics in the series include:

- Engineering Design
- Machinery and Machine Elements
- Mechanical Structures and Stress Analysis
- Automotive Engineering
- Engine Technology
- Aerospace Technology and Astronautics
- Nanotechnology and Microengineering
- Control, Robotics, Mechatronics
- MEMS
- Theoretical and Applied Mechanics
- Dynamical Systems, Control
- Fluid Mechanics
- Engineering Thermodynamics, Heat and Mass Transfer
- Manufacturing
- Precision Engineering, Instrumentation, Measurement
- Materials Engineering
- Tribology and Surface Technology

To submit a proposal or request further information, please contact the Springer Editor in your country:

China: Li Shen at li.shen@springer.com
India: Dr. Akash Chakraborty at akash.chakraborty@springernature.com
Rest of Asia, Australia, New Zealand: Swati Meherishi at swati.meherishi@springer.com
All other countries: Dr. Leontina Di Cecco at Leontina.dicecco@springer.com

To submit a proposal for a monograph, please check our Springer Tracts in Mechanical Engineering at http://www.springer.com/series/11693 or contact Leontina.dicecco@springer.com

Indexed by SCOPUS. The books of the series are submitted for indexing to Web of Science.

More information about this series at http://www.springer.com/series/11236

Chun Lin Saw · Tze Keong Woo ·
Salvinder Singh a/l Karam Singh ·
Didi Asmara Bin Salim
Editors

Advancement in Emerging Technologies and Engineering Applications

Springer

Editors
Chun Lin Saw
Department of Mechanical Engineering
Politeknik Ungku Omar
Ipoh, Perak, Malaysia

Tze Keong Woo
Department of Mechanical Engineering
Politeknik Ungku Omar
Ipoh, Perak, Malaysia

Salvinder Singh a/l Karam Singh
Department of Mechanical and Materials Engineering
Universiti Kebangsaan Malaysia (UKM)
Bangi, Selangor, Malaysia

Didi Asmara Bin Salim
Department of Mechanical Engineering
Politeknik Ungku Omar
Ipoh, Perak, Malaysia

ISSN 2195-4356 ISSN 2195-4364 (electronic)
Lecture Notes in Mechanical Engineering
ISBN 978-981-15-0001-5 ISBN 978-981-15-0002-2 (eBook)
https://doi.org/10.1007/978-981-15-0002-2

This Springer imprint is published by the registered company Springer Nature Singapore Pte Ltd.
The registered company address is: 152 Beach Road, #21-01/04 Gateway East, Singapore 189721, Singapore

Contents

A Study of Oil Spill at Marine Companies: Factors and Effects

Ismila Che Ishak, Wan Muhammad Hafiz Wan Ab Rani, Shaiful Bakri Ismail and Norazimah Mazlan

Abstract Shipping activity is growing rapidly due to the effectiveness of marine transports in fulfilling the high demand of a global trade. The oil spill in the ocean is a serious issue due to its adverse effects, especially in the marine environment and the economy of the country. The oil pollution was caused by several factors, including human error, negligence, technical error, equipment failure, system failure, failure in implementation of required safety procedures, incompetency of crews and natural disaster. The aim of this research is to identify the main factor contributing to the oil spill as a preliminary study to prevent an unnecessary oil spill incident. Questionnaires survey was conducted on staff of oil and gas company and a maritime government agency in Lumut, Perak, Malaysia. The result shows the most significant factor of the oil spill is from human error which is resulted from in compliance of the standard operating procedure such as lack of technical skill (ship handling, operating system), communication failure between staff, fatigue and poor to follow rules and regulation at work place. Meanwhile the second part of this research is to identify which of these three factors of independent variables can be assumed as the most effect variables towards the dependent variables (environmental and economic) by measuring the strength and relationship between the variables. The result indicates natural disaster the highest positive relationship ($r = 0.778$) towards environmental effect and, human error indicates the strongest relationship ($r = 0.660$) for economic factors that effect to the oil spill.

I. C. Ishak (✉) · W. M. H. W. A. Rani · N. Mazlan
Maritime Management Section, Universiti Kuala Lumpur Malaysian Institute of Marine Engineering Technology, UniKL MIMET, Lumut, Perak, Malaysia
e-mail: ismila@unikl.edu.my

W. M. H. W. A. Rani
e-mail: wanmuhammadhafizrani@gmail.com

N. Mazlan
e-mail: norazimah@unikl.edu.my

S. B. Ismail
Technical Foundation Section, Universiti Kuala Lumpur Malaysian Institute Marine Engineering Technology, UniKL MIMET, Lumut, Perak, Malaysia
e-mail: shaifulbakri@unikl.edu.my

C. L. Saw et al. (eds.), *Advancement in Emerging Technologies and Engineering Applications*, Lecture Notes in Mechanical Engineering,
https://doi.org/10.1007/978-981-15-0002-2_1

Keywords Maritime industry · Oil spill factors · Oil spill effects

1 Introduction

The shipping industry is the earliest industry in the world with a diversification of exchange of goods of an international trade. Sea transport is a cheap transport cost as compared to transport by air as collectively states by Luoma [35], Chang et al. [9], but the sea transport also has a great risk especially in the event of an oil spill. Some countries depend on oil production and trade in generating national income. The hydrocarbon spill in the marine environment is very dangerous to the marine life, affects the human health, and bring down the economy of a country as state by Teal and Howarth [55], Aguilera et al. [3] and Fingas [20]. Water resources will be tainted, and contamination of the soil and the operating beach and port industry will be delayed because of the oil spill, as the industries require clean sea water supply to resume operation. The biggest catastrophic threat to the sea is the issue of the oil spill and the release of harmful chemicals into the sea. Malaysia is a member of several international conventions concern on marine pollution such as United Nations Convention on the Law of the Sea (UNCLOS) 1972 [56] and International Convention for the Prevention of Pollution from Ship (MARPOL) 1973/78 [36]. Annex 1 of the MARPOL specifically mentions on the prevention of pollution by oil at sea.

This research focuses to analyze the most significant factor contributes to the oil spill and, to determine the effects of the oil spill. This research was conducted at three different organizations which are; oil & gas company, maritime operator organization, and maritime government agency. The selected oil and gas company involves in providing maintenance and repairing the floating oil and gas facilities, piping the oil and gas pipes, as well as the construction of platforms offshore. The maritime operator organization involves in providing a transportation service to passengers. Meanwhile, the maritime government agency is establishing to monitor and manage all matters relating to port, shipping activities, responsible in ensuring the sailing ships sail safely, conducting inspection on vessels on license, ship identification and compliance of regulations, monitoring, providing an assistance to ships and responsible for supervising the seafarers' examination.

2 Methodology

2.1 Theoretical Framework

A theoretical framework is a structure supports a theory of the research and is used to determine the relationship of the study. An independent variable is a variable change in a scientific experiment to study on the effect of the dependent variables.

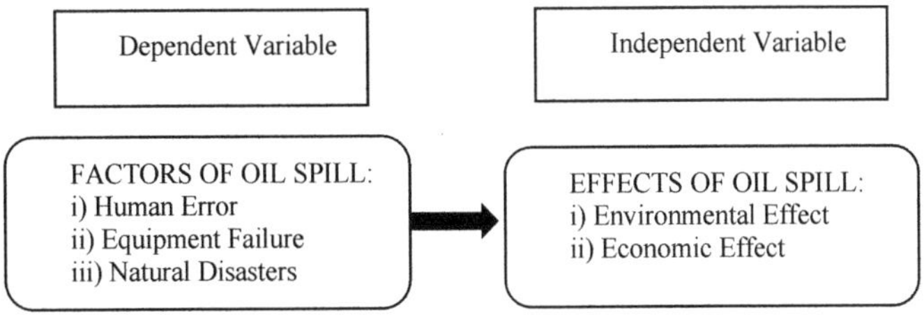

Fig. 1 The research theoretical framework

The dependent variable is the variable which measures in a scientific experiment as mention by Hogreve et al. [28]. The independent variable focuses on the element of factors of an oil spill such as; human error, equipment failure and natural disaster. The human error has collectively state by Chauvin et al. [11]. Chicago: National Safety Council (2008), Rothblum [48], Anderson and LaBelle [7], Kontovas et al. [30], Gasparotti [22] and Abdulrazaq and Kader [1]. Meanwhile, for the equipment failure supports by Okoye and Okunrobo [41], Okechukwu Anyanwu [41], Lee et al. [32], Akpomuvie [6], Omofonmwan and Odia [42], and for natural disaster has agree by Sholeye et al. [52], Akpofure et al. [5], Hogarth [27], Pine [46], Cruz and Krausmann [14], Grimaldi et al. [25] and Wang et al. [58]. On the other hand, the elements in the dependent variables are environmental and economic effects. The environmental effect has support by Teal and Howarth [55], Aguilera et al. [3] and Fingas [20], Farrington [19], Siliman et al. [53], Sumaila et al. [54], Corn [13], Demopoulos and Strom [17], Ajide and Isaac [4], Adelana et al. [2], Ordinioha and Brisibe [44], Barry [8], EPA [18], Klemas [29], DeLeo et al. [16], Sagerup et al. [50], Shigenaka [51] and Giri et al. [24]. Finally, for the economic effect has collectively agree by Montewka et al. [37], Zock et al. [62], Verma [57], Hayworth et al. [26], Opukri and Ibaba [43], National Commission [39], Gill et al. [23], Barry [8], Opukri and Ibaba [43] and Ajide and Isaac [4] (Fig. 1).

2.2 *Population, Sample and Respondents*

The population of this research involves employees in the oil and gas company, maritime operator organization, and maritime government agency. The 92 sample is taken at 40% of the total population. 65 respondents participated as shown in Table 1, and the response rate is at 70.65%.

Design of Questionnaire. The questionnaire consists of five sections which are; Section A: Respondent's background, Section B: Company's Background, Section C: Factors to oil spills, Section D: Effects of the oil spill and, Section D: Suggestion and Opinion from respondents.

Table 1 Population, sample and respondents

Breakdown	Population	Sample	Respondents
Oil and gas company	140	42	30
Maritime operator organization	40	20	15
Maritime government agency	50	30	20
Total	230	92	65

3 Results Analysis

This section discusses three main analysis which are demographic analysis, factor analysis and regression analysis. The demography analysis is a statistical and mathematical analysis use to study respondents' size, composition, distribution of human populations and, how these features change over time. Data are obtained from the record of the events such as gender, age, marital status, educational background and working experiences as state by Lawrence [31], Neubert and Caswell [40], Xie [61] and Li [34]. In addition, descriptive statistic is a summarize table from the research result and reflect population, sample, respondents, and the data has split into mode, mean, median, standard deviation and variance. It is applied to compress and summarizes the data clearly for report preparation as mention by Data [15], Runyon [49], Wiersma [59] and Leech [33]. Moreover, a regression analysis is used as it is one of the most widely used technique for analysing multifactor data and, is apply to express the relationship between a variable of interest (the response) and a set of related predictor variables as agree by Pedhazur [45], Fox [21], Montgomery [38] and Chatterjee [10].

3.1 Demographic Analysis

Demographic study shows the background data from each 65 respondents that randomly selected for this survey. This analysis is important to learn more about a sample characteristic to determine the suitability of their response towards this survey and further explain the result. Table 2 shows list of selected respondents' background, such as gender, age, highest education and working experiences. This selected background is according to their relevancy towards oil spill factor and effects. From 65 respondents, there are 50 male and 15 female respondents. There is a difference in gender as the oil and gas company, maritime government agency and maritime operator organization are involved in a heavy engineering industry, which requires more male manpower in general, and female staff involve in administrative matters. The range of the age is between five years, which refer to the age range of the staff of the selected companies. The group of the age is considering several factors such as their position and working experiences. In the age range between 31 and 35 years

Table 2 Respondents demography results

Descriptions	Results	Percentage
Gender	Frequency	
Male	50	76.9
Female	15	23.1
Age	Frequency	Percentage
20–25 years	4	6.2
26–30 years	11	16.9
31–35 years	32	49.2
36–40 years	18	27.7
Marital status	Frequency	Percentage
Single	24	36.9
Married	35	53.8
Divorce	6	9.2
Highest education	Frequency	Percentage
SPM/PMR	1	1.5
Diploma	5	7.7
Degree	26	40.0
Master	33	50.8
Working experiences	Frequency	Percentage
Less 5 years	29	44.6
5.1–10 years	12	18.5
10.1–15 years	19	29.2
15.1–20 years	3	4.6
20.1–25 years	2	3.1

old, shows the highest feedback which is at 49.2%, and followed by the age range between 36 and 40 is at 27.7%. This value is enough to represent age range for all the samples as a majority. The majority staff who is included in this age are holding positions from middle management to top management and, it is related to the position and experience to counter the issue towards the factor and effects of oil spill. In terms of marital status, the majority are married which is at 53.8%. There is not much difference for the academic qualification among respondents for Bachelor holding is at 40.0 and 50.8% for Masters. Only 7.7% respondents holding Diploma. In the scope of the research required at least staff with Diploma to carry marine operation daily tasks at ports. Besides that, this research shows that most of the respondents with working experiences are less than 5 years at 44.6%. The working experiences staff who are especially involved in marine operation are eligible to the organization due in covering issues in the oil spill. It is because experiences staff is highly knowledgeable and manageable to handle big incidents and the employers have taken an

Table 3 Descriptive statistic

Factor	Results			
	N	Mean	Std. Deviation	Std. Error mean
Human error	65	4.0484	0.34835	0.04321
Equipment failure	65	3.8879	0.37669	0.04672
Natural Disaster	65	3.8484	0.41717	0.05174

appropriate decision in recruiting them and placed at the relevant working experiences in combating the oil spill issue. The duration of more than 5 years to 15 working experiences are enough to measure the capability of their work and this is supported by Wiesenfeld et al. [60] states that, experiences workers gaining a lot of experiences on working situation and meet the standard quality of work parallel with the salary gained.

3.2 *Descriptive Statistics (Factors)*

There are three main selected factors that contribute to oil spill at selected marine company which are human error, equipment failure and natural disaster. These factors can be simply measured using descriptive statistics. The analysis in Table 3 shows the descriptive statistics for all the variables in the research. Typically, the mean, standard deviation and number of respondents (N) who participated in the research are given. Looking at the mean, conclude that the factor of human error is the most important variable that influences and contribute to oil spill at marine company as agreed by respondents and it has the highest mean of 4.05. This is supports by Rothblum [48], claim that 80% of oil tanker accidents that lead to oil spill are from the human activities and errors. Based on the survey to the respondents, most common sub-factors are significant and represented to the human error are; lack of technical skill (ship handling, operating system), communication failure between staff, fatigue and poor to follow rules and regulation at work place.

3.3 *Correlation (Effect)*

From this analysis, the relationship between independent variables and dependents variables are evaluates. In this section, contain two results of correlation between independent variables with two variables of dependent variables which are environmental effect and economy effect respectively. Ideally, the aim for the second part of this research is to identify which of these three factors of independent variables can be assumed as mostly effected to the oil spill by measuring the strength and relationship between the variables either tend to increase and decrease the direction of the lines.

Correlation Effect on Environmental. The reason for the environmental and economic are chosen as the dependent variables due to the collective severe effects on ecosystems and the environmental where the oil spill is released, it is also involve oil toxic effect on the habitats, seafood, marine mammals, plankton, coral and marine habitat by Siliman et al. (2012), effect ocean and bring the marine life to extinction either quickly or slowly by Farrington [19], takes a long time to restore oil spill as affects habitat, mangroves, food chain structure in the area by Sumaila et al. [54], effect flora and fauna on the seafloor, fish that eat plants on the seafloor are also affected by the inundation of plants that are stuck with oil by Corn [13], Demopoulos and Strom [17], effect people near oil spills and cause their crops to be damaged due to mixing with the spill oil and underground contaminated water by Ajide and Isaac [4], Adelana et al. [2], effect the reduction of nutrients and proteins in foods such as vegetables, cassava, and fish which are mixed with toxic oils and can harm humans and animals by Ordinioha and Brisibe [44] and Verma [57], effect life and health of the people in the immediate area of the oil spill by Barry [8], Ordinioha and Brisibe [44], effect on respiratory problems, skin inflammation, difficult to get clean food and water by Adelana et al. [2] and Zock et al. [62], effect DNA damage which can cause cancer and multigenerational birth defects by EPA [18], effect the marine life by DeLeo et al. [16], Sagerup et al. [50] and, effect wetland and estuarine habitat by Klemas [29].

Correlation Effect on Economic. In addition, the process of cleaning up the oil spill require quite a while and the cost is expensive. The oil spill causes the coastal areas to be contaminated and disrupted the recreational activities such as camping, fishing, and swimming. The cleaning up is expensive and need to consider the types of oil spill, quantity spill, size of the spill, spill distance of the oil with the coast by Montewka et al. [37]. It effects income to decline and causing rural population move to the urban area to find additional sources of income in sustaining life, affects the tourist activities and cause the number of tourists to decrease by Hayworth et al. [26], effect the closure of the fishery sector and causing fishermen to lose the source of income by National Commission [39]. The tourism industry that rely on the water activities such as skating, kayaking, swimming, and rafting are also affected and cause the hotels, restaurant owners, tenant boats, dive tour operators and others who have income from the recreational activities in the coastal zone will experience a significant economic loss by Gill et al. [23]. The activities such as farming, agriculture, and fishing will be affected and forced into unemployed or to change to another jobs for survival by Barry [8], Opukri and Ibaba [43] and effect economics and human health through the contamination and death of fishes by Ajide and Isaac [4].

Correlation analysis describe the strength of the relationship between two or more variables based on Table 4. Pearson correlation coefficient (r) is a measure to determine the strength of the alliance between the two variables as propose by Coakes et al. [12] and, two variables are related to each other then, the variables are

Table 4 Range of values for the correlation coefficient

Correlation coefficient	(r)
Perfect negative	−1
No correlation	0.00
Very weak	0.00 to 0.19–0.00 to 0.19
Weak	0.20 to 0.39–0.20 to 0.39
Moderate	0.40 to 0.59–0.40 to 0.59
Strong	0.60 to 0.79–0.60 to 0.79
Very strong	0.80 to 1.0–0.80 to 1.0

said to be correlated as a claim by Puth et al. [47]. This analysis helps to derive the degree and the direction of such relationships both variables.

Tables 5 and 6 indicates the result of relationship between independent variables (human error, equipment failure and natural disaster) with dependent variables (environmental and economy). Ideally, this analysis is conducted to identify which one of the factor mostly effect towards the oil spill by measuring the strength of the coefficients. According to Table 5, it shows a strong association between all three independents (factors) variables with dependents variable (environmental) with the range of coefficient $r = 0.683$ to $r = 0.778$. The result shows that natural disaster indicates the highest correlation coefficient that gives positive and strong relationship with environmental effect ($r = 0.778$) as the natural disasters are a natural event, and is out of human influence which cause phenomena such as earthquakes, hurricanes, adverse weather conditions as collectively claim by Sholeye et al. [52], Ajide and Isaac [4], Adelana et al. [2], Ordinioha and Brisibe [44] and Verma [57].

Meanwhile, the second result from Table 6, shows a strong association between only two independents (factors) variables with dependents variable (economic) with the range of coefficient $r = 0.642$ to $r = 0.660$ which are human error shows the highest coefficient correlation value ($r = 0.660$). It is true that the human error factor

Table 5 Correlation independents variables and environmental effect

Pearson correlation (Sig. 2-tailed = 0.000, N = 65)	
	Environmental effect
Human error	0.694**
Equipment failure	0.683**
Natural disaster	0.778**

Table 6 Correlation independents variables and economy effect

Pearson correlation (Sig. 2-tailed = 0.000–0.002, N = 65)	
	Economy effect
Human error	0.660**
Equipment failure	0.378**
Natural disaster	0.642**

in contributing to the oil spill leads to economic efficiency. This is because the people nearby area of the oil spill tends to lose income from affected fish with the mixing chemical from the oil spills and the cost to clean up the spill oil is expensive as collectively agree by Akpofure [5], Opukri and Ibaba [43], Gill et al. [23], Montewka et al. [37] and Ajide and Isaac [4].

4 Conclusion and Recommendation

As a conclusion for the first objective, it was found out that human error is the main factor of the oil spill as agree by Abdulrazaq and Kader [1], Chauvin et al. [11], Adelana et al. [2] and Rothblum [48]. Meanwhile, the conclusion for the second objective is that most of the independent variables have a strong relationship with the dependent variables. Human error, equipment failure, and natural disaster have a strong relationship towards the environmental and economic effect. The presence of the oil spill in the environment contributes significantly to the degradation of the environment and affects the economy of that area. It is undoubtedly that the oil spill is a serious issue, which can lead to the destruction of nature and the organism, whether it is immediate or long-term as agreed by Verma [57], Farrington [19], and Ordinioha and Brisibe [44].

As a recommendation, an efficient cleanup and remediation programs need to be conducted as the oil remains in the water. The human error can be reduced if one takes an extreme precaution and proper safety assessments. The related authorities such as Department of Environment (DOE) and Marine Department (MARDEP) need to ensure the person handling vessels or jobs related to carrying oil in the sea, especially oil tankers, must have a valid certificate, not under the influence of alcohol and prohibit the use of illegal drugs. In minimizing the pollution of the oil spill by human error to the marine environment, the workers or crews need an adequate and enough series of training. Secondly, the rule on double hull requirement needs to be complied with by tanker ships. Double layers of a watertight hull surface are a precaution in minimizing the pollution, which normally occurred during a collision, grounding or any other unexpected accident. Finally, sea pollution due to oil spills can also be reduced by using environment-friendly technology, by replacing the petroleum power to a solar power consumption.

Acknowledgements This research paper is conducted by the collaboration gained from the oil and gas company, maritime operator organization and maritime government agency located in Lumut, Perak, Malaysia. This research is financially supported by Universiti Kuala Lumpur Malaysian Institute Marine Engineering Technology (UniKL MIMET). Certainly, without the cooperation granted from all team members, this research cannot be successfully implemented.

References

1. Abdulrazaq AO, Kader SZSA (2014) Vessel-sourced pollution: a security threat in Malaysian waters. J Sustain Dev Law Policy 3(1):22–36
2. Adelana SO, Adeosun T, Adesina AO, Ojuroye MO (2011) Environmental pollution and remediation: challenges and management of oil Spillage in the Nigerian coastal areas. Am J Sci Ind Res 2(6):834–845
3. Aguilera F, Méndez J, Pásaro E, Laffon B (2010) Review on the effects of exposure to spilled oils on human health. J Appl Toxicol 30(4):291–301
4. Ajide OM, Isaac OO (2013) An assessment of the physical impact of oil spillage using GIS and Remote Sensing technologies: Empirical evidence from Jesse town, Delta State, Nigeria. Br J Arts Soc Sci 12:235–252
5. Akpofure RR, Grace O, Israel O, Okokoyo IE (2006) Job satisfaction among educators in colleges of education in Southern Nigeria. J Appl Sci 6:1094–1098
6. Akpomuvie OB (2011) Tragedy of commons: analysis of oil spillage, gas flaring and sustainable development of the niger delta of Nigeria. J Sustain Dev 4(2):200–209
7. Anderson CM, LaBelle RP (2011) Update of comparative occurrence rates for offshore oil spills. Spill Sci Technol Bull 6(5):303–321
8. Barry FB (2010) Environmental Injustices: conflict and Health Hazards in the Niger delta. Substantial Research Paper, Washington, DC, pp 1–73
9. Chang S, Stone J, Demes K, Piscitelli M (2014) Consequences of oil spills: a review and framework for informing planning. Ecol Soc 19(2)
10. Chatterjee S (2015) & Hadi. Regression analysis by example. Wiley, A. S.
11. Chauvin C, Lardjane S, Morel G, Clostermann JP, Langard B (2013) Human and organisational factors in maritime accidents: analysis of collisions at sea using the HFACS. Accid Anal Prev 59:26–37
12. Coakes, SJ, Steed, L Ong, C (2009) SPSS version 16.0 for windows: analysis without anguish. Wiley Australia, Ltd, Australia
13. Corn ML (2010) Deepwater Horizon oil spill: coastal wetland and wildlife impacts and response. DIANE Publishing
14. Cruz AM, Krausmann E (2009) Damage to offshore oil and gas facilities following hurricanes katrina and rita: an overview. J Loss Prevent Proc 21:620–626
15. Data S (1988) Using descriptive statistics In: Devore J, Peck, Bartz AE (eds) basic statistical concepts. Macmillan, New York
16. DeLeo DM, Ruiz-Ramos DV, Baums IB, Cordes EE (2016) Response of deep-water corals to oil and chemical dispersant exposure. Deep Sea Res Part II 129:137–147
17. Demopoulos AW, Strom DG (2012) Benthic community structure and composition in sediment from the northern Gulf of Mexico shoreline, Texas to Florida (No. 2012–1153). US Geological Survey (2012)
18. EPA A (2011) Inventory of US greenhouse gas emissions and sinks: 1990–2009. Environmental Protection Agency
19. Farrington JW (2014) Oil pollution in the marine environment II: fates and effects of oil spills. environment: science and policy for sustainable development 56(4):16–31
20. Fingas M (2013) The basics of oil spill clean-up. CRC Press (2013)
21. Fox J (1997) Applied regression analysis, linear models, and related methods. Sage Publications, Inc.
22. Gasparotti C (2010) Risk assessment of marine oil spills. Environ Eng & Manag J (EEMJ) 9(4)
23. Gill DA, Picou JS, Ritchie LA (2012) The exxon valdez and BP oil spills: a comparison of initial social and psychological impacts. Am Behav Sci 56(1):3–23
24. Giri C, Long J, Tieszen L (2011) Mapping and monitoring Louisiana's mangroves in the aftermath of the 2010 Gulf of Mexico oil spill. J Coastal Res 27(6):1059–1064
25. Grimaldi CSL, Casciello D, Coviello I, Lacava T, Pergola N, Tramutoli V (2011) An improved RST approach for timely alert and near real time monitoring of oil spill disasters by using AVHRR data. Nat Hazards Earth Syst Sci 11(5):1281–1291

26. Hayworth JS, Clement TP Valentine JF (2011) Deepwater Horizon oil spill impacts on Alabama beaches. Hydrol Earth Syst Sci 15(12):3639 (2011)
27. Hogarth W (2015) Testimony on the effects of hurricanes katrina and rita on the fishing industry and fishing communities in the Gulf of Mexico before the House committee on resources, Subcommittee on Fisheries and Oceans, available at: http://www.ogc.doc.gov/ogc/legreg/testimon/109f/hogarth1215.htm
28. Hogreve J, Iseke A, Derfuss K, Eller T (2017) The service-profit Chain: a meta-analytic test of a comprehensive theoretical framework. J Mark 81(3):41–61
29. Klemas V (2010) Remote sensing techniques for studying coastal ecosystems: an overview. J Coastal Res 27(1):2–17
30. Kontovas CA, Psaraftis HN, Ventikos NP (2010) An empirical analysis of IOPCF oil spill cost data. Mar Pollut Bull 60:1455–1466. https://doi.org/10.1016/j.marpolbul.2010.05.010
31. Lawrence BS (1997) Perspective—The black box of organizational demography. Organ Sci 8(1):1–22
32. Lee GH, Pouraria H, Seo JK, Paik JK (2015) Burst strength behaviour of an aging subsea gas pipeline elbow in different external and internal corrosion-damaged positions. Int J Nav Arch Ocean Eng 7(3):435–451
33. Leech N, Barrett K Morgan, GA (2013) SPSS for intermediate statistics: use and interpretation. Routledge
34. Li J, Li H, Jakobsson M, Li SEN, SjÖDin PER, Lascoux M (2012) Joint analysis of demography and selection in population genetics: where do we stand and where could we go? Mol Ecol 21(1):28–44
35. Luoma E (2009) Oil Spills and Safety Legislation. Publications from the centre for maritime studies, University of Turku, Finland
36. MARPOL 73/78. International Maritime Organization (1973/1978)
37. Montewka J, Weckström M, Kujala P (2013) A probabilistic model estimating oil spill clean-up costs–a case study for the Gulf of Finland. Mar Pollut Bull 76(1):61–71
38. Montgomery DC, Peck EA, Vining GG (2012) Introduction to linear regression analysis (Vol. 821). Wiley (2012)
39. National Commission (2011) Financial crisis inquiry commission, & United States. Financial crisis inquiry commission. The financial crisis inquiry report, authorized edition: Final report of the national commission on the causes of the financial and economic crisis in the United States. Public Affairs,
40. Neubert MG, Caswell H (2000) Demography and dispersal: calculation and sensitivity analysis of invasion speed for structured populations. Ecology 81(6):1613–1628
41. Okoye C, Okunrobo L (2014) Impact of oil spill on land and water and its health implications in odu-gboro community, sagamu, ogun state, nigeria. World J Environ Sci Eng 1:1–21
42. Omofonmwan SI, Odia LO (2009) Oil exploration and conflict in the niger-delta region of Nigeria, Kamla-Raj. J Hum Ecol 26(1):25–30
43. Opukri CO, Ibaba IS (2008) Oil induced environmental degradation and internal population displacement in the Nigeria's Niger Delta. J Sustain Dev Afr 10(1):173–193
44. Ordinioha B, Brisibe S (2013) The human health implications of crude oil spills in the Niger Delta, Nigeria: an interpretation of published studies. Niger Med J 54(1): 10–6 (2013)
45. Pedhazur EJ, Kerlinger FN (1973) Multiple regression in behavioral research. Holt, Rinehart and Winston, New York
46. Pine J (2006) Hurricane katrina and oil spills: impact on coastal and ocean environments. Oceanography 19(2):37–39
47. Puth MT, Neuhäuser M, Ruxton GD (2014) Effective use of Pearson's product–moment correlation coefficient. Anim Behav 93:183–189
48. Rothblum MA (2006) Human error and marine safety, vol 4. US coast guard risk-based decision-making guidelines, US Coast Guard Research and Development Center
49. Runyon RP, Coleman KA (2000) & Pittenger. Fundamentals of behavioral statistics. McGraw-Hill, D. J.

50. Sagerup K, Nahrgang J, Frantzen M, Larsen LH, Geraudie P (2016) Biological effects of marine diesel oil exposure in red king crab (Paralithodes camtschaticus) assessed through a water and foodborne exposure experiment. Mar Environ Res 119:126–135
51. Shigenaka G (2002) Oil toxicity. In Oil spills in mangroves. Office of Response and Restoration, NOAA Ocean Service, NOAA, Seattle, Wash, pp 23–35
52. Sholeye O, Salako A, Ayankoya S (2012) Oil spills and community health: Implications for resource limited settings. J Toxicol Environ Health Sci 4(9):145–150
53. Silliman BR, van de Koppel, J, McCoy MW, Diller J, Kasozi GN, Earl K, … Zimmerman AR (2012) Degradation and resilience in Louisiana salt marshes after the BP–Deepwater Horizon oil spill. Proc Natl Acad Sci 109(28):11234–11239
54. Sumaila UR, Cisneros-Montemayor AM, Dyck A, Huang L, Cheung W, Jacquet J, … Watson R (2012) Impact of the deepwater horizon well blowout on the economics of US Gulf fisheries. Can J Fish Aquat Sci 69(3):499–510
55. Teal JM, Howarth RW (1984) Oil spill studies: a review of ecological effects. Environ Manage 8(1):27–44
56. UNCLOS (1972)
57. Verma A (2016) Application of computational transport analysis: oil spill dynamics (Doctoral dissertation, Faculty of the Graduate School of the University at Buffalo, State University of New York)
58. Wang Q, Zhang S, Li Y, Klassen W (2011) Potential approaches to improving biodegradation of hydrocarbons for bioremediation of crude oil pollution. J Environ Prot 2(01):47
59. Wiersma W, Jurs SG (2005) Research methods in education: an introduction
60. Wiesenfeld BM, Reyt JN, Brockner J, Trope Y (2017) Construal level theory in organizational research. Annu Rev Organ Psychol Organ Behav 4:367–400
61. Xie Y (2000) Demography: Past, present, and future. J Am Stat Assoc 95(450):670–673
62. Zock JP Rodríguez-Trigo G, Rodríguez-Rodríguez E, Espinosa A, Pozo-Rodríguez F, Gómez F, … Barberà J. A. (2012) Persistent respiratory symptoms in clean-up workers 5 years after the Prestige oil spill. Occup Environ Med, 69(7), pp, 508–513,

Dynamics Analysis of Single Buoy Mooring in Malaysian Waters

Mazlan Muslim, Md Salim Kamil and Asmalina Mohamed Saat

Abstract The research is carried out to study the behaviour of the Deepwater Offloading Single Buoy Mooring (SBM) as well as the mooring line after combining polyester to the existing mooring line design. The first mooring line is consisting of chain and spiral strand wire which is becomes current practice in oil and gas industry. The second mooring line is designed by introducing polyester into the system. Static analysis is carried out by using lump mass method which is proposed by Wingerie (2008). There are 2 mathematical software have used for static analysis which are Mathcad and Matlab. Both of the mooring designs are then be analysed in ANSYS AQWA for dynamic analysis.

Keywords Single buoy mooring · Mooring designs · Dynamic analysis

1 Introduction

In the paper a catenary anchor leg mooring (CALM) buoy moored in a water depth of 1,200 m is considered. The purpose of this CALM buoy is to provide an offloading point for transferring oil from an floating production storage offloading (FPSO), which is moored at a distance of approximately 1 nautical mile, to a shuttle tanker. In this analysis, the total length required is calculated using static analysis and the overall performance of mooring system will be analysed in ANSYS. Comparison between 2 types of mooring line will be analysed. The first type consists of chain and spiral strand rope whilst the second type will be combination among mooring chain, polyester and spiral strand.

M. Muslim (✉) · M. S. Kamil · A. M. Saat
UniKL MIMET, 32200 Lumut, Perak, Malaysia
e-mail: mazlanm@unikl.edu.my

C. L. Saw et al. (eds.), *Advancement in Emerging Technologies and Engineering Applications*, Lecture Notes in Mechanical Engineering,
https://doi.org/10.1007/978-981-15-0002-2_2

2 Objectives

i. To design the mooring line which is a combination of chain, spiral strand and polyester in a mooring line to reduce the risk of mooring failure due to corrosion effect.
ii. To determine the horizontal offset of Single Buoy Mooring after polyester rope is added to the mooring line segment.
iii. To compare the performance between both of the mooring line design using results gained from dynamic analysis in ANSYS AQWA.

3 Environmental Factors

In this study the Single Buoy Mooring would be used at Malaysian sea water and the water depth chosen was 1200 m which is located at Kikeh, Sabah (see Fig. 1). The significant wave height is considered as 6.31 m and wave period equals to 13.1 s.

Fig. 1 Kikeh field location, Sabah, Malaysia waters

4 Problem Statement

Steel wire rope and chain have been used for mooring floating offshore production systems since their introduction nearly 30 years ago. A survey carried out by Noble Denton Europe Limited (2006) has shown that a lot of serious incidents associated with floating production system have occurred in the past, including loss of station. According to Chapline et al. (2008), service life of mooring line can be influenced by the local conditions. Warmer tropical sea water has become the main factor that deteriorated the service life of mooring line due to corrosion.

5 Synthetic Fibre

The key advantage of using the synthetic fibre ropes over chains and wires is their relatively low in cost. The comparative cost and mass analysis of several single-line catenary and taut-moored arrangements including several different anchor configurations has been carried by Ridge et al. (2010). The comparison highlight about the advantages to replace conventional mooring chains by using lightweight but durable mooring components. For example, the ability of nylon to reduce peak loadings would allow smaller gauge mooring chain to be used in anchor-chain- surface buoy-rope-device configurations. The study noted that the reduced by 88 kg/m and the cost by over RM450,000 per mooring line (estimated) can be achieved by adopting this design to the overall mass of the system. These reductions would partly be attributable to the specification of lower capacity components, such as anchors. Clearly in practice the feasibility of using a particular mooring system and actual cost savings that are achieved will depend on the case in question.

6 Mooring Configurations

According to Joshua (2013), there are many types of mooring configurations used for offshore structures, and they range from simple passive moorings to complex active systems. The three main configuration categories are single-point moorings, spread moorings, and dynamic positioning systems. Single point moorings utilize one mooring line, and can have one or more anchors. They are often used for deep-water meteorological buoys or small floats. They offer a large amount of compliance for dynamic wave environments, but they have large watch circles and do not provide directional control. To support a wide range of offshore structures, spread mooring is used because of their design to use multiple mooring lines and anchors. Their complexity can be greatly varied and the designs of the line are either catenary lines, tensioned lines or the combination of both lines. They usually have less compliance and a larger underwater footprint however; they offer directional control and typically have much smaller watch circles than single points mooring. To change and control mooring configurations, dynamic positioning systems such as winches or thrusters is used to utilize active controls. Single point or spread mooring is used and they are often very complex. The system is used for the large offshore structures, such as oil-rigs or floating wind turbine.

7 Analytical Calculations of Static Forces

Usually, mooring line consist of different types of material especially for deep water application. It may be combined with one another to withstand environmental force and secure the floating structure. In this study, there are 2 types of mooring line would be analysed. The first type is chain—wire and the latter is chain-polyester-wire connection. In static analysis, 2 methods were used to find the initial parameter before dynamic analysis is carried out. These 2 methods are analytical solution and element solution which were proposed by Wingerie (2008). The results from both solutions are compared to validate the data. As referred to Fig. 2, the line is segmented into 2 parts, 1 being chain part whilst being the wire part.

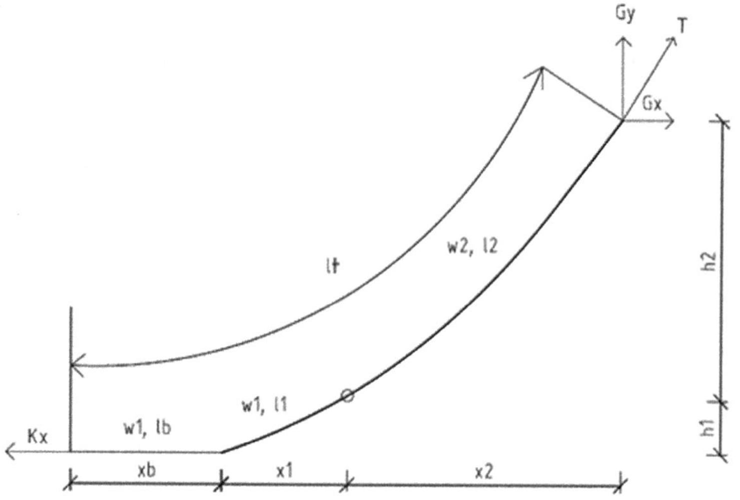

Fig. 2 Mooring line segment

Table 1 Buoy parameters

Parameter	Value (m)
Buoy hull diameter	17
Skirt diameter	21
Buoy height	7.65
Draft	5.65
KG	3.40
K*xx*	4.39
K*yy*	4.39
K*zz*	6.01
Fairlead radius	9.5

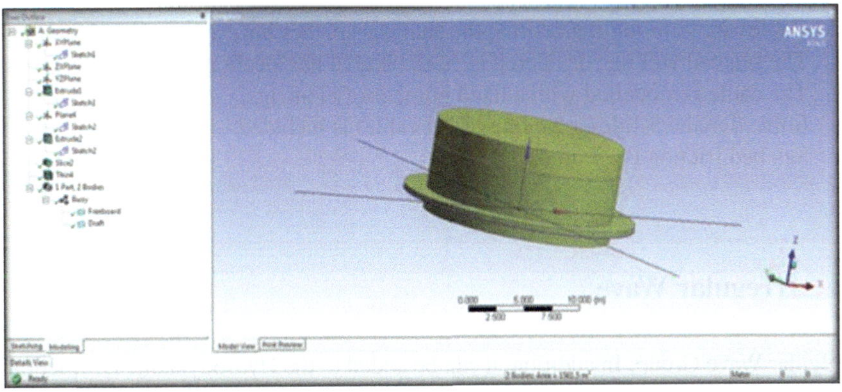

Fig. 3 Buoy construction completed

8 Buoy Construction

The buoy is divided into 2 part which consist of freeboard and underwater part. The buoy has cylinder shape for both parts that make it easy to construct. It was constructed by using Ansys modeler. The parameter of the buoy is as in Table 1. Figure 3 shows the buoy design after the construction process was completed.

9 Equation of Motion

The response X of a structure in waves is calculated by solving the equation of motion in the frequency domain for unit wave amplitude, in Eq. (1):

$$[-\omega^2(M_s + M_a(\omega)) - i\omega B(\omega) + C]X(\omega) = F(\omega) \tag{1}$$

Where;

M_s is structure mass
M_a is added mass (frequency dependent) B is damping (frequency dependent)
C is hydrostatic stiffness
F is wave force (incident and diffracting forces).

10 Mooring System Simulation

Estimation of mooring system is found from AQWA program as well. The procedure has the following simplification:

i. Cables are semi-taut/taut during the analysis i.e. they have a defined pre-tension.
ii. The sea bed will be considered as horizontal at the location of the anchors.
iii. The cable is modelled with a fixed number of elements.
iv. Inline dynamics (along the line of the cable) is included.
v. Sea bed friction is ignored.

11 Irregular Wave

Irregular Wave Group: In addition, it can be added to the Hydrodynamic Diffraction system when Linearized Tube Drag is specified for the analysis (only a single irregular wave can be used in the Hydrodynamic Diffraction system, and no Cross Swell should be specified).

12 Dynamics Analysis of Mooring Lines

In this study, dynamic analysis is conducted by using ANSYS AQWA. The buoy is constructed by using ANSYS MODELER and all the input parameters are defined. This is including the environmental condition and material properties for all the mooring line. The buoy is analysis by using DNV requirement for mooring design. There are 3 types of mooring design analysis which are:

i. An ultimate limit state (ULS) to ensure that the individual mooring lines have adequate strength to withstand the load effects imposed by extreme environmental actions.
ii. An accidental limit state (ALS) to ensure that the mooring system has adequate capacity to withstand the failure of one mooring line, failure of one thruster or one failure in the thrusters' control or power systems for unknown reasons. A

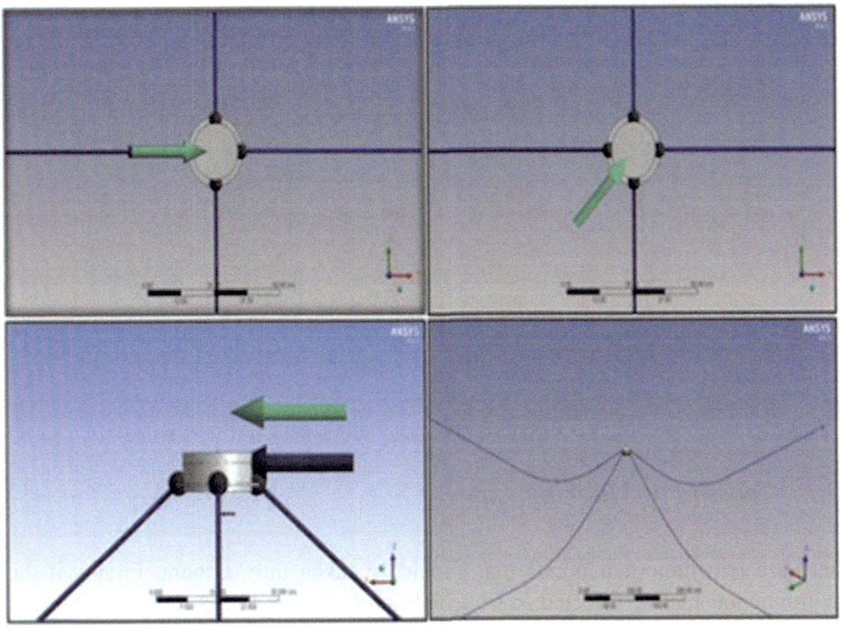

Fig. 4 Forces applied to buoy and mooring lines

single failure in the control or power systems may cause that several thrusters are not working.

iii. A fatigue limit state (FLS) to ensure that the individual mooring lines have adequate capacity to withstand cyclic loading.

iv. Since there is time and information constraint during this study, the student has focused to only the ultimate limit state design. Generally, the force applied to the buoy and mooring line is directed from different angle as shown in Fig. 4.

13 Mooring Simulation

Usual options for the mooring simulation analysis are taken into consideration:

i. Head Seas;
ii. Quartering Seas;
iii. Beam Seas.

In this analysis, the researcher had focused on heads seas and quartering seas only because the effect on beam seas will be similar as in head seas. This is due to the geometry shape of the buoy. The buoy has cylinder shape so that it would give constant projection area for any load of direction.

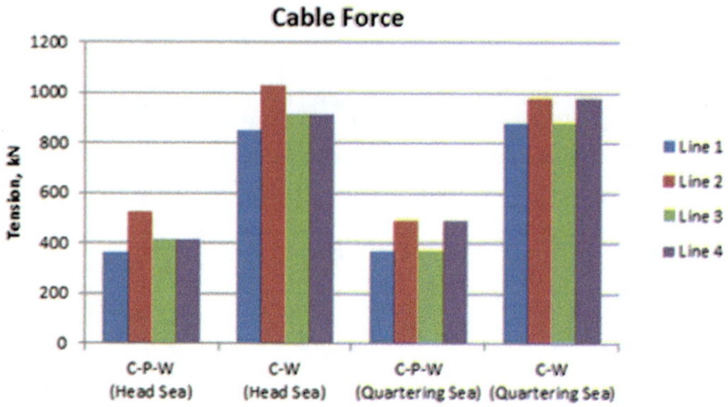

Fig. 5 Maximum tension on each mooring line

13.1 Tension of Mooring Lines

There are two important points that should be taken into account. Firstly, if the environmental load is directed to 0° which is head seas, line 2 will incur highest tension because the line is in line with the load direction. Secondly, if the load direction is 45°, line 4 will experience highest tension but the value is lower than the first case. The tensions on all 4 lines are distributed quite even in quartering seas. Refer to the bar chart (Fig. 5); the tension of each line for each case is presented in order to get clear perspective before further comparison is made.

14 Hydrodynamics Performance

Whereas the total length and tension of the mooring line are analyzed, hydrodynamic motion of the buoy is compared for both design. The performance of mooring line is translated by the rotational and translational motion of the buoy since couple dynamic analysis between mooring line and buoy are carried out by using ANSYS. The comparison is made only for head seas direction because the characteristics from other angles are quite similar due to the symmetrical shape of the buoy.

15 Results and Discussions

The metacentric height (GM) for the buoy is 2.4 m and restoring moment is about 9.872 kN.m/degree. Metacentric height is measured from the centre of gravity to the metacentre of the buoy. Positive value of metacentric height is required to show the stability of the buoy when external force is applied to the body. There are 3 types of GM characteristics of a floating structure or body including:

i. GM less than zero, which means it has negative value. In this condition, the buoy is considered as unstable equilibrium. The uprighting moment produced becomes less than zero. In other words, the buoy doesn't have initial stability and when external force is applied onto the buoy, it will incline to a certain angle and then it will tend to heel over and finally it will capsize. There is no restoring moment which will restore the buoy to an upright position.
ii. GM is equal to zero, thus the uprighting moment is equal to zero. In this case, the buoy is considered has initial stability but after environment load is applied to the buoy, it will remain in the same angle of heel until another external load is applied. This happens when the KM equals to KG. The position of the buoy (angle of heel) can be determined by the amount of load applied to the buoy since there is no countering force from the buoy.
iii. GM more than zero. Thus the uprighting moment would be positive value. The buoy is considered in stable equilibrium. The buoy would able to return to its initial state although external load is applied to the buoy. The restoring moment will restore the buoy its original position. However, the amount of load should be less than the restoring moment, otherwise, the buoy might capsize. GM should always become positive to provide stable condition. When the GM is comparatively high, thus the righting moment at small angle of heel will also be comparatively high. It means that, larger or higher moment is required to incline the buoy.

As stated in DNV Offshore Standard for offshore loading buoys, the GM shall be positive. There is no specific value mentioned in the standard since there is no person working on board for 24 h like offshore platform.

16 Conclusions

From this research, several important conclusions can be stated. Polyester has generated a big impact toward dynamic performance of the buoy. Polyester has achieved a better performance compared to the chain—spiral strand mooring line after couple dynamic analysis had been carried out.

It provides shorter horizontal excursion as well as less rotational and translation motion. It is an important criterion in mooring design because aggressive hydrodynamic motion would lead to lost host connection. Besides that, excesses horizontal offset would generate extra tension to the riser connection that could result in riser failure, thus big disaster might be happen.

It also generates lower total cable force that result in lower equipment specification. These reductions would partly be attributable to the specification of lower capacity components, such as hawser, anchor, shackle, etc.

The polyester should be considered for deep water mooring because of its excellent performance and it would be able to reduce the risk of failure due to corrosion. Although it requires longer total mooring line, the overall benefits that we got is considerably worth it.

17 Recommendations

From this research, there are several characteristics that can be improved and enhanced in order to achieve a better result and analysis. In addition, it is also crucial so that any improvement towards the design and configurations may be developed in a way to reduce the surge responses.

First and foremost, to improve the validity of the hydrodynamic response calculated and estimated using both of the analysis (statics and dynamics), it is vital that a model analysis is made in the wave tank. Its function to test the designs of the buoy and mooring line in a real life manner where researcher can observe and analyze the hydrodynamic characteristics thoroughly.

Moreover, couple dynamic analysis between FSO, SBM and mooring line should be made to get a better mooring design. However, FSO's dynamic motion was not included due to the higher processor memory required to run additional floating structure model in the analysis.

According to OS E-301, there are 3 types of mooring design analysis. In this analysis, the researcher had focused on the ultimate limit state analysis. The remaining 2 analysis; accidental limit state and fatigue limit state analysis should be carried out to determine the overall performance of mooring line.

Finally, environmental data such as seabed or soil condition can be added into the analysis. It would enable the friction effect between the laid mooring line and seabed may be calculated. However, during the research, seabed frictions have been neglected due to difficulties to gain the information and time constraint during the research.

References

1. AQWA (2012) user manual
2. Barltrop N (1998) Floating structures: a guide for design and analysis. Oilfield Publications Limited
3. Bea M (2004) Conceptual design of a semi-submersible floating oil and gas production system for offshore Malaysia. Chicago
4. Chakrabarti S (2005) handbook of offshore engineering. Elsevier, Amsterdam
5. Calvert J (2014) The catenary. http://mysite.du.edu/~jcalvert/math/catenary.htm. Last accessed 12 Aug 2014
6. Chaplin C, Potts A, Curtis A (2008) Degradation of wire rope mooring lines in SE Asian waters. Offshore Asia
7. Childers M (1973) Mooring systems for hostile waters. Petroleum Engineer 45(5):58–70

8. Chopra A (2007) Upper saddle river. In: Dynamics of structures, 3rd ed. Pearson education
9. Cozijn J, Bunnik T (2004) Coupled mooring analysis for a deep water CALM buoy. Proceedings international conference on offshore mechanics arctic engineering
10. Faltinsen O (1990) Sea loads on ships and offshore structures. Cambridge University Press, Cambridge
11. Flannery C, Choo B (2007) Kikeh batch setting: case study. In: Proceedings of SPE/IADC drilling conference
12. Harris R, Johanning L, Wolfram J (2004) Mooring systems for wave energy converters: a review of design issues and choices
13. Low Y, Langley R (2006) A comparison of time domain and frequency domain approaches for the fully coupled analysis of deepwater floating systems. In: Proceedings international conference on offshore mechanics and arctic engineering
14. Montasir OA, Yenduri A, Kurian VJ (2014) Effect of mooring line properties and fairlead slopes on the restoring behavior of offshore mooring system. Research journal of applied sciences, engineering and technology 8(3):346–353
15. Nakajima T, Motora S, Fujino M (1982) On the dynamic analysis of multi-component mooring lines. In: Offshore Technology Conference, OTC 4309, pp 105–120

Forensic Investigation on the Propulsion Train System for Vibration Induced Phenomenon Onboard a Naval Vessel Using Fast Fourier Transform

Hamdan Nuruddin, Muhammad Hazwan Azizi, Muhammad Nasuha Mansor, Shaiful Bakri Ismail and Iwan Zamil Mustaffa Kamal

Abstract Vibration is a phenomenon that produces sound waves. For a naval vessel reducing the acoustic signature was vital in order for her to fulfill her role as an effective weapon platform because sound waves can reveal the vessel's position to enemy warship especially submarines which in turn will deploy acoustic seeking torpedoes to destroy the vessel. The research was to investigate the source of excessive vibration onboard a naval vessel. This abnormally was probably caused by the major repair on the vessel during the scheduled maintenance. Data were taken at various locations in the machinery room among them were at the propeller shaft, main bearing, gear box, plummer block, main engine and auxiliary engines using an accelerometer and SKF MICROLOG analyzer. All together thirty-eight points were taken. The arrival of advanced signal processing tools such as MATLAB greatly change the sphere of development of vibration study. The spectral analysis, time synchronous signal averaging, high frequency envelop analysis and time-frequency analysis can be analyzed by using MATLAB software. The time domain data were converted to frequency domain using Fast Fourier Transform (FFT). The frequency data was analyzed for each item of the propulsion train system and compared against the Vibration Criteria Chart from classification society. It was found that the propeller and propeller shaft were having excessive reading beyond the allowable limits. It was recommended that the propeller be balanced accordingly and the propeller shaft to be straightened and realigned.

Keywords Vibration · Propulsion train system · Propeller and propeller shaft · Fast fourier transform

---

H. Nuruddin (✉) · M. N. Mansor · S. B. Ismail · I. Z. M. Kamal
Universiti Kuala Lumpur, Malaysian Institute of Marine Engineering Technology, Lumut, Malaysia
e-mail: hamdann@unikl.edu.my

M. H. Azizi
DALAC Marine Engineering and Services Sdn.Bhd, Johor, Malaysia

C. L. Saw et al. (eds.), *Advancement in Emerging Technologies and Engineering Applications*, Lecture Notes in Mechanical Engineering,
https://doi.org/10.1007/978-981-15-0002-2_3

1 Introduction

Vibration is a mechanical phenomenon that gives adverse impact on the machinery, ship structure and cause discomfort to the passengers and crew. Any physical bodies that has mass and mechanical properties will vibrate. This body when disturbed from its equilibrium position will cause vibratory motion. The vibratory forces from the mechanical motion of the body will restore the body to it equilibrium state. The damping force will reduce the amplitude of vibration with time so that the body stops vibrating. Vibration problem might be caused by auxiliary equipment also not just the primary equipment.

There are four major causes of vibration by machineries that are imbalance, misalignment, wear and looseness. Imbalance is a heavy spot in a rotating component when the unbalance weight rotates around the machine's axis will create a centrifugal force. For the angular misalignment, this happen when the axes of a motor and pump are not parallel. Wear can occurs when a roller bearing becomes pitted for instance the bearing rollers will cause vibration each time they travel over the damage area. Looseness of bearings also can cause vibration [1]. The marine classification society, such as Det Norske Veritas—Germanischer Lloyd (DNV GL) has made a Vibration requirement for all vessel that used DNV GL classification.

A vibration analysis was done to a patrolcraft which belongs to the Royal Malaysian Navy. After major repair, the vessel under gone a sea trial and it was experiencing heavy vibration at the stern area. It was decided that it was necessary to measure the vibration by using SKF MICROLOG equipment. The time series data was automatically transform to frequency domain data by the SKF MICROLOG. The data was analyzed to identify the source of vibration. It is highly likely the source of vibration is coming from the propeller but it is important to identify other source of vibration as well. In general, the possible sources of vibration include shaft line bearing, engine, turbocharger, turbine and gears. Every source of vibration has its own limitation, measurement method as well as velocity [2].

2 Methodology

Several significant points were identified in the engine room for data measurement locations using SKF MICROLOG equipment as shown in Fig. 1. The locations were:

i. PORT Gearbox Output Shaft—Radial and Axial
ii. PORT FORWARD Plummer Block Bearing Input Shaft—Radial and Axial
iii. PORT FORWARD Plummer Block Bearing Output Shaft—Radial and Axial
iv. PORT AFT Plummer Block Bearing Input Shaft—Radial and Axial
v. PORT AFT Plummer Block Bearing Output Shaft—Radial and Axial.

A built-in Fast Fourier Transform (FFT) function in MATLAB was used to convert the data from time domain to frequency domain. From the graph, the maximum

Fig. 1 SKF Microlog device

amplitude at corresponding frequency was determined. The amplitude was compared by using Military Standard TYPE III to determine whether the amplitudes were in acceptable range. A recommendation to minimize the vibration was applied to ensure the vibration produced were acceptable for the naval vessel.

3 Result and Discussion

The primary outputs of the result were represented by an amplitude-frequency graph. Each graph was analyzed using specific formula to determine which frequency belongs to a specific machinery.

3.1 Analyzed Frequency Data

Several readings at different points were selected to determine the trend of the frequency data. The locations selected were engine room, engine room output plummer bearing, aft auxiliary input plummer bearing and output gearbox for different range of RPMs.

These graphs were produced for the above machineries. Figure 1a showed that the vibration reading was taken from auxiliary engine room at 84.9 RPM. The highest vibration amplitude was 0.24167 mm/s RMS produced from the propeller shaft. Figure 1b is the vibration reading taken from output plummer bearing at 84.9 RPM. The highest vibration amplitude was 0.21286 mm/s RMS produced from propeller shaft.

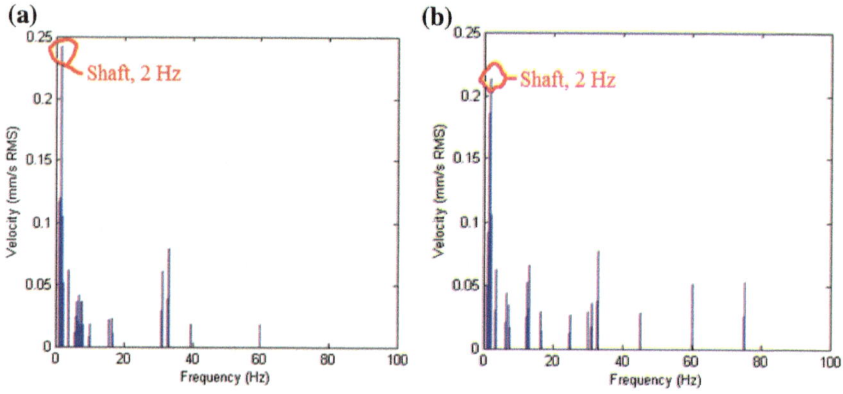

Fig. 1 **a** Auxiliary room at 84.9 RPM **b** engine room output plummer bearing at 84.9 RPM

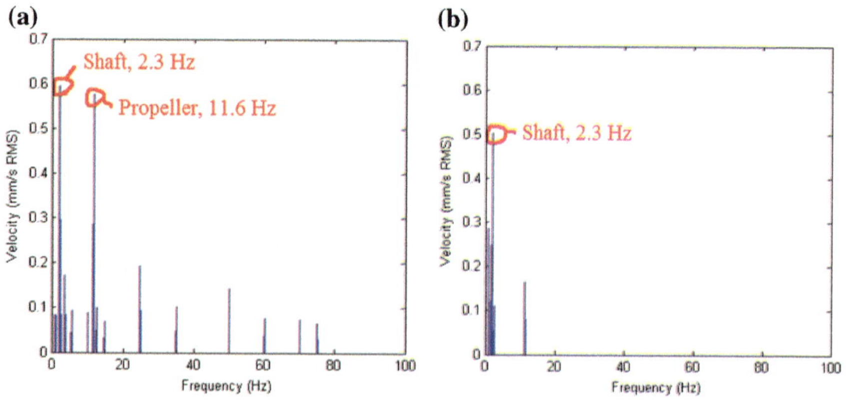

Fig. 2 **a** aft engine room output plummer bearing at 141.9 RPM **b** Gearbox inner at 141.9 RPM

Table 1 Highest vibration

No.	Frequency (Hz)	Amplitude (mm/s RMS)	Amplitude (mm)	Vibration sources
1	2.3	0.93144	0.09116	Shaft
2	2.3	0.59483	0.0582	Shaft
3	2.3	0.68498	0.06704	Shaft
4	2.1	0.50105	0.04904	Shaft
5	2.5	0.57766	0.05201	Shaft
6	11.5	0.35109	0.00687	Propeller
7	11.2	0.39679	0.00798	Propeller
8	11.6	0.57549	0.01117	Propeller

Figure 2a showed that the vibration reading was taken from aft engine room output plummer bearing at 141.9 RPM. The highest vibration amplitude was 0.59483 mm/s RMS produced from the propeller shaft while 0.57549 mm/s RMS was produced by the propeller. Figure 2b was the data taken from the gearbox inner at 141.9 RPM. The highest vibration amplitude was 0.50105 mm/s RMS produced from the propeller shaft. From the result, the highest vibration amplitude was produced from the propeller shaft and the propeller. The highest reading was selected from the propeller shaft and propeller blades. The amplitudes were compared against the Military Standard TYPE III to determine whether the amplitudes were in acceptable range (Table 1).

3.2 Acceptance Vibration Limit

For the propeller it was acceptable because its amplitude was just below the allowable limit. However a recommendation to minimize the amplitude was needed because the vibration can still reach the limits when the RPM was increased [3].

3.3 Recommendation and Remedial Measure

(i) It was recommended that shaft straightness check to be carried out. This will enable the ship's Design Authority to determine if either propeller or shaft was deformed probably due to the collision with the sea bed. The Main and Intermediate 'A' Bracket Bearing Clearances should be taken and recorded as part of the straightness check. A total of six dial gauges will be required for each shaft in order that simultaneous readings can be obtained. The 'A' Bracket Journal readings are expected to be within 0.051 mm Total Indicator Reading and within 0.127 mm at the Stern Gland Position. Any reading that fall outside

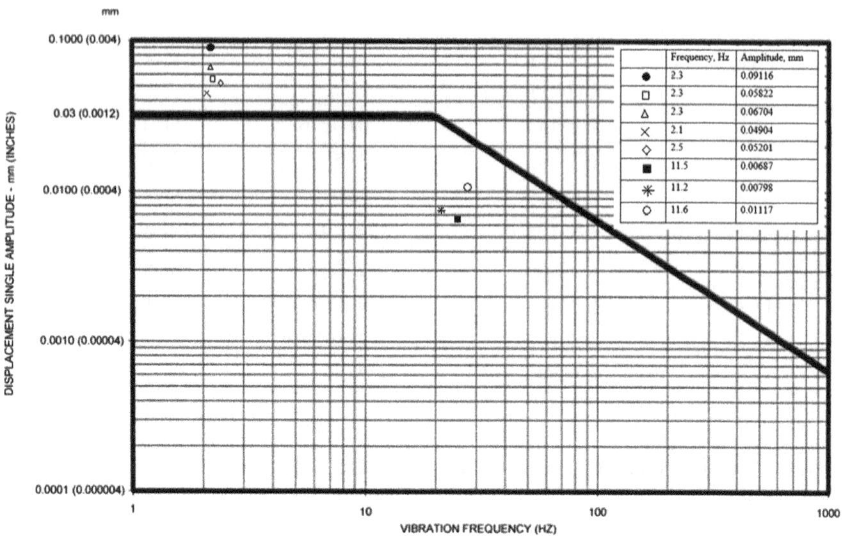

Fig. 3 Result for acceptance vibration criteria

of the tolerances will be referred back to the Design Authority who may require additional readings to be collected and action taken accordingly.

(ii) It was also recommended that the balancing of the propeller to be carried out. The aim was to achieve interchangeability of blades from different manufacturers. The basis for the allowable unbalance was the ISO R484 criteria [4]. For blade with large Expanded Area Ratio (EAR) and medium to high rpm, an extra balancing around the propeller blade pitch setting axis ("Y" axis) is necessary to minimize the risk for dynamic unbalance (Fig. 3).

4 Conclusion

It was concluded that the vibration problem onboard the patrolcraft was successfully investigated. The highest vibration occurred was detected from the propeller shaft with the highest amplitude (mm/s RMS was 0.93144, 0.59483, 0.68498, 0.50105 and 0.57766). While increasing the RPM will increase the amplitude of propeller blades (mm/s RMS' was 0.57549, 0.39679 and 0.35109). By referring the Standard Military TYPE-II vibration acceptance criteria, these vibration amplitudes produced by the shaft had exceeded the limit. Although the amplitudes produced by the propeller was still within the range, a recommendation need to be applied. This is because the amplitudes were nearly reaching the limit. It was recommended that remedial measures need to be applied to minimize the vibration such as 'Shaft Straightness Check' for propeller shaft and 'Balancing of Propeller Blade' for propeller.

Appendix: Formulas and Calculation

See Tables 2 and 3.

Table 2 Shaft rotation calculation

Shaft rotation speed (RPM)	Running speed/second (Hz)
141.9	2.36
139.9	2.33
84.9	1.42

Table 3 Rolling element bearing frequency

Shaft rotation speed (RPM)	Pinion gear speed (RPM)	BPFI* (Hz)	BPFO** (Hz)
141.9	549.56	110.98	87.2
139.9	586.18	109.42	85.97
84.9	335.73	66.4	52.17

$$*BPFI = \left(\frac{N_b}{2} + 1.2\right) \times (Pinion\,Gear\,Speed)$$

$$**BPFI = \left(\frac{N_b}{2} - 1.2\right) \times (Pinion\,Gear\,Speed)$$

See Table 4.

Table 4 Propeller blade frequency

Shaft rotation speed (RPM)	Propeller blade (Hz)
141.9	11.83
139.9	11.65
84.9	7.08

*Number of blade = 5

$$**\ Blade\ Frequency = \frac{N \times Shaft\ Speed}{60}$$

References

1. Adams LM (2001) Rotating machinery vibration: from analysis to troubleshooting. Second Edition, New York
2. Chakraborty S (2016) Types of vibrations on ships—machinery vibrations. Marine insight. Retrieved from http://www.marineinsight.com
3. Defence DO (1999) Mechanical vibration of shipboard equipment. USA: USA. Gockenbach MS (2005b) A Practical Introduction to Matlab. Retrieved from Howard I (1995) Vibration signal using MATLAB. Retrieved from Australia
4. Lewis EV (1988) Principles of naval architecture vol 2. The society of naval architects and marine engineers. Pavonia Avenue, Jersey City, NJ, p 601

Thermal Characteristics of Electrically Conductive Adhesives (ECAs): Study on Micro- and Nano-Sized Silver Particle Effect

M. Zulkarnain, M. Mariatti and Zulzamri Salleh

Abstract The electrically conductive adhesives (ECAs) composite of micro- (2–3.5 μm of diameter) and nano-sized (80 nm in diameter) silver (Ag) particle effect were prepared by using the epoxy polymer. The characteristics of particles size are analysed on the synergistic effect in thermal conductivity that respect to volume fraction (vol.%). In both micro- and nano-particles study, the volume fraction factor which is set in the range of 2–8 vol.%. The hybrid-sized technique is proposed by using the volume ratios of the micro- and nano-sized particles. During hybrid process, the combination size is varied in three sets: 75:25, 50:50, and 25:75 (the number on the left indicates the micro-sized particle content, while nano-sized on the right side). A significant improvement in the optimization of the particles dispersion synergistic effect on thermal conductivity has been obtained with new techniques in the experimental data. The influence of the hybrid filler composition on the thermal properties of the hybrid system was studied. A positive effect was observed in the thermal conductivity result when the composition of micro- and nano-sized Ag particles reached a 75:25 weight ratio. The nano-sized Ag particles became interconnecting particles in the interstitial spaces between micro-sized particles.

Keywords Thermal conductivity · Electrically conductive adhesives (ECAs) · Particle contact resistance · Particle structure

---

M. Zulkarnain (✉) · Z. Salleh
Universiti Kuala Lumpur Malaysian Institute of Marine Engineering Technology (UniKL MIMET), Sitiawan, Malaysia
e-mail: m.zulkarnain@unikl.edu.my

M. Mariatti
School of Materials and Mineral Resources Engineering, Universiti Sains Malaysia, Nibong Tebal, 14300 Seberang Perai Selatan, Penang, Malaysia

C. L. Saw et al. (eds.), *Advancement in Emerging Technologies and Engineering Applications*, Lecture Notes in Mechanical Engineering,
https://doi.org/10.1007/978-981-15-0002-2_4

1 Introduction

In the past two decades, the alternative electrically conductive adhesives (ECAs) have been investigated to replace lead-free solder for interconnection material. The interconnecting material for electronic application such as die attach adhesives, flip-chip interconnection, and surface-mount electronic component have some advantage when applied the ECAs such as good electrical conductivity, low cost, extendibility to fine pitch interconnect and environmental friendly [1, 2].

The major technological problem in the electronic application is contributed by head dissipation such as chips because it has a fundamental maximum operating temperature. In addition their reliability rapidly declines at increased operating temperatures. For these reasons the thermal conductivity of the ECAs composite is a parameter of major importance. Several works have been done to improve thermal conductivity; Yuan et al. [3] employed epoxy resin that insulating by various particle fillers, such as aluminium oxide (Al_2O_3), graphite, copper (Cu), aluminium (Al), zinc oxide (ZnO), diamond, boron nitride and silver (Ag). They reported that the thermal conductivity resulted in eight kind of variation where graphite filler showed the higher thermal conductivity. Meanwhile, the investigation found that flaky-shape filler tend to perform particles network than spherical-shape. In addition, the sharp-corner-shape filler generated the higher thermal conductivity. Another researcher generated Boron nitride (BN)/polymer composite to investigate thermal conductivity value [4]. Hong Jun Ahn et al. reported the orientation and particles size of BN with an amphiphilic agent can improve the performance of thermal conductivity. The results showed that the BN/PVB composite with amphiphilic agents such as $C_{14}H_6O_8$ and $C_{27}H_{27}N_3O_2$ given a higher thermal conductivity than the PVB composite with pristine BN only. Other finding reported that good dispersion and interfacial adhesion with $C_{14}H_6O_8$ on the $C_{14}H_6O_8$-treated BN/PVB composite can improve thermal conductivity compare to $C_{27}H_{27}N_3O_2$-treated composite.

Particle dispersion efficiency brings to isotropic composite in such thermally conducting network. Some analysis developed on different fillers with distinct shapes and dimensions were combined on the basis of complementary structural and property advantages to generated conducting network. Yuan et al. reported that the synergistic effect on thermal conductivity by using different filler. The composite was generated by using the combination of tetrapod-shaped zinc oxide (T-ZnO) whiskers and boron nitride (BN) flakes onto polymer phenolic formaldehyde resin (PF) [5]. The characteristics of thermal conductivity are investigated that synergistic effect by adding the particles. Thermal conductivity value was measured at 30 wt% BN and 30 wt% T-ZnO, the thermal conductivity reached at 1.96 W/mK that higher 6.8 times than pure PF. Another researcher developed the composite by using graphene nanosheets particles that embedded onto epoxy resin to improve the thermal conductivity of anisotropic conductive adhesives (ACAs) [6]. In preparing of composite, graphene oxide (GO) and Al(OH)3-coated graphene (Al-GO) fillers were produced using the Hummers method and a simple sol–gel method, with aluminium isopropoxide as the aluminium precursor. As the results, particles dispersion of functions graphene fillers showed well distributed within the epoxy resins. That synergistic on the thermal conductivity of composite was increased compared to those without fillers.

Silver-filled epoxies have some advantages when used as electrically conductive adhesives application in the interconnection material electronic devices component [7]. One probably reason is high thermal conductivity as well as electrical conductivity properties especially for attachment of chips to substrates application. Both thermal and electrical conductivity give big impact for ECAs performance, for this reason on the Ag particle size effect still have challenges to investigate due to lack information on thermal conductivity of Ag particle on micro- and nano-sized effect.

2 Experimental Method

2.1 Materials

The epoxy product used in this investigation was bisphenol-A-(epichlorohydrin) with another name was EPON™ Resin 8281 (EPON 8281) that supplied by Hexion Specialty Chemical, USA. The curing agent of Polyetheramine D230 (PEA; density = 0.946 g/mL at 298 K), which was used for EPON 8281, it was manufactured by BASF Corporation. The condition mixture between EPON 8281 and PEA was fixed at 100:32. The Ag particles used in this study were purchased from Sigma–Aldrich, Inc. The sheet data from the supplier shows that the average particle sizes were 80 nm and 2–3.5 μm, and the density was 10.49 g/cm^3. The Chloroform (99.0–99.4%) purity was used in the study to lubricate the particle distributions in the epoxy resin with mixing ratio is 1:1 by weight and were sonicated in ultrasonic (model NeyALTRAsonik) bath for 10 min.

2.2 Preparation of ECA Composite

Single-filler (micro- and nano-sized) conductive adhesives were prepared using 2–8 vol.% fillers. For the hybrid system, the total Ag micro- and nano-sized filler loading was fixed at 6 vol.%. The volume ratios of the hybrid system using micro- and nano-sized particles were varied in three sets: 75:25, 50:50, and 25:75 (the number on the left indicates the micro-sized particle content; that on the right indicates the nano-sized particle content). The epoxy and Ag filler mixture was stirred for approximately 10 min using a homogenizer, and then the mixture was sonicated for 30 min to facilitate filler dispersion. The mixture was subsequently vacuumed for approximately 0.5 h at room temperature (35 °C) to remove bubbles. The curing agent (32 parts by weight) was added into EPON 8281 (100 parts), and the mixture was stirred for another 10 min. Finally, the mixture was stored in a vacuum oven for 2 h at room temperature before it was cured in the oven at 100 °C for 1 h and post-cured at 125 °C for 3 h.

2.3 Characterization

Thermal conductivities of all the samples were measured by the Hot Disk TPS-2500 thermal constants analyser at room temperature. Specimens with 1 cm × 1 cm dimensions and 3 mm of thickness were prepared for the thermal conductivity tests. This technique has been previously used for investigation of thermal properties of other ECAs such as the phase change materials and industrial thermal greases [4, 8]. For these measurements we sandwiched an electrically insulated flat nickel sensor with the radius 1 cm between two identical samples of the same composition. The sensor acted as the heat source and temperature monitor simultaneously. The surfaces of the specimens were flattened and cleaned to reduce resistance at the sensor-sample surfaces.

3 Results and Discussion

3.1 Thermal Conductivity of Micro- and Nano-Sized Particles

The thermal conductivity of ECAs was generally characterized by the varied with the volume fraction of the Ag filler loading. Figure 1 illustrates the relationship between the experimentally measured thermal conductivity and Ag volume fraction in the epoxy composites for both micro- and nano-sized particle. Both epoxy composites filled with micro- and nano-sized Ag particles showed similar conductivity trends

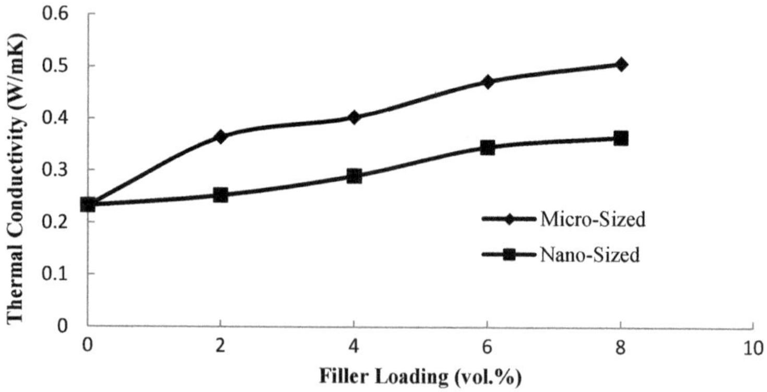

Fig. 1 Thermal conductivity of Ag-filled epoxy composites versus the volume fraction of Ag fillers

that characterized by filler loading. The thermal conductivity of the epoxy composite showed increasing significantly by adding several of Ag particles. According to previous study [3–5], the thermal conductivity is dependent on the density, specific heat capacity, and thermal diffusivity of the composites. The measured data suggest that the through thermal conductivity is strongly influenced by the inclusion of high thermally conductive Ag-fillers. At low contents of Ag, the thermal conductivity of the Ag/epoxy composites is slightly lower than the higher filler loading of Ag. The difference becomes larger at high filler Ag particles contents. For example, with 2 vol.% of Ag filler particles, the thermal conductivity of the micro-sized composite is 0.3647 W/mK, while 8 vol.% of Ag filler particles showed at 0.5082 W/mK that increased almost 2 times higher.

From the results, the thermal conductivity values determined for the Ag micro-sized particle is significantly higher than that measured for the nano-sized particle (Fig. 1). This condition showed that by improved the thermal conductivity of micro-Ag particle from 0.3647 W/mK at the filling load of 2 vol.% Ag particles, which is higher than nano-sized that mentioned at 0.2527 W/mK. However, the thermal conductivity of the epoxy filled with the Ag nano-sized particle at 8 vol.% is 0.3658 W/mK, increased by almost 2 times that of the pure epoxy. Nevertheless, is still much smaller than that of the epoxy filled with the Ag micro-sized. For comparison, the thermal conductivity of micro-sized at 8 vol.% of Ag particles is 0.5082 W/mK that increased by more than 2 times of the pure epoxy. From this observation, it seems that Ag particle size affected the thermal conductivity of ECAs which micro-sized has higher the thermal conductivity. This phenomenon could be attributed to the fact that the thermal conductivity of the composites slightly increased with increasing the Ag particle size. A similar phenomenon has been reported by previous work that the thermal conductivity of the composites was affected by particle size [8].

3.2 *Thermal Conductivity of Hybrid-Sized Systems*

In hybrid system study, the mixing ratio between micro- and nano-sized of Ag filler was fixed at 6 vol.% of filler loading. Three ratios were considered in this study which was conducted on 75:25, 50:50, and 25:75. The thermal conductivity analysis of hybrid-sized system was shown in Fig. 2. The thermal conductivity results showed varied depending on the particle size combination ratio. It were presented on 0.4367, 0.4092, and 0.3477 W/mK for the volume loading ratios of 75:25, 50:50, and 25:75, respectively. The thermal conductivity is decreasing with increasing nano-particles in the hybrid system. This phenomenon may be due to the density of the composites slightly decreased with decreasing the Ag particle size. In addition, when each particle size reached 25:75 volume ratio of filler loading, the thermal conductivity shows an almost similar value that obtained on nano-sized particles at 6 vol.%. This phenomenon proved that the decrease in particle size caused the thermal conductivity decreased. The thermal conductivity value of hybrid system decreased as increasing

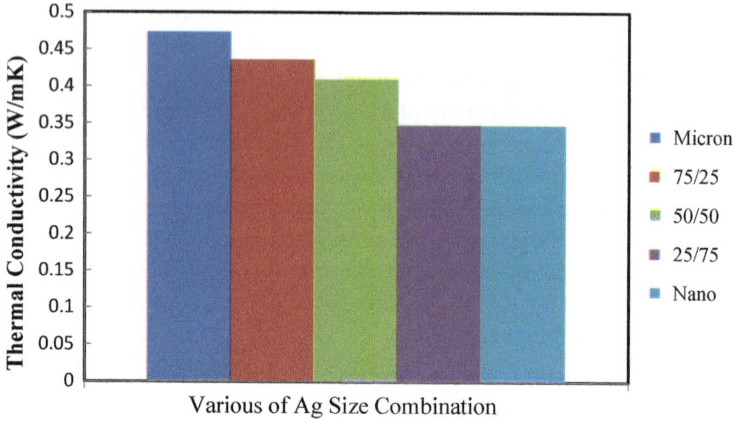

Fig. 2 Thermal conductivity of ECA with 6 vol.% Ag filler loading versus the ratio of micro- and nano-Ag particles

the nano-particle ratio. The decreasing the thermal conductivity value was denoted by adding nano-particles in the system.

4 Conclusions

The study of thermal conductivity characteristics of ECAs by proposing micro- and nano-sized Ag particle effect has been successfully presented. The thermal conductivity of ECAs with micro-sized Ag particle showed positive effects for both volume fraction loading trend and in hybrid-sized system analysis. The ECAs filled with Ag nano-sized particles bring a decreasing on thermal conductivity and also on their distribution, which can be attributed to the higher resistance between nano-sized particle contacts. Since the nano-sized Ag particles generate a decreasing effect on thermal conductivity, in the hybrid system observation, a 75:25 volume ratio gives the highest thermal conductivity where micro-sized content is higher. The simply adding nano-particles have a negative effect on the overall conductivity because of the increased number of contact points.

The volume loading ratios of 75/25, 50/50, and 25/75 had promoted the effect of formation on the thermally conductive pathway. It proofed by three models of micrograph scanning of the hybrid-sized composite. The Ag particles were not well distributed and resulted in agglomeration. However, the pathway of particle networks of nano-sized Ag was obvious in the system, which led to a thermal transfer. The negative effects of the hybrid system are shown by composite 50/50 and 25/75. The typical distribution was dominated by the nano-sized Ag particle distribution where a large number of agglomerates and too much space among particles existed. Consequently, a segregated network of particles formed at a much lower filler loading, and the resistivity increased upon the addition of nano-sized particles. The filler particles were more separated by the epoxy matrix and at this condition, and the thermal conductivity values were lower than other systems. Nano-particle led to introduce the higher contact resistance in the composite system. The total resistance between two particles in resin is the sum of contact resistance and tunnel resistivity of resin.

Acknowledgements The author acknowledges the financial support of the FRGS grant (203/P.Bahan/6071125) and post-graduate research fund (PGRU), as well as the facilities provided by the School of Materials & Mineral Resources Engineering and Post Graduate Research Grant Scheme (1001/PMEKANIK/8042005) of Mechanical Engineering, Universiti Sains, Malaysia.

References

1. Li Y, Wong CP (2004) Recent advances on electrically conductive adhesives (ECAs). In: IEEE polytronic, pp 1–7
2. Kang SK, Rai RS, Purushothaman S (1998) Development of high conductivity lead (pb)-free conducting adhesives. Packag, Manuf Technol A 21(1):18–22
3. Fu Y-X, He Z-X, Mo D-C, Lu S-S (2014) Thermal conductivity enhancement with different fillers for epoxy resin adhesives. Appl Therm Eng 66:493–498
4. Ahn HJ, Cha S-H, Lee WS, Kim ES (2014) Effects of amphiphilic agent on thermal conductivity of boron nitride/poly(vinyl butyral) composites. Thermochim Acta 591:96–100
5. Yuan F-Y, Zhang H-B, Li X, Li X-Z, Yu Z-Z (2013) Synergistic effect of boron nitride flakes and tetrapod-shaped ZnO whiskers on the thermal conductivity of electrically insulating phenol formaldehyde composites. Composites: Part A 53:137–144
6. Kim J, Yim B, Kim J, Kim J (2012) The effects of functionalized graphene nanosheets on the thermal and mechanical properties of epoxy composites for anisotropic conductive adhesives (ACAs). Microelectron Reliab 52:595–602
7. Amoli BM, Trinidad J, Hu A, Zhou YN, Zhao B (2015) Highly electrically conductive adhesives using silver nanoparticle (Ag NP)-decorated graphene: the effect of NPs sintering on the electrical conductivity improvement. J Mater Sci: Mater Electron 26:590–600
8. Zhang S, Cao XY, Ma YM, Ke YC, Zhang JK, Wang FS (2011) The effects of particle size and content on the thermal conductivity and mechanical properties of Al2O3/high density polyethylene (HDPE) composites. eXPRESS Polym Lett 5(7):581–590

Flexural Analysis for Syntactic Foam Sandwich Panels

Zulzamri Salleh, M. Islam, J. Epaarachchi, M. Zulkarnain and Yaseen Adenan Ahmed

Abstract The investigation on the strain value for flexural sandwich panel syntactic foam were also carried out using the FEA approach to predict the properties' behaviour in this study. It was found that the micro strain for SG1 for FEA was 17% higher than the experimental value, even though they were at the same loading setting. However, the prediction for the micro strain of SG2 was only 2.7% different, which was considered a good agreement to predict the properties of syntactic foam core sandwich panel for different loading values.

Keywords Syntactic foam · Strain gauge · Finite element analysis

1 Introduction

The syntactic foams known as closed foams possess a higher density than open cell foams, and it is difficult to synthesis the existing porosities in syntactic foams. However, syntactic foams have considerable superior mechanical properties that make it possible for them to be used in load bearing structural applications. The presence of porosity inside hollow particles, called microballoons, leads to lower moisture absorption and lower thermal expansion, resulting in better dimensional stability [1, 2]. The size and distribution of porosity can be controlled very closely in these foams by means of microballoon volume fraction and wall thickness. A previous study also indicated that mechanical properties are varied with different filler content [3]. A comprehensive understanding of the influence of microballoons/matrix adhesion, and wall thickness of the matrix on the compressive failure mechanisms

Z. Salleh (✉) · M. Zulkarnain · Y. A. Ahmed
Malaysian Institute of Marine Engineering Technology, Universiti Kuala Lumpur, 32200 Lumut, Perak, Malaysia
e-mail: zulzamri@unikl.edu.my

M. Islam · J. Epaarachchi
Centre for Future Materials and School of Mechanical and Electrical Engineering, Faculty of Health, Engineering and Sciences, University of Southern Queensland, Toowoomba, QLD 4350, Australia

C. L. Saw et al. (eds.), *Advancement in Emerging Technologies and Engineering Applications*, Lecture Notes in Mechanical Engineering,
https://doi.org/10.1007/978-981-15-0002-2_5

of these foams is still lacking [4]. This is also supported by Gupta et al. (2004) who noted that to achieve better mechanical properties wall thickness and volume fraction need to be considered [5]. These parameters always correlate with the density of glass microballoons and filler content. Previous reports show that the void content or porosities are less when the density of glass microballoons is increased [6]. Therefore, all these main parameters need to be considered in this study to ensure a better understanding for exploring marine applications. Thus, increasingly demanding for lightweight materials had gained their usage not limited due to higher good mechanical strength for aerospace [7], marine, thermal insulation and packaging [8] but also environmentally user friendly.

This foam is also known as low density foam, which is used in buoyancy aid facilities for offshore applications [9]. Many studies have reported that the presence of porosity inside a thin wall thickness shell promises to give a better result during environmental test conditions such as moisture absorption and thermal analysis [1, 2]. At the same time, the stress concentration factor (SCF) around the hole for homogenous material, such as syntactic foam, can be determined using a strain gauge (SG) unit.

In this case, mechanical properties such as the modulus elasticity and Poisson's ratio are dependent on the composition of the composite material is observed, highlighting the need for developing the composites material. Poveda et al. (2010) found that different types of glass microballoon with different sizes cause the Poisson's ratio to vary [10]. Other experimental work also showed that the modulus of epoxy or vinyl ester matrix syntactic foams can be tailored within a range of 1–3 GPa while their strength can be controlled in the range of 30–110 GPa [5, 11]. Beside of experimental work, theoretical studies also tend to assume the Poisson's ratio using the model particularly in FEA. Previous report also mentioned that lack of availability of Poisson's ratio values is a limitation for modelling efforts [12]. The determination of both the modulus of elasticity and Poisson's ratio from the experiment can be used in FEA modelling to finalise the SCF value of syntactic foam. Desirable to composite plate subjected to various transverse static loading conditions [13]. They found, that the peak stress concentration factor occurs on the boundary of the hole and that the stress concentration factor was also strongly influenced by the uniformly distributed normal and equivalent stresses. Limitation of availability the characteristic such as modulus elasticity and Poisson's ratio in functional graded material like syntactic foam lead to focus in this study using the modelling FEA methodology.

This paper presents the characterisation of the tensile properties of a dog bone specimen glass microballoon/vinyl ester syntactic foam. The main objective of this work is to investigate the behaviour of syntactic foam under tensile loading by using FEA. Tests on coupons and full-scale specimens were undertaken to determine the tensile properties of syntactic foam. The tensile tests on coupons were conducted. On the other hand, the tests on full scale specimen were performed using the procedures available in previous studies. The details of these tests are presented in the next sections. Aside from these tests, a finite element analysis (FEA) particularly stress concentration factor (SCF) was carried to simulate the tensile behaviours of full-scale

Fig. 1 Strain gauges attched to the specimen with indication SG1 and SG2

specimen by using the Strand7 software. The results obtained from the experiment were compared with those of FEA.

2 Methodology

2.1 The Flexural Syntactic Foam Core Sandwich Panel

The flexural syntactic foam sandwich panel has been used for the evaluation, using the FEA modelling method. The strain gauges, also supplied by Bestech Australia Pty. Ltd, were attached with glue in the middle of the top (SG1) and bottom (SG2) of the skin sandwich panel as shown in Fig. 1. The strain gauge was manufactured by Tokyo Sokki Kenkyujo Co. Ltd. with the general specification: type FLG-02-11, gage factor 2.05, length 0.2 mm and width 1.4 mm, resistance 120 Ω (Bestech 2015).

2.2 The FEA Finite Element Analysis(FEA) Modelling

The study of the mechanical properties, particularly the tensile specimens, assumed that the panels were constructed with one homogeneous layer with diameter hole, diameter: 3 mm, as indicated in Fig. 2a. The flexural specimen, taken from the sandwich syntactic foam, was assumed to be constructed with two homogeneous layers, which is shown as two skin layers for top- bottom and core in the middle, as shown in Fig. 2b. The diagram was illustrated and simulated using CREO 3.0 Parametric software, which includes simulation analysis and refined meshing methods. The considered sandwich panels were symmetric, i.e., its skins had an identical thickness t. The thickness of the core was donated by the symbol c as the syntactic foam. With reference to the terminology used by Allen, the sandwich panels can be classified as thick skins and non-antiplane core (Allen 1969). The deflection of a thick-skinned and non-antiplane core can be computed using solutions. The first solution was the

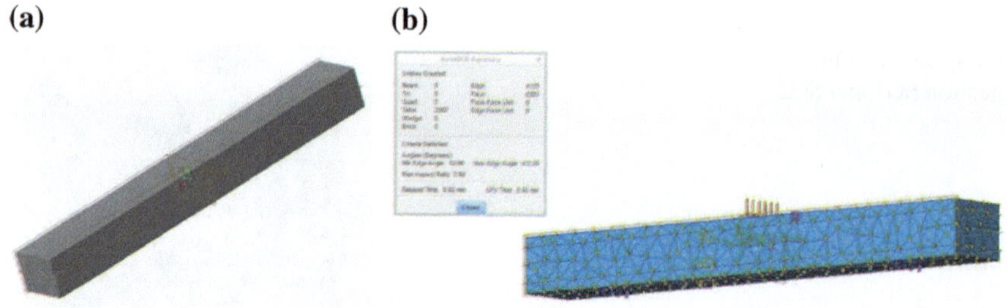

Fig. 2 Typical flexural 3-point bending specimen is illustrated using CREO 3.0 parametric software for **a** Flexural dimensioned **b** Redefined automesh

approximation that the field along the sandwich core is linear, and the second solution used the Total Potential Energy theorem. Because this study was most likely to use the actual size of the specimens, it was beneficial to continue with this solution. The constitutive model utilised for the foam was considered as a crushable foams plasticity model (Deshpande and Fleck 2000).

3 Result and Discussion

The failure mode pattern of flexure specimens, tested under a 3-points bending test and a FEA analysis using CREO simulation software, is revealed in Fig. 3. The failure mode observed in the flexural test, which was characterised by compressive failure, was also revealed to be due to de-bonding unsymmetrical shear failure between the skin and core areas that were in direct contact with the loading ramps. The crack formation under the flexure testing, due to compression, could clearly be seen in the simulated failure mode shown in Fig. 3a. It was apparent from the simulated failure that the side area (the initial compression zone) was imminent. In the figure, the cracked portion is represented by a blue and red-coloured strip at the side edge area. It is worth noting that whilst the surface contact with the loading rams provided a concave green-coloured shape, the middle area produced a small convex line at the side of the sandwich panels. This simulation confirmed the results obtained from the load-strain relationship that whilst this region is compressed during the initial loading, the increase of the loading until failure shifted the surface into tension mode, as revealed in Fig. 3b, for the support beam condition.

Figure 4 shows the load-μstrain relationships obtained from both the flexural testing and the FEA simulation. It should be noted that the micro strain values indicated in the figure are the values at the top of the skin (SG1), while another strain gauge (SG2) was located exactly perpendicular with the loading ramp at the bottom mid-span section of the sandwich panel. As can be seen in Fig. 4a, the linear fitting line represented as the FEA simulation was correlated with the experimental micro strain SG1, and then with tabulated data. The peak load obtained from the experiment

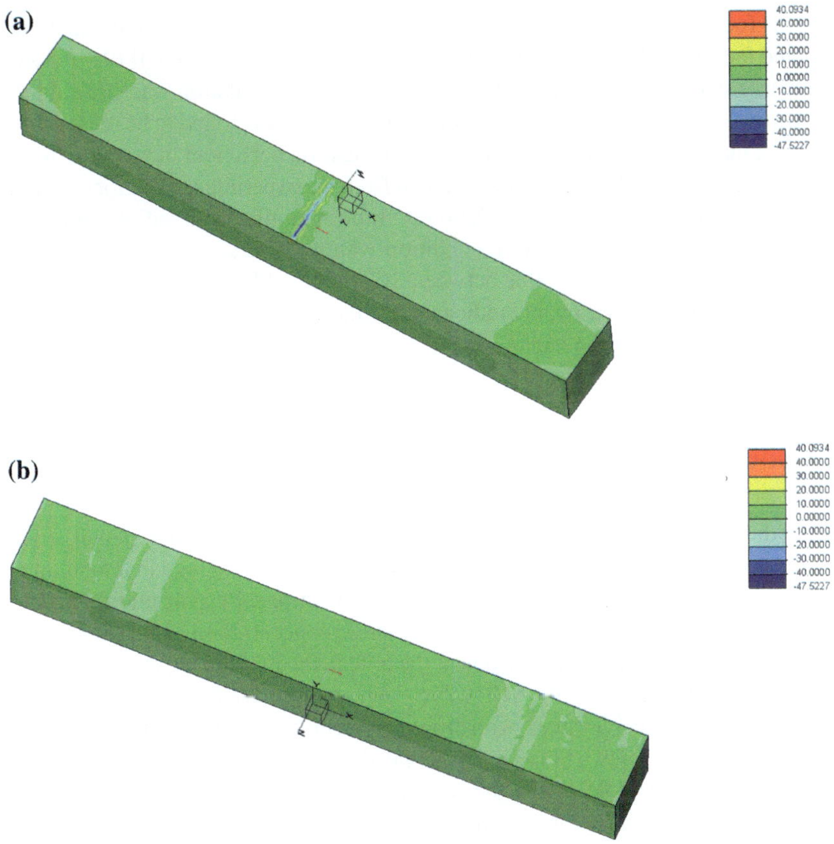

Fig. 3 Comparison of the flexural failure mode of syntactic foam core sandwich panel **a** FEA flexural simulation **b** Support beam dented

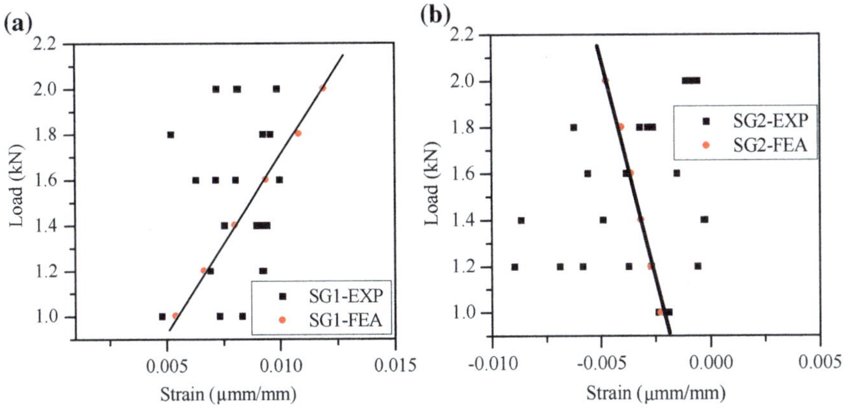

Fig. 4 Comparison of **a** Flexural testing and **b** FEA simulation for SG1 and SG2

SG1 was found to be 2 kN at failure strain 0.00989 μstrain. On the other hand, the predicted failure load using the FEA simulation at 2 kN is showed a 0.01191 μstrain. In this case, the μstrain value predicted from the FEA simulation was 17% higher than the experimental value. This difference of the value was found to be reasonable, indicating that the FEA simulation predicted the flexural behaviour of syntactic foam sandwich panels well. The peak load obtained from the experiment SG2, for example at 1.2 kN, was at failure strain −0.002665 μstrain. A similar observation applied to the FEA simulation at this loading condition where the strain failure could be −0.002743. The differences in value between SG2 and the FEA simulation was about 2.7%. Again this value reasonably indicated that the FEA was in good agreement for another comparison μstrain value.

4 Conclusion

The prediction of strain value between local strains from the experimental strain gauge was compared with the FEA simulation when their varied load in longitudinal and transverse axes was applied to flexural sandwich panel's syntactic foam. For the tensile specimen, the determination of the stress concentration factor (SCF) used one strain gauge, which was attached near the hole in the middle of the extensometer length. The results show that the SCF values were comparable between experiments with extensometer and SG values, with different percentages from 0.40 to 1.36%. The investigation of SCF for two SG were investigated using a specimen of 10 wt%, which was attached near the hole area at the same position as previous tensile specimens. The comparison and prediction were made between experimental values and the FEA analysis results. It can be estimated that the experimental values of around 90–70% followed the FEA values for SG1 and SG2, respectively. The investigation on the strain value for flexural sandwich panel syntactic foam were also carried out using the FEA approach to predict the properties' behaviour in this study. It was found that the micro strain for SG1 for FEA was 17% higher than the experimental value, even though they were at the same loading setting. However, the prediction for the micro strain of SG2 was only 2.7% different, which was considered a good agreement to predict the properties of syntactic foam core sandwich panel for different loading values.

References

1. Gupta N, Woldesenbet E (2003) Hygrothermal studies on syntactic foams and compressive strength determination. Compos Struct 61:311–320
2. Sauvant-Moynot V, Gimenez N, Sautereau H (2006) Hydrolytic ageing of syntactic foams for thermal insulation in deep water: degradation mechanisms and water uptake model. J Mater Sci 41:4047–4054

3. Vasanth C, Dinesh P, Gupta N (2012) Thermal expansion behaviour of hollow glass particle/vinyl ester composites. J Mater Sci 47:5596–5604
4. Swetha C, Kumar R (2011) Quasi-static uni-axial compression behaviour of hollow glass microspheres/epoxy based syntactic foams. Mater Des 32:4152–4163
5. Gupta N, Woldesenbet E, Mensah P (2004) Compression properties of syntactic foams: Effect of cenosphere radius ratio and specimen aspect ratio. Composites: Part A 35:103–111
6. Gupta N, Ye R, Porfiri M (2010) Comparison of tensile and compressive characteristics of vinyl ester/glass microballoon syntactic foams. Compos B Eng 41:236–245
7. Devi K, John B, Ninan CNK (2007) Effect of low-density filler on mechanical properties of syntactic foams of cyanate ester. J Appl Polym Sci
8. Salleh Z, Islam M, Ku H (2014) Study on compressive properties of syntactic foams for marine applications. J Multifunct Compos 21–27
9. Tien C, Gupta N, Talalayev A (2009) Thermoanalytical characterization of epoxy matrix-glass microballoon syntactic foams. J Mater Sci 44:1520–1527
10. Poveda R, Gupta N, Porfiri M (2010) Poisson's ratio of hollow particle filled composites. Mater Lett 64:2360–2362
11. Gladysz G, Perry B, McEachen G, Lula J (2006) Three-phase syntactic foams: structure property relationships. J Mater Sci 41:4085–4092
12. Jain N, Mittal N (2008) Finite element analysis for stress concentration and deflection in isotropic orthotropic and laminated composite plates with central circular hole under transverse static loading. Mater Sci Eng 498:115–124
13. Gupta N, Woldesenbet E, Kishore, Sankaran S (2001) Studies on compressive failure features in syntactic foam material. JS Sandw Struct Mater 4:249–272
14. Deshpande V, Fleck N (2000) Isotropic constitutive models for metallic foams. J Mech Phys Solid 48:1253–1283

Single Order Multiple Regression Model on Existing Vessel Design Index (EVDI)

Aminatul Hawa Yahaya and Muhamad Hazim Muksan

Abstract This paper utilized a multivariate technique to examine the relationship between the dependent variable with independent variables. The coefficients (weights) are the marginal impacts on each variable, and the size of the weight can be directly interpreted using multiple regressions (MR). Parameter tests (Global test, Multicollinearity test, Coefficient test and the Wald test) were carried out on all the sixty-three possible models. The best model was obtained using the Eight Selection Criteria (8SC). The goodness-of-fit tests were carried out to validate the best model obtained. MR was used to determine the best model that related. The model was used to estimate the value of EVDI™ based on selected significant variables.

Keywords Existing vessel design index (EVDI™) · Multiple regressions (MR) · Eight selection criteria (8SC) · Best model

1 Introduction

Multiple regressions (MR) is used to estimate or predict the variance in an interval dependent, based on linear combinations of interval, dichotomous (binary) or dummy independent variables. Multiple regression analysis (MRA) treats all independent variables (IV) in the analysis as numerical. Numerical variables are interval or ratio scale variables whose values are directly comparable. However, researcher wants to include an attribute or nominal scale variable in MRA.

The Existing Vessel Design Index (EVDI™) is the tools to measure the Carbon Dioxide (CO_2) that vessel emitted during their journey and by using these tools we can measure the energy efficiency of the vessel so that we can increase information flows [1]. The EVDI™ calculated the CO_2 emission per nautical miles that the ship or vessel travel across the ocean.

A. H. Yahaya (✉) · M. H. Muksan
Universiti Kuala Lumpur, Malaysian Institute of Marine Engineering Technology, Lumut, Perak, Malaysia
e-mail: aminatulhawa@unikl.edu.my

C. L. Saw et al. (eds.), *Advancement in Emerging Technologies and Engineering Applications*, Lecture Notes in Mechanical Engineering,
https://doi.org/10.1007/978-981-15-0002-2_6

2 Methodology

2.1 *Multiple Regression (MR) Models*

Multiple regression analysis, a form of general linear modelling is a statistical technique that can be used to analyze the relationship between a single dependent (criterion) variable and several independent (predictor) variables. The objective of regression analysis is to predict a single DV from the knowledge of one or more IV's. In regression equation, intercept is represented by β0. The amount of change in the DV due to the IV is represented by the βj, and also known as a regression coefficient. Using a mathematical procedure known as least squares estimation, β0 and βj can be made such that the sum of squared errors (SSE) of prediction is minimized [2]. The prediction error, the difference between actual and predicted values of the IV, is termed the residual (ε). The specific MR model can be stated as follows:

$$Y_i = \beta_0 + \beta_1 X_{1i} + \beta_2 X_{2i} + \ldots + \beta_k X_{ki} + \varepsilon_i \tag{1}$$

where Yi is the ith value of the considered DV, βj is the jth coefficient that describes the corresponding variable Xj for j = 1, 2,..., k and Xji is the ith value of IV, where i = 1, 2, ..., n and j = 1, 2, ..., k. The εi is the corresponding error term of the ith observation.

According to Gujarati [3] there are eight basic assumptions of MR models as listed below:

- Regression analysis is a linear procedure
- At least one observation in the IV is different
- Mean population error is zero
- Absence of autocorrelation
- Homoscedasticity
- Given variable X and ε are independently distributed
- Sample size > number of IV's
- Errors should be normally distributed for each set of values of the independents

In the development of the MR Models, the four phases of model building procedure are involved [4, 5]. Explanations on these phases are as follows:

Phase 1: All Possible Models

All possible models, N can be calculated by using the formula:

$$N = \sum_{j=1}^{q} \left(C_j^q \right) \tag{2}$$

Where N is the number of possible models generated and q is the number of variables. For example if q = 6, the possible model are

Table 1 Eight Selection Criteria (8SC) for best model identification

AIC: $\left(\frac{SSE}{n}\right)e^{\frac{2(k+1)}{n}}$	RICE: $\left(\frac{SSE}{n}\right)\left(1-\frac{2(k+1)}{n}\right)^{-1}$
FPE: $\left(\frac{SSE}{n}\right)\frac{n+k+1}{n-(k+1)}$	SCHWARZ: $\left(\frac{SSE}{n}\right)(n)^{\frac{2(k+1)}{n}}$
GCV: $\left(\frac{SSE}{n}\right)\left(1-\frac{k+1}{n}\right)^{-2}$	SGMASQ: $\left(\frac{SSE}{n}\right)\left(1-\frac{k+1}{n}\right)^{-1}$
HQ: $\left(\frac{SSE}{n}\right)(\ln n)^{\frac{2(k+1)}{n}}$	SHIBATA: $\left(\frac{SSE}{n}\right)\frac{n+2(k+1)}{n}$

$$\left(C_1^6\right)+\left(C_2^6\right)+\left(C_3^6\right)+\left(C_4^6\right)+\left(C_5^6\right)+\left(C_6^6\right)=63 \quad (3)$$

Phase 2: Selected Models

This phase will be divided into two sub-phases, where two different tests should be carried out to determine the selected model.

Phase 3: Best Model

Identification of the best model should be based on Eight Selection Criteria (8SC) as shown by Noraini et al. [6]. The objective is to determine a model with the lowest value of a criterion statistic. The calculation of the criterion statistics will be based on the Sum of Square Error (SSE), number of estimated parameters and the sample size. Table 1 shows the details of each model selection criteria.

Where, n would be the number of observations, (k + 1): the number of model's parameters and SSE the sum of square of error. The Akaike Information Criterion (AIC) and Finite Prediction Error (FPE) are developed by Akaike [7]. The Generalised Cross Validation (GCV) is developed by Golub et al. [8] while the HQ criterion is suggested by Hannan and Quinn [9]. The RICE criterion is discussed by Rice [10] and the SCHWARZ criterion is discussed by Schwarz [11]. The SGMASQ is developed by Ramanathan [12] and the SHIBATA criterion is suggested by Shibata [13].

Phase 4: Model's Goodness-of-fit

Tests were carried out to investigate the randomness and normality of the models' residuals. Randomness test (Run Test) would be implemented in order to verify the randomness of the residual distribution. For verification of normality, Kolmogrov-Smirnov is used. Besides this quantitative test, graphical output can be used as supporting evidences.

Mean Absolute Percentage Error (MAPE)

MAPE is commonly used in model verification because it produces a measure of relative overall fit. The absolute values of all the percentage errors are summed up

Table 2 The description of variable shows in the model

Variables	Description
Y	Existing vessel design index (EVDI™)
X1	Current vessel age (AGE)
X2	Speed (SPD)
X3	Maximum continuous rating (MCR) of engine
X4	Deadweight tonnage (DWT)
X5	Main engine element (MEE)
X6	Auxiliary engine element (AEE)

and the average is computed. In this study, MAPE is used to verify the best model obtain. It usually expresses accuracy as a percentage, and is defined by the formula:

$$MAPE = \frac{1}{a}\sum_{t=1}^{q}\left|\frac{A_t - F_t}{A_t}\right| \times 100 \tag{3}$$

where A_t is the actual value and F_t is the forecast (estimated) value.

The difference between A_t and F_t is divided by the actual value A_t again. The absolute value of this calculation is summed for every fitted or forecast point in time and divided again by the total number of fitted point's a.

3 Source of Material (Data)

This dataset was a published dataset, previously used in previous research. The ownership of the dataset will be remain anonymous due to mutual agreement by the data owner and researcher. This study includes 200 samples of ship registered in Malaysia, data including vessels information and EVDI^{TM}. Details on some variables will be remain anonymous due to private and confidential terms on the dataset. Few factors which affecting the calculation of EVDI^{TM} of the vessel was chosen and will be analysed. Table 2 shows all of these data in quantitative variables and listed as follows.

4 Analysis and Result

In the development of the MR models for this datasets, Existing Vessel Design Index (EVDI^{TM}) would be the dependent variable (DV) noted by Y, whereas AGE (X_1), SPD (X_2), MCR (X_3), DWT (X_4), MEE (X_5) and AEE (X_6) would be the independent variables (IV). Based on six single IVs, the number of models would be 63.

Pearson Correlation analysis verifies that EVDI™ have a negative correlation with all IV's except the AGE (X_1). There is no existence of multicollinearity between IV's. The Pearson Correlation Analysis also has been carried out to the other 62 possible models and no multicollinearity detected in the models. Thus, no elimination should be made among the independent variables. The next step is to proceed with the Global Test and Coefficient Test to all 63 possible models at significant level less than 0.05. In this paper, model M63 is selected for the illustration purpose.

From ANOVA test indicates F-statistics value as 31.7395, p-value is 1.907 X 10^{-16}, less than 0.05. The decision is to reject the null hypotheses where all regression coefficients are zero. The next the step is to perform the coefficient test for evaluating the insignificant variable for all the coefficients in model M63. For M63, the final model named as M63.0.2, with six variables eliminated.

The Coefficient Test also being applied to all possible models and from the result, the list of selected models can be obtained. From 63 possible models, after Phase 2, only 17 models from each group have been selected. The best model was then chosen from the selected models by using the Eight Selection Criteria (8SC). Noted that M49.0.0 is similar to M63.0.2. By using the Wald test, the complete model (M63) was taken as initial possible model and M63.0.2 as the reduced model.

The complete (C) model (M63):

$$Y = \beta_0 + \beta_1 X_1 + \beta_2 X_2 + \beta_3 X_3 + \beta_4 X_4 + \beta_5 X_5 + \beta_6 X_6 + \varepsilon \tag{4}$$

SSE (C) = 0.5784 and df (C) = 53

The reduced (R) model (M63.0.2):

$$Y = \beta_0 + \beta_1 X_1 + \beta_3 X_3 + \beta_4 X_4 + \beta_6 X_6 + \varepsilon \tag{5}$$

SSE (R) = 0.5178 and df (R) = 47

The value of F critical value from F Distribution Curve is $F_{0.05,6,47} = 2.34$ and the calculated value of F is 0.9182. Since the calculated value of F is less than $F_{0.05,6,47}$, the decision is to accept H_0. The removal of insignificant variables in coefficient test is justified. The final phase of model building is applying the Goodness-of-Fit test on the final best model. Since the significant value = 0.077 is > than 0.05, therefore H_0 is accepted. There is enough evidence at 0.05 significant levels that the standardized residual is normal as in Fig. 1. Finally, the best multiple regression models were given as:

$$Y = \beta_0 + \beta_1 X_1 + \beta_3 X_3 + \beta_4 X_4 + \beta_6 X_6 + \varepsilon \tag{6}$$

By substituting the defined variable, the final model that explained the relationship between EVDI™ and contributing factor is:

$$Y = 20.474 + 0.258\,X_1 - 0.000046\,X_3 - 0.000081\,X_4 + 0.002X_6 + \varepsilon \tag{7}$$

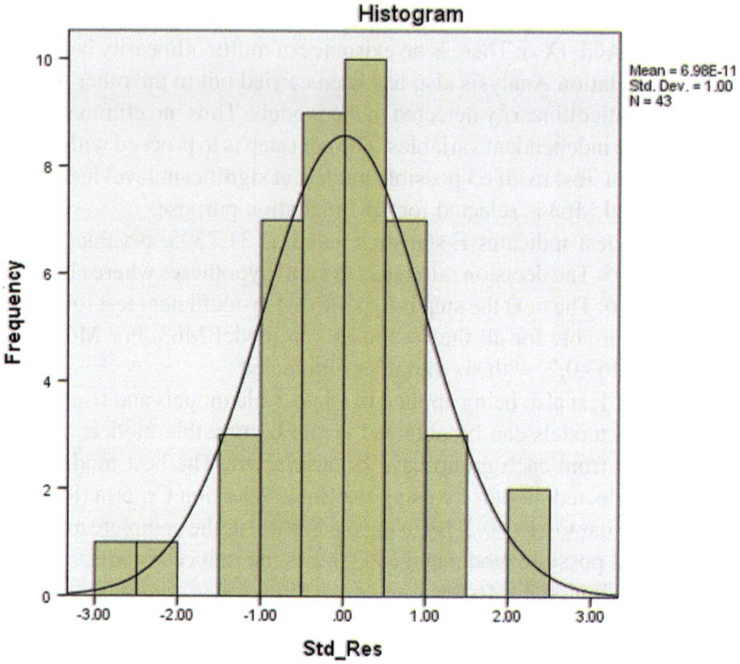

Fig. 1 Histogram with normal curve

The next step is to calculate the MAPE. By taking the remaining observation that has not been included in the model building analysis, the value of MAPE obtained is 6.61%. This value indicates that this model could be best used for estimation of missing value or forecasting.

5 Discussions

In this study, MR is used to investigate the relationships between EVDI^{TM} and other contributing factors in predicting the Green House Gas emission by vessel. By knowing contributing factors in predicting the Green House Gas emission, the model for estimation of EVDI^{TM} can be developed. From the model, it is clear that the vessel age (AGE) giving the highest impact in determining the EVDI^{TM} value. The correlation coefficient between EVDI^{TM} and AGE exhibit a strong positive relationship. AGE is dominant factor that may indicate a significant effect of EVDI^{TM} value. Another factor that contributed in estimation of EVDI^{TM} value are the Maximum continuous rating (MCR), Deadweight Tonnage (DWT) and Auxiliary Engine Element (AEE).

6 Conclusion

The process of model building involves a search for the best way to specify a relationship between a dependent variable and a set of independent variables. In this study, the estimation of the missing value in this work cannot be completed as there are more than one parameter missing in the dataset. Out of six independent parameters, only four parameters give the reading and for this observation the value of the dependent variable is also undefined. For the comparison of this analysis, future work on MRA with variable interaction should be carried out.

References

1. Paolo AT (2014) Energy efficiency and time charter rates: energy efficiency savings recovered by ship owners in the Panamax market. Transp Res Part A: Policy Pract 173–184 (2014)
2. Kutner MH, Nachtsheim CJ, Neter J (2008) Applied Linear Regressions Models. McGraw-Hill, New York
3. Gujarati DN (2006) Essentials of econometrics, 3rd edn. McGraw-Hill Inc., New York
4. Noraini A, Zainodin HJ, Amran A (2011) Improved stem volume estimation using P-Value approach in polynomial regression models. Res J For 5(2):50–65
5. Zainodin HJ, Noraini A, Yap SJ (2011) An alternative multicollinearity approach in solving multiple regression problem. (Academic Journals Inc) Trends Appl Sci Res 6(11):1241–1255
6. Noraini A, Zainodin HJ, Nigel Jonney JB (2008) Multiple regression models of the volumetric biomass. Wseas Trans Math 7(7):492–502
7. Akaike H (1974) A new look at statistical model identification. IEEE Trans Auto Control 19:716–723
8. Golub GH, Heath M, Wahba G (1979) Generalized cross-validation as a method for choosing a good ridge parameter. Technometrics 21:215–223
9. Hannan EJ, Quinn B (1979) The determination of the order of an auto regression. J Royal Stat Society 41(B):190–195
10. Rice J (1984) Bandwidth choice for nonparametric kernel regression. Annals of Stat 12:1215–1230
11. Schwarz G (1978) Estimating the dimension of a model. Ann Stat 6:461–464
12. Ramanathan R (2002) Introductory econometrics with applications, 5th edn. Thomson Learning, South-Western Ohio
13. Shibata R (1981) An Optimal Selection of Regression Variables. Biometrika 68:45–54

A Preliminary Design of an Adjustable Double-Outrigger Stabilizer for Fishing Boat

Norfadhlina Khalid, Muhammad Haziq Zulkifli, Nor Hisham Mustafa and Z. Z. Mukhtar

Abstract An outrigger stabilizer is a developed product attached to a floating structure by means of increasing the stability of the floating structure. The presence of outrigger has a definite effect on the heave, pitch, and roll motion of the craft as compared to the hull without an outrigger. Stabilization is used to counteract the forces of a boat's rolling motion. During fishing activities such as netting, casting or fishing if the wind or waves are strong at the coastal area, it will increase the probabilities of rolling motions on the boat. While using the outrigger, the extra stability comes at the cost of increased weight complication in the setup and the drags are increased with the installation of device. The aims of this study are to design the adjustable double-outrigger stabilizer for the small fishing boat and to determine the stability of the outrigger. The AutoCAD software is used to design a mould of the double-outrigger and the Maxsurf software is used to analyze the stability with the outrigger. As a result, the adjustable double-outrigger stabilizer for the small fishing boat is produced with adjustable arms providing the boat with necessary buoyancy and stability.

Keywords Stabilizer · Outrigger · Boat stability · Maxsurf

1 Introduction

An outrigger stabilizer is a floating mechanism or device that is attached to the side of the boat or canoe which helps to stabilize the boat or canoe from flipping over [1]. Doran (1974) stated that there are only three types of multi-hull water craft defined as hull form which are double canoes, single outrigger and double outrigger [2]. Figure 1 shows the types of outrigger stabilizer.

N. Khalid (✉) · M. H. Zulkifli · N. H. Mustafa · Z. Z. Mukhtar
Universiti Kuala Lumpur Malaysia Institute of Marine Engineering Technology (UniKL MIMET), Lumut, Perak, Malaysia
e-mail: norfadhlina@unikl.edu.my

C. L. Saw et al. (eds.), *Advancement in Emerging Technologies and Engineering Applications*, Lecture Notes in Mechanical Engineering,
https://doi.org/10.1007/978-981-15-0002-2_7

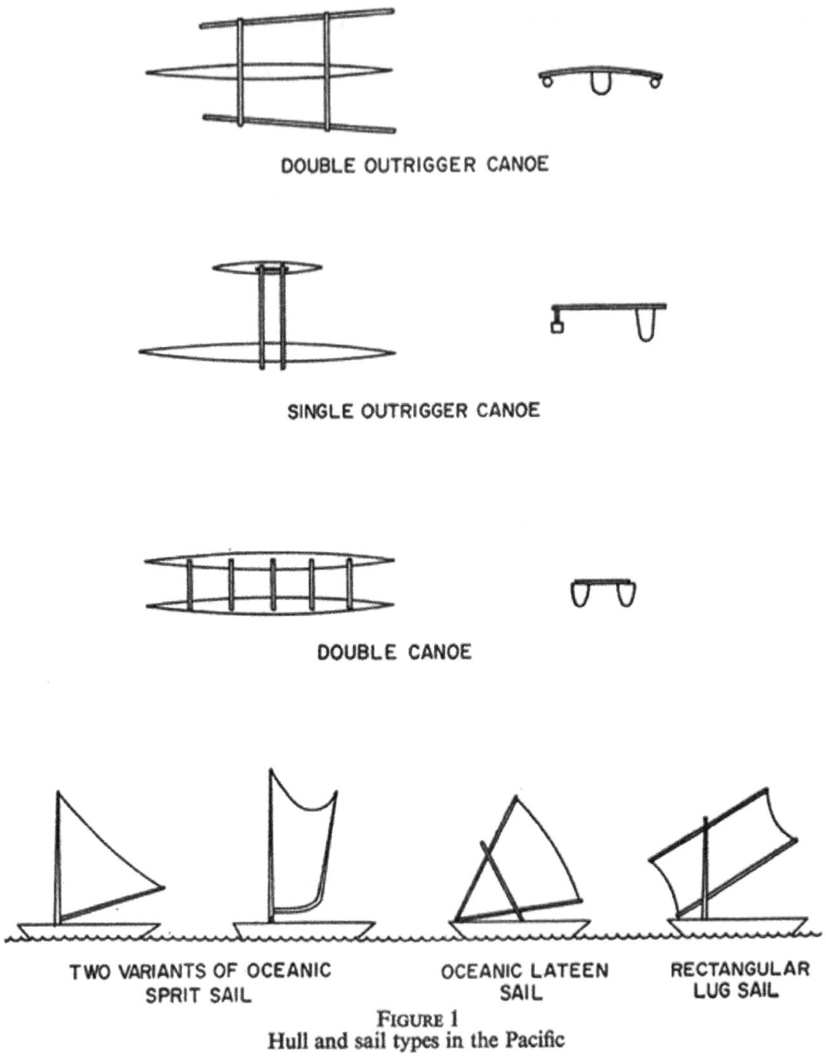

Fig. 1 Types of outrigger stabilizer

The stabilizer is required for small boats. Small boats should be installed with outrigger stabilizers to prevent any accidents such as yawing or rolling [3]. This is important as many accidents involve capsized boats or small fishing boats in the river and the sea due to lack of stability. Most of the current designs of outrigger only have one function which is stability although it can be improved to give more benefits to fishermen. Some of the designs are not multi-functional which pose difficulties to fishermen in performing maintenance on their outrigger for example its fixture on boat, not adjustable, inability to assemble and re-assemble. The aims of this study are thus to design a double-outrigger with adjustable function outrigger for small fishing boats and to determine the stability of the boat using the Maxsurf software.

2 Methodology

The preliminary design of the double-outrigger was designed according to the characteristics that were required for the outrigger which was developed with the dimensions suitable for small fishing boats with a length of around 8 ft. The design of the hull must be narrow at the aft and forward to reach the objective of it being multipurpose where the design must reduce the drag effect. Figure 2 shows the detailed drawing of a double-outrigger stabilizer using AutoCAD.

The next stage is developing the design concept of stabilizer using the Maxsurf Modeller. The input of the surface size of the model such as the length, breadth and depth of the stabilizer were required to develop the actual scale of outrigger structure. Then, the design was exported into the AutoCAD drawing before printing, which was transforming to the actual model scale. The printed of actual scale-body plan scale was needed to develop a part of the outrigger and cut all the parts by pieces on the plastic board. The plugs were made by using plastic board and epoxy which was light, easy to cut and strong enough to keep the structure assembled together [4]. Figure 3 shows the plugging process of outrigger model using plastic board. The glass with resin was used to cover all the plugs, before sanding it. The grey coat

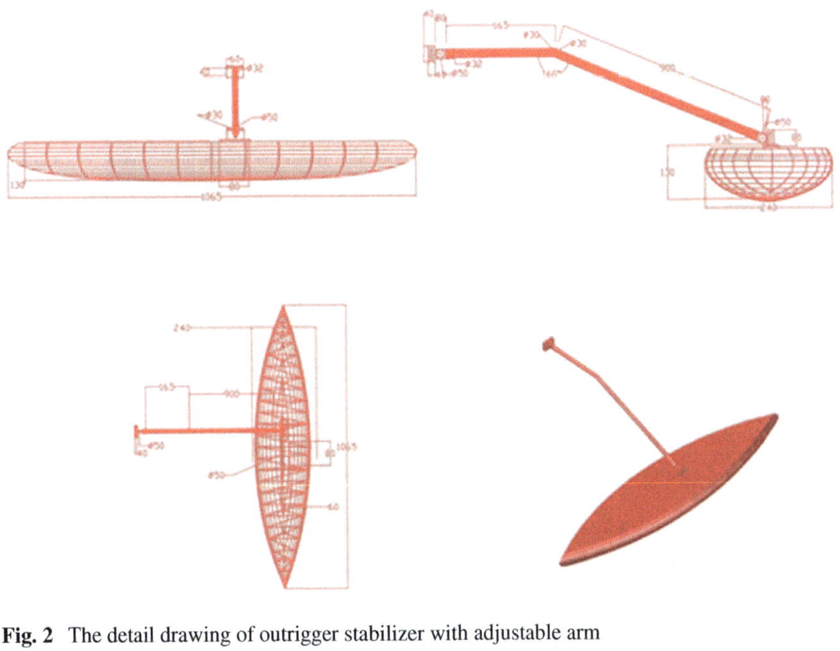

Fig. 2 The detail drawing of outrigger stabilizer with adjustable arm

Fig. 3 Plugging process of the outrigger model structure

was applied before sanding it again. Next the black gelcoat was used and the mould was polished for finishing.

After polishing the mould, the fibre-glass work started by putting a wax inside the mould before applying the resin and fibre mat. The mixing resin with hardener ratio was 2:1 where 2 layers of the fibre mat were applied to produce a strong and light stabilizer hull [5]. Figure 4 shows the lamination process of the mould. The topside decks were developed using a sketch on the board and laminated on the drawing on board. The deck and hull form were different parts. By using the lamination method both of the part can be combined before sanding all over the component.

Fig. 4 Lamination process using fibre and resin

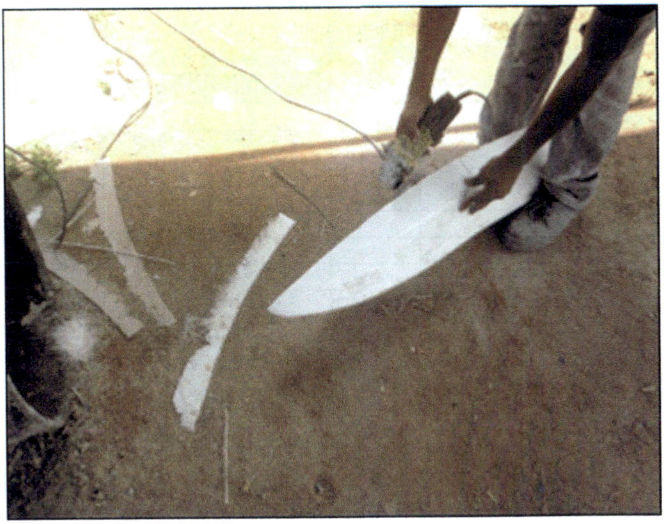

Fig. 5 Sanding the hull form

This was to produce smooth surface before applying the paint. Figure 5 shows the sanding process of the hull form.

3 Results and Discussion

The design has been developed from the Maxsurf Modeller that can be opened straight on Maxsurf stability without inserting all the input data of the structure because the stabilizer itself did not have any structure inside the hull. The information data of the lightship weight was required to get a manual weight scale. Precise calculations for stability were performed directly by using the Maxsurf Stability. All the data were displayed in graphical and tabular form for easy updates of any changes made by running the analysis progress. The integrated load case function was convenient to set up for any situation of the loading condition. The loadcase was used to define a stability result. The result of the boat stability shows load case condition was accepted according to the requirement of the design criteria applicable to boats. Table 1 shows the stability result of the outrigger.

The outriggers have been tested by doing sea trial to determine the boat stability and outrigger. The sea trial consisted of the outrigger stabilizer being attached to the side-hull of the boat using clamp. The adjustable arms were assembled on both the outrigger and these arms are able to pull in/out for adjustment and are also easy to assemble and reassemble in order to reduce space of the jetty area. Figure 6 shows the final product of the outrigger stabilizer with an adjustable arm.

Table 1 The stability result of double-outrigger stabilizer using Maxsurf software

Particulars	units	Loadcase Condition 100%	Requirements
Lightship	kg	2.4	
Deadweight	kg	2.4	
Displacement	kg	2.4	
Draft FP	meter	0.043	
Draft AP	meter	0.043	
Angle of Trim (+ve stern -ve bow)	degree	0	
LCB from zero pt. (+ve fwd)	meter	0.55	
LCF from zero pt. (+ve fwd)	meter	0.552	
Area from 0 to 30 degree	m.degree	0.4071	shall not less than 0.1513
Area from 0 to 40 degree	m.degree	0.4371	shall not less than 0.1566
Area from 30 to 40 degree	m.degree	0.0233	shall not less than 0.0189
Max. GZ at 30 degree or greater initial GMt	meter	0.026	shall not less than 0.020
Initial GMt	meter	0.055	shall not less than 0.050

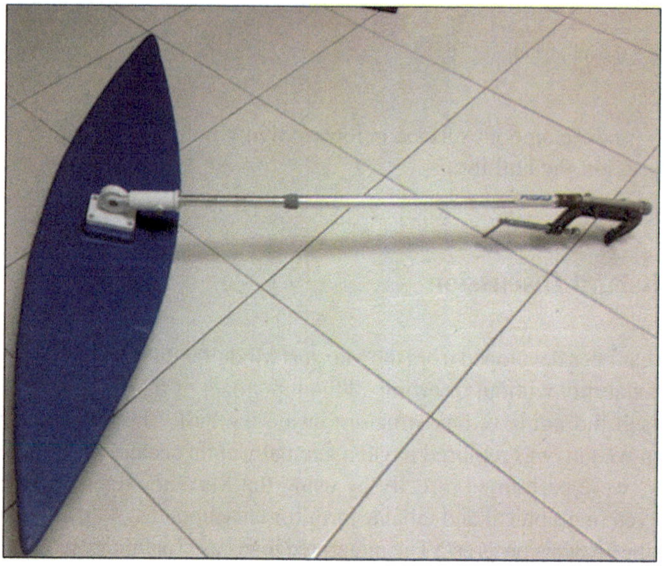

Fig. 6 The final prototype of outrigger stabilizer with an adjustable arm

4 Conclusion

In this study, a preliminary design of an adjustable outrigger stabilizer for fishing boat has been explained in order to develop the outrigger prototype and determine the boat stability. Small fishing boats can avoid the rolling effect motion by attaching the stabilizer outrigger device. This study was mooted by numerous cases about tipping accidents on small boats which caused the boat to capsize or flipover. The design of the stabilizer outrigger provides an adjustable function which is easy to assemble and disassemble with the adjustable rod. The user can simply adjust the length of stabilizer on the side from their boat hull. As a result, an adjustable double-outrigger stabilizer for small fishing boat is produced with an adjustable arms and the result of the boat stability shows load case condition is accepted according to the requirements of the design criteria which are applicable to the boats.

References

1. Abramovitch D (2003) The outrigger: a prehistoric feedback mechanism. In: Proceedings of the 42nd IEEE conference on decision and control, Maui, Hawaii, USA, December 2003, p 2000
2. Edwin JD (1974) Outrigger Ages. J Polyn Soc 83:130–140
3. Hornell J, Outrigger devices: distribution and origin. F.L.S., F.R.A.I. director of fisheries, Madras Government (Retired). J Polyn Soc 52(3):91–100
4. Tarvainen K, Pfäffli P, Estlander T, Suuronen K (2012) Boat builders. In: Kanerva's occupational dermatology. Springer, Berlin, Heidelberg, pp 1303–1307
5. Shamsuddin MZ, Ragnarsson E (2003) A conceptual design of a fibre reinforced plastic fishing boat for traditional fisheries in Malaysia. Published in Fisheries Training Programme, The United Nations University, pp 1–53

Experimental Analysis of FRP Confined Concrete for Underwater Application

Z. Z. Mukhtar, A. Abu Bakar, A. Fitriadhy, M. S. Abdul Majid and Asmalina Mohamed Saat

Abstract In marine industry, current practice shows that composite materials already being used in a number of marine structures such as high and low-pressure tubing, bridge and jetty as well as accommodation modules for offshore structures. Fiber Reinforced Plastic (FRP) confined concrete has been widely accepted in the inland construction technology as a way to reduce cost. This method seems feasible as steel structure can be filled by concrete and confined by FRP for underwater application. This study concentrates on FRP confined concrete cylindrical column specimen under axial compressive loading and Vacuum In Fusion method has been applied for FRP confinement process. The testing results showed that cylindrical column which is properly confined by FRP confinement can achieve high levels of strength and ductility if compared to those of plain concrete. Results confirmed that external confinement produced by FRP can significantly enhanced compressive strength, ductility and energy absorption capacity. The highest compressive strength is 29.32 Mpa for 1.5 mm FRP confinement and it is about 56% higher than compressive strength of specimen without FRP confinement. Stress-strain relationship, ultimate strength and ductility of specimens are analysed in detail based on experimental results.

Keywords FRP confined concrete · Marine structure · Axial compressive · Ductility · Ultimate strength · Stress Strain

Z. Z. Mukhtar (✉) · A. M. Saat
Malaysian Institute of Marine Engineering Technology, Universiti Kuala Lumpur, 32200 Lumut, Perak, Malaysia
e-mail: zaimi@unikl.edu.my

A. Abu Bakar · A. Fitriadhy
School of Ocean Engineering, Universiti Malaysia Terengganu, 21030 Kuala Nerus, Terengganu, Malaysia

M. S. Abdul Majid
School of Mechatronic, Universiti Malaysia Perlis (Pauh Campus), 02600 Arau, Perlis, Malaysia

C. L. Saw et al. (eds.), *Advancement in Emerging Technologies and Engineering Applications*, Lecture Notes in Mechanical Engineering,
https://doi.org/10.1007/978-981-15-0002-2_8

1 Introduction

In the past few decades, concrete is most frequent element that was used in the construction industries. It is known with its ability to withstand a huge amount of compression and a very flexible element to fight against most of weather condition. Fibre Reinforced Polymer has been recently popular in concrete strengthening. During this era the undertaking of FRP composite as a common construction material has rapidly rise and including the studies. Due to a significant result in this topic, the uses of FRP is gaining more frequent and popular in order to strengthening a structure. FRP is a composite material that made of the combination of polymer elements reinforced with fibre. The polymer usually used in construction are thermosetting plastic, vinylester and epoxy. Another reason for FRP keeps growing over the recent years due to increase application of composite. This includes the research of high performance resin system and new types reinforcement. It's even being considered as enhancement or upgrading to alternative for infrastructure components.

Composite material is an element that was engineered together or a combination of two elements to form a new types of elements by using chemical or physical fusion. Composite structure mostly have strong ductility and durability. Commercial material commonly has glass or carbon fibres in matrices based on thermosetting polymers, such as epoxy or polyester resins. Sometimes, thermoplastic polymers may be preferred, since they are moldable after initial production. There are further classes of composite in which the matrix is a metal or a ceramic. For the most part, these are still in a developmental stage, with problems of high manufacturing costs yet to be overcome. Furthermore, in these composites the reasons for adding the fibres (or, in some cases, particles) are often rather complex; for example, improvements may be sought in creep, wear, fracture toughness, thermal stability, etc.

In marine industry, current practice shows that composite materials already being used in a number of marine structures such as high and low-pressure tubing, bridge and jetty as well as accommodation modules for offshore structures. Hence, it is important to prolong the age of the marine structure by using proven technology. Fiber Reinforced Plastic (FRP) confined concrete has been widely accepted in the inland construction technology as a way to reduce cost. This method seems feasible as offshore steel structure space can be filled by concrete and confined by FRP.

Seica and Parker [1] discussed the experimental results to investigate Carbon-FRP strengthening of an artificially degraded steel beam of circular cross-section (Fig. 1a) under four-point loading. Finally, the steel beam in bending, although it was expected to reach the plastic moment, it only reach 94% of it. However, all the CFRP wrapped specimens did reach the plastic moment and also showing increased ductility. It was concluded that the use of CFRP composites to enhance the strength of tubular steel in underwater application is perfectly feasible.

Marwan et al. [2] presented an experimental study on the axial compressive strength of square and circular concrete column strengthen with FRP wrap (Fig. 1b). According to the test result, it was shown that the FRP wrap increase the strength

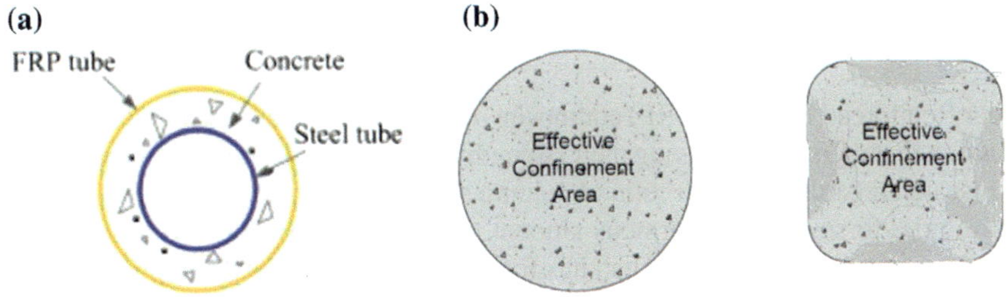

Fig. 1 **a** Bending test specimen. **b** Compression test specimen

and ductility significantly for circular concrete column. However, the FRP wrap did not increase the strength of square concrete column.

In FRP confined concrete column design, the main objective is to develop a confinement model to predict the ultimate strength and strain of the confined columns. Nowadays, the confinement models can be divided into 2 categories. The first category is design oriented model where the model is directly based on interpretation of test results. The second one is analysis-oriented model where the model is based on incremental numerical procedures by accounting explicitly for equilibrium and radial displacement compatibility. The design-oriented model can be expressed as [3–15]:

$$\frac{F'_{cc}}{F'_{co}} = \left[1 + k\frac{(f_r)^m}{(f'_{co})^m}\right]' \tag{1}$$

where F'_{cc} is the compressive strength of the confined concrete, F'_{co} is the compressive strength of unconfined concrete, f_r is the lateral confining pressure, $\left(F'_{cc}/F'_{co}\right)$ is the confinement effectiveness, $\left(F'_r/F'_{co}\right)$ is the confinement ratio, k is the effectiveness coefficient and m is the power coefficient of the confinement ratio. The f_r can be found by using the following formula;

$$f_r = \frac{2f_{frp}t}{D} \tag{2}$$

where f_{frp}, t and D are the hoop tensile strength and wall thickness of the FRP shell and diameter of the concrete core itself. It is noted that almost all the design oriented confinement models are based on normal strength concrete with sufficient confinement and there is a needs to determine the structural behaviour of FRP confinement for concrete cylinders. There are some important factors also need to be evaluated as follows:

1.1 Shape of Effects

Fiber Reinforced Polymer (FRP) composite materials are widely used to wrap compression members for improving their strength and ductility. There were tremendous of research regarding the selection of column shape to be wrapped by FRP composite in order to sustain an enormous amount of load to be distribute to the ground. A research from University of Utah has found that the use of FRP composite straps or hoops is beneficial for columns with a circular cross-section but not for columns with a square or rectangular cross section. In the case of using FRP composite prefabricated shells, it is advantageous to use expansive cement concrete rather than non-shrink grout to achieve the shape modification. Square columns, which were modified to circular using FRP composite prefabricated shells and expansive cement concrete. It showed a higher increase in axial strength capacity and ductility, compared to square columns, which were modified to circular using FRP composite prefabricated shells and non-shrink grout [16].

The behavior of FRP-confined concrete in circular cylinder has been extensively studied, and therefore researchers have upturned their attention to FRP-confined concrete in rectangular columns. The ultimate strength and ductility of FRP-confined concrete increase significantly with the enhancement of the number of composite layers and the confinement level. Existing strength models of FRP-confined concrete in rectangular columns were reviewed and evaluated by data collected from literatures. And then improved model was proposed for more accurate estimation of the ultimate strength of the FRP-confined concrete in rectangular columns. An improved model has been presented and shown to provide satisfactory predictions of test results, which can serves as a reference for retrofitting rectangular concrete columns with FRP composite jackets [17].

1.2 Fiber Properties

Almost all the columns are subjected to an eccentric axial load, which can be solved into a co-axial compressive load and a bending moment. Because of this, almost all the columns should be treated as beam-columns. Fiber orientation is an important variable in the structural design of FRP wrapped concrete columns. The fiber was oriented to 0°, 45°, and 90° from the loading axial load direction. It is found that the strength, ductility, and failure mode of FRP wrapped concrete cylinders depend on the fiber orientation and wall thickness. Fibers oriented at a certain angle in between the hoop direction and axial direction may result in strength lower than fibers along hoop or axial direction. In hypothesis, the lower the degree of FRP orientation wrapping from the loading direction, the higher its ability to withstand load [18].

Some researchers have done a factorial study upon tensile strength of coir reinforced epoxy composite. In their study, the volume fraction, curing time and compression load during the solidification of composites were taken as parameters. From the results, they concluded that volume fraction influences the tensile strength of the composites [19]. Authors also increased the percentage volume fraction of fiber and found that the tensile properties of composites increased to some extent. Curing time also showed some effects on the characteristics of composites meanwhile the influence of compression load on the properties of composites were not revealed properly.

1.3 Vacuum In-fusion Process

Vacuum infusion process (VIP) takes a different approach, in that a vacuum is drawn while the materials are still dry. From that point, resin is infused using vacuum pressure. Ideally, any excess resin that is introduced will eventually be sucked out into the vacuum line. As a result, only the minimum amount of resin is introduced. This lowers weight, increase strength, and maximizes the properties of fibre and resin. Due to the nature of VIP, resin usage becomes very predictable. While a standard lay-up varies in resin content due to the human variable, VIP is remarkably consistent. Even when creating a large product, resin usage will be predictably similar upon repeated attempt.

Vacuum infusion provides another valuable benefit which is time. Many resins have a pot-life of about 30 min, though there are certainly some that offer work times of up to 2 h. Large projects can easily approach the 2 h mark, and even small, seemingly simple projects can quickly turn frantic when e pesky leak in the vacuum seal cannot be found. Also, depending on when the bag is applied, the amount of resin removed can vary from part to part. Vacuum infusion offers unlimited set-up time. Because the vacuum is applied while reinforcements are still dry, there is no resin clock to work against [20].

After the bag is applied, leaks can be patiently sought out. If something is not sitting properly, simply release vacuum and readjust. No time constraints are introduced until it is decided that it is time to infuse the resin. Until that moment, changes can be made again and again. Finally, vacuum infusion is a much cleaner process. There are no brushes or rollers, and therefore no splashing or spattering. No one will be required to hover over an open mould, saturating a laminate by hand, trying not to drip on himself. In addition, the VIP provides a cleaner, safer, and friendlier work environment. Though it is still important to work in a well-ventilated area and use other appropriate safety equipment.

2 Test Specimen and Experimental Setup

2.1 Test Specimen

This experimental exercise has been divided into two phases; (1) To prepare the concrete block complete with FRP confinement and steel reinforcement, (2) To conduct the mechanical compressive testing for all specimen in order to determine their mechanical properties. The preparation of concrete cylinder is based on ACI Standard 211.1 and the grade of concrete cylinder block produced was A20 with size of 150 mm in diameter and 300 mm in length. The concrete was cast, compacted, finished, demolded, and cured for 28 days in an ASTM standard curing room with 100% relative humidity. Two (2) bared concrete cylinder have been prepared without any confinement to ensure that the standard compressive strength value for each concrete mixture for all specimen cylinder is similar for A20 grade. After 28 days of curing time, the cylinder blocks are confined with Glass-Fiber Reinforced Plastic (GFRP) by using Vacuum In Fusion method.

The general setup of resin in fusion method are shown in Figs. 2 and 3 which are using the same principle of vacuum pressure, used to drive resin into a laminated layer of fiber mat. For this study, CSM450 fibre mat has been used and thicknesses of GFRP confinement are 0.3, 0.5, 0.7, 1.0 and 1.5 mm. For FRP in-fusion process, vacuum packing method was applied. Vacuum packing is a method of packaging that removes air from the packaging prior to sealing of concrete cylinder. This will ensure there is no air bubble inside the plastic vacuum packaging. Once the process was completed, inlet and outlet hoses were clamped and resin in-fusion process will begin. It will take about 8 h to complete the GFRP confinement process.

2.2 Mechanical-Compression Test

Mechanical-Compression Test were done by using KENCO 3 MN machine at Concrete Lab of Civil Engineering Technology Department, Universiti Malaysia Perlis (UniMAP). The test was performed according to ASTM C39 standard. This is a testing standard for concrete cylinder to determine compression strength, stress-strain relationship, force-displacement relationship, energy generation and etc. The standard test specimen with size of Ø150 mm × 300 mm has been used for the test. Strain gauge was set at the body of concrete cylinder to determine the displacement of the specimen during testing. Please refer to the Fig. 4 for preparation of specimen of mechanical-compression test.

Fig. 2 Concrete mixture preparation and moulding

3 Result and Discussion

From the experimental data, there were 6 number of specimens have been tested. One (1) specimen was tagged as a 'control' concrete cylinder and another 5 specimens were infused for GFRP confinement with thickness of 0.3, 0.5, 0.7, 1.0 and 1.5 mm. Data for specimen with 0.7 mm did not take into account due to insufficient confined cylinder due to incomplete curing time during in-fusion process. The sufficient or insufficient confinement can be quantified by Spoelstra and Monti's confinement ratio criterion [21]. According to Spoelstra and Monti's, if the confinement ratio is smaller than 0.07, the confinement is insufficient and it will generate unaccurate test result. The confinement ratio can be determined as f_r/f'_{co}, and f_r can be calculated by using Eq (2).

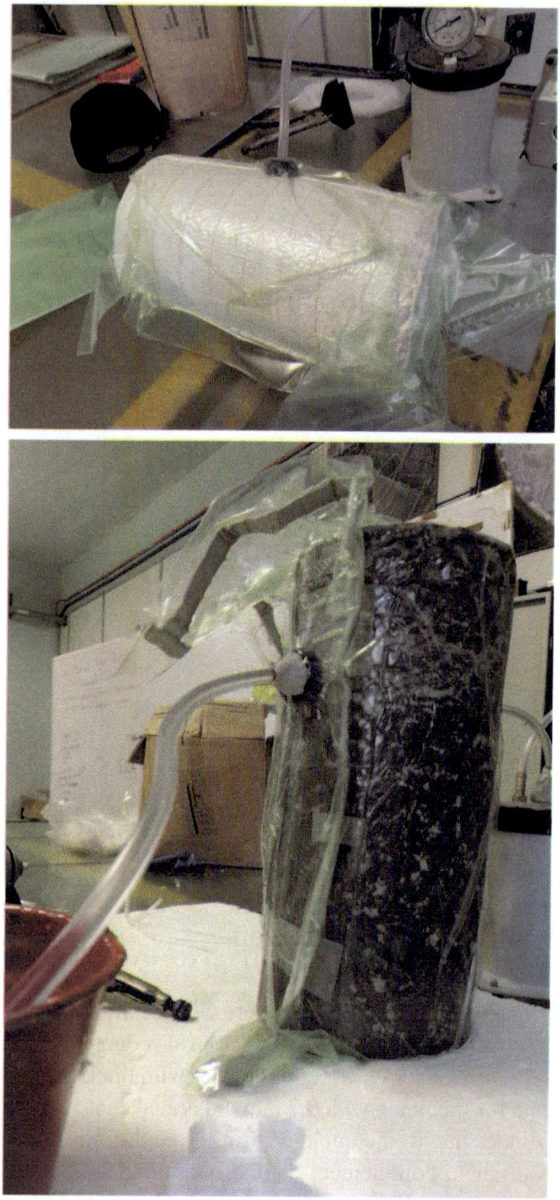

Fig. 3 FRP confinement by using in-fusion process

Fig. 4 Experimental setup

3.1 Stress Versus Strain Relationship

The result above shows the comparison of all thickness in grade A20 concrete in one graph of stress versus strain. By combining all the graph together, the trend of the graph can be easily noticed which shown the strength of the specimens are differs by the thickness of the confinement. The graph shows the highest compressive strength of the specimen is the specimen with 1.5 mm of thickness confinement at the top ultimate tensile strength of 29.32 Mpa. It follows by thickness of 1 mm at top compressive strength of 24.47 Mpa. The blue line indicates the compressive strength for the thickness of 0.5 mm at the value of 22.98 MPa and finally the thinnest confinement which is 0.3 mm where the ultimate compressive strength is only at 20.65 Mpa. The highest strain value can be determined at 0.0149 for concrete cylinder with 1.5 mm thickness and the lowest strain is 0.0136 for specimen with 0.3 mm confinement thickness as shown in Fig. 5.

3.2 Test Database of FRP-Confined Concrete Cylinders

A total number of 5 specimens were tested and the data from the experiment is presented as follows:

From Table 1, we can see the trend of incremental of compressive strength for FRP-confined concrete proportionate to GFRP confinement thickness. With confinement

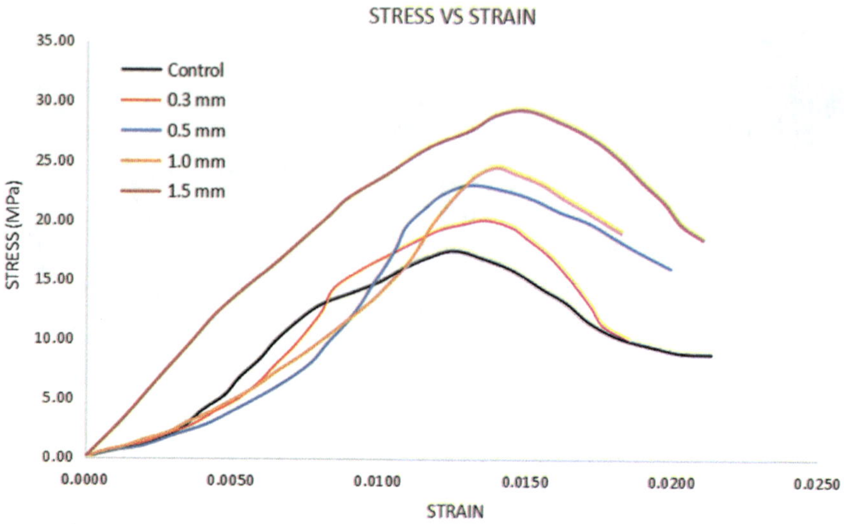

Fig. 5 Stress versus strain graph for concrete grade A20

thickness of 0.3 mm, the compressive strength of concrete column is increased from 18.79 to 20.9 Mpa and it almost 10% increment and it also happens for 0.5 mm thickness, where the compressive strength increases from 18.79 to 22.98 MPa. The incremental percentage of strength is about 22%. The highest incremental of strength due to FRP confinement is involving 1.5 mm thickness of GFRP confinement. The compressive strength is increased from 18.79 Mpa to 29.32 Mps and it is about 56% increase from it's original strength. The application of vacuum in fusion method during FRP confinement process for concrete cylinder specimen is very effective and it also can reduce manufacturing defect such as uneven laminating thickness and surface finish. For strain enhancement ratio, Ɛcu/Ɛco, the range of incremental is about between 1.088 and 1.192 for all thicknesses of GFRP confinement from 0.3 to 1.5 mm. The average strain value recorded for all FRP confined concrete specimens is 0.0139.

4 Conclusion

This study was focused on the compression test result and analysis, the following conclusion are obtained. Firstly, insufficient confinement must be avoided in practice during in-fusion confining process. Otherwise it may lead to premature structural failure during experimental testing and inaccurate result will be produced. Application of vacuum infusion method will reduce manufacturing failure during mechanical testing.

Table 1 Summary of experimental result

Spec No.	Specimen dimension		FRP		Concrete properties		Measured ultimate condition		Strength enhancement ratio	Strain enhancement ratio
	D (mm)	H (mm)	T frp (mm)	F frp (MPa)	F'_{co} (MPa)	Ɛco (%)	F'_{cc} (MPa)	Ɛcu (%)	F'_{cc}/F'_{co} (MPa)	Ɛcu/Ɛco
1	150	300	0	–	18.79	0.0125	–	–	–	–
2	150	300	0.3	363	18.79	0.0125	20.6	0.0136	1.096	1.088
3	150	300	0.5	363	18.79	0.0125	22.98	0.0131	1.222	1.048
4	150	300	1.0	363	18.79	0.0125	24.47	0.014	1.302	1.12
5	150	300	1.5	363	18.79	0.0125	29.32	0.0149	1.560	1.192

Secondly, by analysing all the specimens for compression testing, it can be concluded that the highest compressive strength for FRP-confined concrete is occured to thickest GFRP confinement at 1.5 mm at 29.32 Mpa. The enhancement strength of this specimen is up to more than 50% from original bare concrete cylinder compressive strength which is 18.79 Mpa.

Thirdly, a specimen with lower thickness such as 0.3 mm and 0.5 mm achieve up to 20.6 MPa and 22.98 Mpa for their compressive strength. It mean that these two (2) specimens have lower toughness and more brittle compared to 1.0 and 1.5 mm specimens.

Finally, the analysis of the results also indicate that the confinement of GFRP on concrete cylinder column is very effective and practical in order to increase its strength and pro-long the life cycle of the material for underwater application. It will enhance the concrete strength, deformability and ductility significantly.

References

1. Seica MV, Parker JA (2007) FRP materials for the rehabilitation of tubular steel structure for underwater application. Compos Struct 80(3):440–450
2. Yousef MN, Feng MQ, Mosallam AS (2007) Stress-strain model for concrete confined by FRP composite. Compos Part B: Eng 38:614–628
3. Lam L, Teng JG (2003) Design-oriented stress-strain model for FRP-confined concrete. Constr Build Mater 17:471–489
4. Lam L, Teng JG (2002) Strength models for fiber-reinforced plastic-confined concrete. J Struct Eng 128:612–623
5. Samaan M, Mirmiran A, Shahawy M (1998) Modeling of concrete confined by fiber composites. J Struct Eng 124:1025–1031
6. Toutanji HA (1999) Stress-strain characteristics of concrete columns externally confined with advanced fiber composite sheets. ACI Mater J 96:397–404
7. Richart FE, Brandtzaeg A, Brown RL (1928) A study of the failure of concrete under combined compressive stresses. Engineering experimental station bull, vol 185. Urbana (IL), Univ. of Illinois
8. Newman K, Newman JB (1974) Failure theories and design criteria for plain concrete. In: Proceeding of international civil engineering mathematics conference. on struct. (Solid mech. And engrg.des). Wiley Interscience, New York, pp 936–995
9. Cusson D, Paultre P (1995) Stress-strain model for confined high strength concrete. J Struct Eng 121:468–477
10. Karbhari VM, Gao Y (1997) Composite jacketed concrete under uniaxial compression-verification of simple design equations. J Mater Civ Eng 9:185–193
11. Fardis MN, Khalili HH (1982) FRP-encased concrete as a structural material. Mag Concr Res 34:191–202
12. Miyauchi K, Inoue S, Kurota T, Kobayashi A (1999) Strengthening effects of concrete columns with carbon fiber sheet. Trans Jpn Concr Inst 21:143–150
13. Cheng H-L, Sotelino ED, Chen W-F (2002) Strength estimation for ERP wrapped reinforced concrete columns. Steel Compos Struct 2:1–20
14. Saafi M (2000) Design and fabrication of FRP grids for aerospace and civil engineering applications. J Aerosp Eng 13:144–149
15. Lin H-J, Liao C-I (2004) Compressive strength of reinforced concrete column confined by composite material. Compos Struct 65:239–250

16. Pantelides CP, Gergely I, Reaveley LD (1999) Composite retrofit design for R/C bridges. In: Proceeding of the transportation research board 78th annual meeting, TRB, Washington, D.C., Jan. 10–14
17. Riad B, Habib M, Nash EC (2010) FRP-confined concrete cylinder: axial compression experiments and strength model. J Reinf Plast Composites 30(16):2469–2488
18. Li G (2006) Experimental study of FRP confined concrete cylinders. Eng Struct JS 28:1001–1008
19. Romli IR, Ahmad NA, Azmin SR (2012) Factorial study on the tensile strength of a coir fiber-reinforced epoxy composite. AASRI Procedia 2:242–247
20. Yuhazri MY, Amirhafizan, MH, Sihombing H (2016) The effect of various weave designs on mechanical behavior of lamina intraply composite made from kenaf fiber yarn. In: IOP conference series. Mater Sci & Eng 160(012021)
21. Spoelstra MR, Monti G (1999) FRP-confined concrete model. J Compos Constr 3:143–150

Scaling of RV Athena's Appendage Drag Using CFD

Iwan Mustaffa Kamal, Muhammad Safwan Rasahidan and Yaseen Adenan Ahmed

Abstract The drag or the resistance contribution of appendages such as brackets, skegs, shafts, rudders and fin stabilizers fitted to R/V Athena were investigated. The issue in estimating the appendage drag is that laminar flows occurred at lower Reynolds number typically in ship model used for ship hydrodynamic testing in a towing tank. In full-scale ship, the flow is usually turbulent. Uncertainties when scaling the model result to full-scale may arise as the flow regime in both cases are not similar, hence the full-scale estimate will be inaccurate. One method to resolve this issue is to scale the appendage drag using a fixed fraction which is known as the 'Beta' approach. The aim of this investigation was to determine the value of beta. R/V Athena was chosen as the case study in this investigation. The results obtained from SHIPFLOW 6.3 shows that the drag in model scale increased at 14% when fully appended. The drag in full-scale increased at 11% when fully appended. It was observed that the skeg contributed to the highest percentage of drag at about 7–8% of the total drag. Finally, the value of beta was estimated to be in between 0.38 and 0.39.

Keywords Reynolds scale effect · Appendage drag · CFD · Beta approach

1 Introduction

1.1 Ship Appendages

Ship appendages are external fittings fitted outside the main hull, such as shaft supporting brackets, struts, bossings, rudders, shaftings, skeg, bilge keels and control surfaces i.e. stabilizer fins etc. These appendages alter the flow around the hull and these result with an increase of drag to the total ship drag. The flow around a body or in this case a ship appendage can have two different flow regimes; it can be laminar

---

I. Mustaffa Kamal (✉) · M. S. Rasahidan · Y. A. Ahmed
Universiti Kuala Lumpur Malaysian Institute of Marine Engineering Technology, Lumut, Malaysia
e-mail: iwanzamil@unikl.edu.my

C. L. Saw et al. (eds.), *Advancement in Emerging Technologies and Engineering Applications*, Lecture Notes in Mechanical Engineering,
https://doi.org/10.1007/978-981-15-0002-2_9

or turbulent depending on the appendage length and the velocity of the flow around the appendages.

For full-scale appliances where in this case of a full-scale ship, the flow is usually fully turbulent. However, when considering model-scale ship used in hydrodynamic testing in a towing tank, the flow around the ship model can be in laminar flow or transitional. Uncertainties when scaling the model test results to full-scale may arise as the flow regime in both cases are not similar, hence the full-scale estimate will be inaccurate. Therefore this study investigated the appropriate appendages scaling method in addressing this issue.

Molland et al. [1] reported that there is only a small amount of actual data on scaling of appendages. One of the published work which deals with the appendage scale effect can be found in Allan [2]. Allan [2] presented the results of a study of both shafting with struts and bossings for three models, 18.70, 22.96, and 33.67 ft. in length, over a range of speed-length ratios from 0.50 to 0.80. Other published work can be found in Lackenby [3], where the British Ship Research Association (BSRA) in the 1950s, conducted a series of resistance experiment using a jet propelled 58 m ship called '*Lucy Ashton*', which is fitted with various appendages. The resistance measurement was made possible by mounting the jet engine to the hull on a load transducer, which allowed the direct measurements of thrust, therefore the resistance of the ship. These sea trial results were compared with six geosim models tested at the National Physics Laboratory (NPL). The results of the appendage drag on '*Lucy Ashton*' can be found in van Manen and Oossanen [4]. From the results it seems that at lower speed, the scale effect on the appendage drag is not apparent as at higher speed. A conclusion was made by Molland et al. [1] that the results from the '*Lucy Ashton*' experiment and other such tests as in Allan [2] tends to be inconclusive.

1.2 Estimating the Appendage Drag

As recommended in ITTC [5] in the ITTC—Recommended Procedures and Guidelines for ITTC1978 Performance Prediction Method, Procedure no. 7.5-02-03-01.4, the full-scale appendage drag coefficient need to be added as C_{AppS} in the full-scale extrapolation of the total drag coefficient C_{TS} as in Eq. 1.

$$C_{TS} = C_F(1 + k) + C_R + \Delta C_F + C_A + C_{AAS} + C_{AppS} \tag{1}$$

where C_F is the frictional drag coefficient of the ship according to the ITTC1957 model-ship correlation line, k is the form factor determined from the resistance test, C_R is the residual drag coefficient calculated by subtracting the frictional drag coefficient from the total drag coefficient C_{TS}, ΔC_F is the roughness allowance, C_A is the correlation allowance, where C_A is determined from comparison of model and sea-trial results and C_{AAS} is the air resistance coefficient in full-scale.

According to Molland et al. [1] there are four primary methods in estimating the appendage drag coefficient C_{AppS}. The four primary methods as listed in Molland et al. [1] are as the followings:

1. Testing the hull model with and without appendages, where the difference of the model total drag coefficient C_{TM}, with and without the appendages represents the appendage drag as shown in Eq. 2.

$$C_{AppM} = C_{TM(appended)} - C_{TM(barehull)} \tag{2}$$

where C_{AppM} is the appendage drag obtained from subtracting the bare-hull total drag coefficient $C_{TM\,(bare\text{-}hull)}$ from the appended total drag coefficient $C_{TM\,(appended)}$. This appendage drag is then scaled up to full-scale using Eq. 3.

$$C_{AppM} = C_{AppS} = \frac{R_{AppM}}{\frac{1}{2}\rho_{fw} S_{AppM} V_M^2} = \frac{R_{AppS}}{\frac{1}{2}\rho_{sw} S_{AppS} V_S^2} \tag{3}$$

where R_{App} is the appendage drag in Newton, ρ_{fw} *and* ρ_{sw} is the specific density of fresh water and seawater respectively, S_{App} is the wetted surface area of the appendages and V is the model or the ship speed in m/s. The subscript M and S refer to model-scale and full-scale respectively. The scaling of the appendage drag can be re-written as in Eq. 4, where λ is the scale ratio of the linear dimensions of the model and ship.

$$\frac{R_{AppS}}{R_{AppM}} = \frac{\rho_{sw}}{\rho_{fw}} \cdot \lambda^3 \tag{4}$$

2. The 'Beta' approach, where in this approach a beta factor is used. The beta factor is derived from a geosim set of appended models of varying scales. The scaling using the 'beta' approach is defined as in Eq. 5.

$$C_{AppS} = (1 - \beta)\, C_{AppM} \tag{5}$$

3. Test a larger separate model of the appendage at higher speeds. With large models of the appendages and high flow speeds, for example in a towing tank, higher Reynolds numbers, closer to full-scale values can be achieved.
4. Use of empirical data and equations derived from earlier model tests which can be found in Lasky [6], Peck [7] and Kirkman and Kloetzli [8]. Some examples of the application of these empirical data and equations can be found in Mustaffa Kamal [9].

There are some issues in testing the hull model with and without appendages as mentioned in (1) above. The assumptions that the appendage drag coefficient for a ship C_{AppS} is similar to its corresponding model C_{AppM} is simply not true as there are scale effects present, where the appendage drag coefficient is smaller for a ship than for its corresponding model.

1.3 Appendage Scaling Using the Beta Approach

One solution in estimating the full-scale appendage drag is to scale the appendage drag coefficient using 'beta' approach as recommended by ITTC [5]. As mentioned above in Sect. 1.2 in (2), this beta factor is derived from a geosim set of appended models of varying scales tested in a towing tank. In the present case, SHIPFLOW 6.3 based on the Reynold's Average Navier Stokes Equation (RANSE) was used to derive the beta factor instead of using a geosim set of appended models tested in a towing tank.

Some recent study on the 'Beta' approach can be found in Oliva Remolà et al. [10]. In Oliva Remolà et al. [10], the scaling of appendages using the 'beta' method was explored. In his work, Oliva Remolà et al. [10] validated the 'beta' method using experimental through towing tank test and computational tools using a RANSE CFD code. However due to the existence of flow separation, reliable numerical results could not be obtained using CFD. The relative difference of the drag in full-scale between extrapolated value and the CFD simulated result is at 38.84%. Oliva Remolà et al. [10] commented that the one of the reason of the discrepancies could be the use of SST turbulence model which is not the most reliable model in computing the Reynold's stress. Therefore, it is the aim of this present work to explore the 'beta' factor using CFD.

2 Methods

2.1 Case Study

This case study used the hullform of R/V Athena, which is a high speed transom stern ship [11]. The hull lines and the particulars of R/V Athena or Model DTMB 5365 are shown in Fig. 1 and Table 1 respectively. Resistance experiments were carried out at the David Taylor Naval Ship Research and Development Center (DTNSRC)

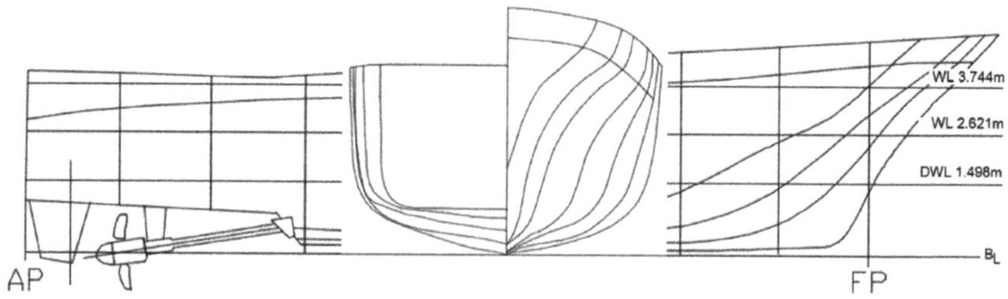

Fig. 1 The sheer and body plan of R/V Athena. note that the appendages of R/V Athena which consists of a pair of rudder, a pair of 'P' bracket, two propeller shafts and a pair of fin stabilizer (not shown in the figure)

Table 1 Ship and model particulars of R/V Athena

Description		Ship	Model
Scale factor	λ	1	8.25
Length between perpendiculars	L_{PP} (m)	46.9	5.69
Length at waterline	L_{WL} (m)	46.9	5.69
Overall length	LOA (m)	47.03	5.7
Breadth	B (m)	6.9	0.84
Draught	T (m)	1.51	0.183
Trim angle	(deg)	0.0	0.0
Displacement	Δ(tonnes)	229.00	0.397
Wetted surface	S_W (m^2)	287.36	4.222
Block coefficient	C_B (−)	0.4775	0.4775

by Jenkins [12]. Jenkins conducted towing tank experiments on a 1/8.25 scale model of R/V Athena over a Froude number range of 0.28–1.00 with the model was made free to sink and trim. This hull form was chosen for this case study because of the extensive amount of experimental evaluations of total, residuary and wave resistance coefficient are available for this hull which can be found in Jenkins [12] and Gadd and Russell [13].

2.2 CFD Computations

The investigation was conducted using a commercial CFD code SHIPFLOW 6.3 which is available from FLOWTECH International AB. The solver used in the SHIPFLOW code is a viscous flow RANSE solver XCHAP. XCHAP is a finite volume computation using the Explicit Algebraic Stress Model (EASM) as the turbulence model. A global approach was chosen for the computation where the computational domain was built using the H-O grid topology. All the grids in this study were created using the SHIPFLOW in-house grid generation module XGRID.

The viscous flow computations were carried out with the computational domain having six boundaries as shown in Fig. 2. The distance between the inlet of the viscous flow and the fore-perpendicular of the ship is at half the length perpendiculars, L_{PP} of the ship. The outlet of the viscous flow is located at 1.5 of length perpendiculars, L_{PP} of the ship behind the aft-perpendicular of the ship. The radius of the cylindrical outer boundary is at 1.0 length perpendiculars of the ship. This is necessary to prevent from any influence of the blockage effect due to shallow depth of the domain boundary.

The main appendages of R/V Athena twin screw hull are two rudders, a pair of bilge keels, two shafts and two shaft brackets as shown in Fig. 3. The simulations in SHIPFLOW were conducted in model scale and in full-scale from bare hull until the hull was fully appended. Overlapping grids were used in the computation of the

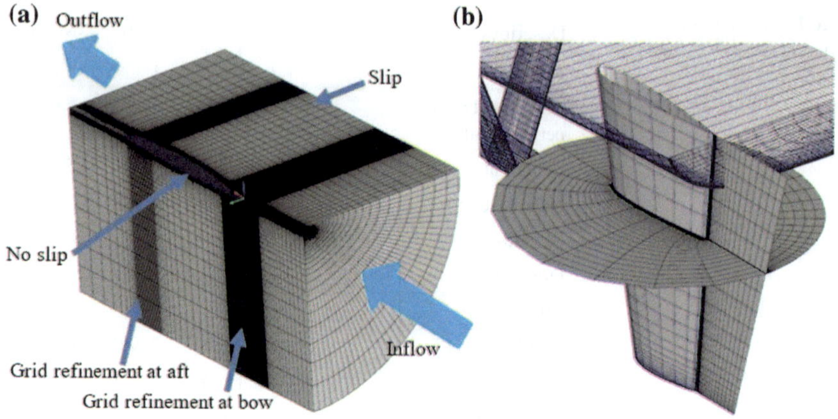

Fig. 2 **a** The computational domain used the H-O grid topology, with the inflow and outflow of the viscous flow computation shown above. No slip condition was applied to the boundary region close to the hull. Note that grid refinements were applied at the bow and the aft region. **b** The overlapping grid of the rudder. A grid clipping of the rudder grids is shown in the figure above

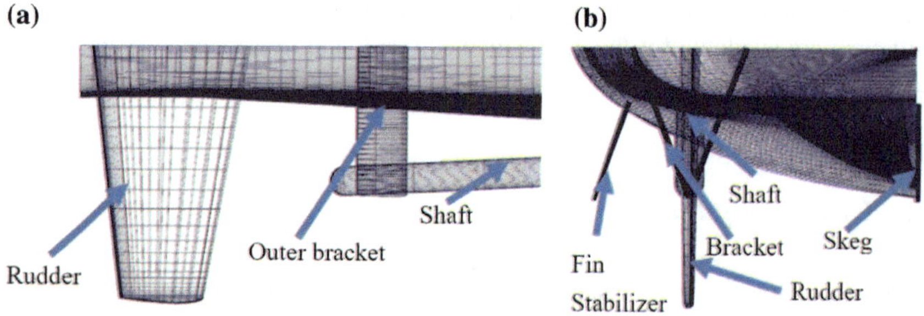

Fig. 3 **a** Profile view of the mesh of R/V Athena showing the mesh of the bare-hull with rudders, shafts and outer brackets. **b** The aft view of the mesh of R/V Athena showing the mesh of the bare-hull with rudders, shafts, fin stabilizer and skeg

appendage drag where the O-O grids of the appendages i.e. rudder overlapped with the cylindrical H-O grid which was meant for the hull geometry as shown in Fig. 2b.

2.3 *Grid Dependence Study*

A grid dependence study was performed in order to define the optimum number of grid cells to be used in the CFD simulations. The study was done using the results of total drag, frictional drag and the viscous drag coefficients of the Athena's bare hull as the main criterion in the study. It was found that the most optimum number of grid cells are at 2.3 million grid cells. Fine grid settings were emphasized at the forward and at the midship region of the Athena's hull.

2.4 Validation Study

A validation study was made based on the experimental results which can be found in Jenkins [12]. In this study, the results of Athena's bare hull drag were compared with the experimental results from Jenkins [12]. The comparison of the simulated CFD total drag with the experimental total drag from Jenkins is shown in Fig. 4. The percentage differences between the CFD computed total drag and the experimental total drag were ranged from −14.8 to 16.3%. There are several reasons

for the disagreement beyond model speed of 5 m/s. The free surface computations was done using inviscid panel method instead of using viscous surface capturing method or Volume of Fluid method (VOF). Therefore the disagreement could be attributed by the inability of the inviscid panel method itself to capture the free surface accurately at higher speed where wave resistance dominates, over-estimating the wave resistance, hence the total resistance of the model.

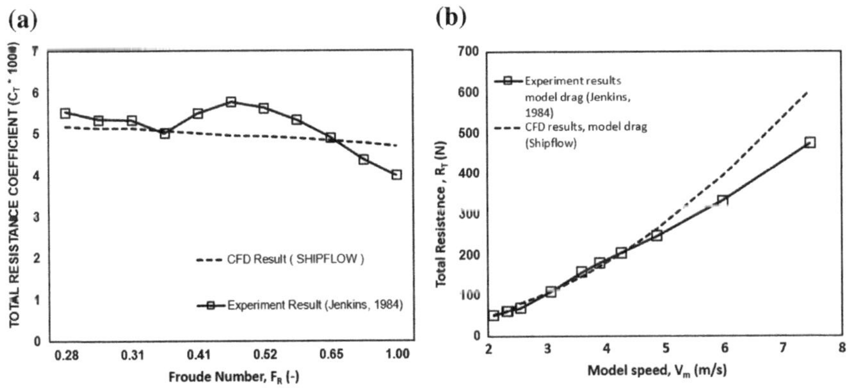

Fig. 4 Validation of simulated CFD results with experimental results from Jenkins (1984). **a** The total drag or resistance coefficient with respect to the Froude number. **b** The total drag or resistance with respect to the model speed in m/s

3 Results

3.1 Drag of Each of the Appendages

The simulations in SHIPFLOW on the appended hull of R/V Athena were done in different stages in order to determine the drag contribution of each of the appendages. The stages of computations were divided into five stages namely (A) Bare-hull with rudders (B) Bare-hull with rudders and shafts (C) Bare-hull with rudders, shafts and brackets (D) Bare-hull with rudders, shafts, brackets and fin stabilizer and (E) Barehull with rudders, shafts, brackets, fin stabilizer and skeg. The computations in stages are necessary in order to quantify the percentage increase in drag contributed by each of the appendages. The percentage increase in drag contributed by each of the appendages in model scale and in full scale is shown in Fig. 5a, b.

Overall, for both model scale and full scale, the drag increased by about 1%. The shafts, brackets and the fin stabilizer contributed to drag increase at all range of speed at about 1% and 2% respectively. The fully appended hull contributed to a bigger margin of appendage drag increase at about 8% in model scale and 7% in full-scale. The fully appended hull i.e. condition E, contributed to a bigger margin of appendage drag increase at about 14% in model scale and 11% in full-scale. It was observed that the skeg contributed to the highest percentage of drag at about 7–8% of the total drag.

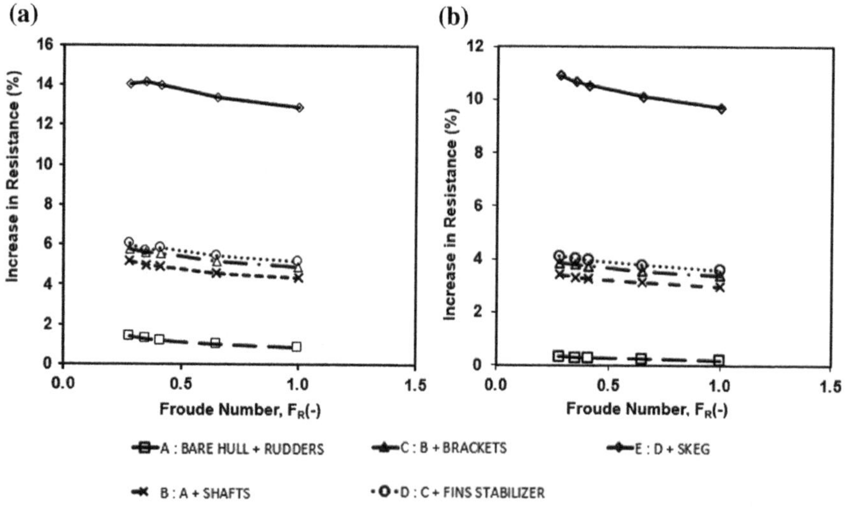

Fig. 5 The plot of accumulated increase in drag or resistance with respect to Froude number contributed by each of the appendages. **a** Model scale. **b** Full scale

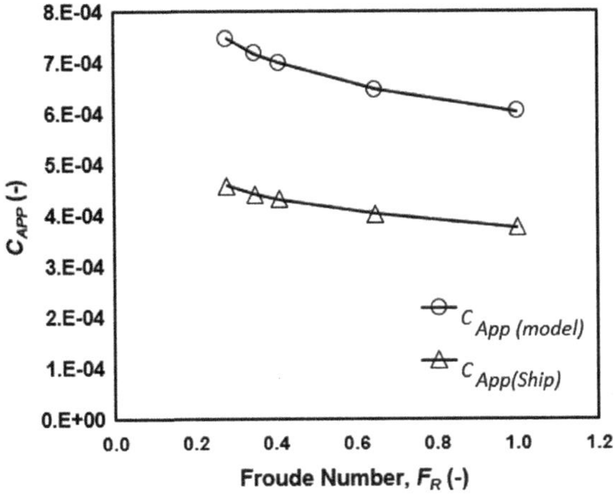

Fig. 6 The comparison of appendage drag coefficients in model scale and in full-scale. It should be noted that the appendage drag coefficient in model scale was higher than the drag coefficient in full-scale

3.2 *Estimation of the Beta Factor*

Figure 6 shows that the appendage drag coefficient of the model is higher than the appendage drag coefficient in full-scale. The values of beta can be obtained using Eq. 2 in the form of $\beta = 1 - C_{AppS}/C_{AppM}$. Table 2 shows the calculated values of beta for Froude number 0.28 up to 1.0. It was found that the values of beta were almost consistent around 0.38–0.39. This agrees well with Oliva Remolà et al. [10] comments that the beta factor depends on the Reynold's number and it is usually in between 0.3 and 0.6. It should be noted that this beta factor found in this study applies strictly to R/V Athena and similar ship form.

4 Conclusion

In this paper, the scaling of appendages of R/V Athena using the beta method was explored. The drag or the resistance contribution of each of the appendages fitted to R/V Athena were investigated. The appendage drag on R/V Athena was highlighted through a comparison study between the model scale and the full-scale drag. The drag of the appendages was simulated using SHIPFLOW. The drag increased up to 14% in model-scale and up to 11% in full-scale. The skeg was the highest contributor of drag with 7–8% increase in drag. The drag coefficient in model-scale was found to be higher that the full-scale drag. This proved that the scale-effects were present in the model-scale which results in higher drag coefficient in model-scale. An equation for force scaling of the drag coefficient was developed. The equation for the scaling is $C_{AppS} = (1 - 0.38)\, C_{AppM}$. It should be noted that this equation only applies strictly to R/V Athena and similar ship form.

Table 2 Appendage scaling results of RV Athena

F_R	$C_{T(appended)\ Ship}$ * 1000	$C_{T(bare\text{-}hull)\ Ship}$ * 1000	$C_{T(appended)\ Model}$ * 1000	$C_{T(bare\text{-}hull)\ Model}$ * 1000	$C_{App\ Ship}$ * 1000	$C_{App\ Model}$ * 1000	β
0.28	4.661	4.20	5.926	5.18	0.458	0.748	0.39
0.35	4.586	4.14	5.808	5.09	0.442	0.719	0.39
0.41	4.536	4.10	5.727	5.03	0.432	0.700	0.38
0.65	4.393	3.99	5.502	4.85	0.403	0.648	0.38
1.00	4.267	3.89	5.307	4.70	0.377	0.604	0.38

References

1. Molland AF, Turnock SR, Hudson AD (2011) Ship resistance and propulsion: practical estimation of propulsive power. Cambridge University Press, England
2. Allan JF (1950) Some results of scale effect experiments on a twin-screw hull series. Trans Inst Eng Shipbuild Scotl 93:353–381
3. Lackenby H (1955) BSRA resistance experiments on the Lucy Ashton. Part III. The ship model correlation for the shaft appendage conditions. Trans R Inst Nav Arch 97:109–166
4. van Manen JD, Oossanen PV (1988) Principles of naval architecture volume II. The Society of Naval Architects and Marine Engineers, New Jersey, USA
5. ITTC (2014) ITTC—Recommended Procedures and Guidelines–1978 ITTC Performance Prediction Method, ed. International Towing Tank Conference
6. Lasky MP (1980) An investigation of appendage drag. David W Taylor Naval Ship Research and Development Center Bethesda MD, Ship Performance Dept
7. Peck RW (1976) The determination of appendage resistance of surface ships. AEW Technical Memorandum 76020
8. Kirkman KL, Kloetzli JW (1980) Scaling problem of model appendages. ATTC, University of Michigan, Ann Arbor
9. Mustaffa Kamal I (2016) The powering performances of large waterjet and propeller driven catamarans at medium-speed, Ph.D., Australian Maritime College, University of Tasmania, Launceston, Tasmania
10. Oliva Remolà A, Pérez Rojas L, Pérez Arribas FL (2013) A contribution to appendage drag extrapolation using computational tools
11. Heffner JA (1989) The athena research ship system—a decade plus of service. David W Taylor Naval Ship Research and Development Center Bethesda, M.D., Ship Hydromechanics Dept
12. Jenkins DS (1984) Resistance characteristics of the high speed transcom stern ship R/V athena in the bare hull condition, represented by DTNSRDC model 5365. David W Taylor Naval ship research and development center bethesda MD
13. Gadd G, Russell MJ (1981) Measurements of the components of resistance of a model of RV. National Maritime Institute, Athena

Effective Control of SO_x and NO_x Release to the Atmosphere from Emissions of Ship Propulsion Engines

Md Salim Kamil, Mazlan Muslim and Asmalina Mohamed Saat

Abstract The reduction of noxious air pollution from ship discharges had been dealt with since the early of 1960s by the world's maritime regulatory body, the International Maritime Organisation under the jurisdiction of the United Nations. The auspicious Annex VI of the Regulations 12–15 of the IMO addresses the environmental pollution issues and emission limits of SO_x and NO_x in particular from ship propulsion engines including ODS and VOC. This first version of the regulation had been adopted by MARPOL convention in 1997 and the subsequent revisions ratified by many nations are highlighted in this paper. As of after the first version of the regulation introduced, continuous efforts were made by the Marine Environment Protection Committee (MEPC) of IMO, there were series of improvement progression on the regulations to ensure more stringent rules adoption and to be complied by all parties concerned. Several researches and initiatives are being carried out currently and also in the past since then the period of combustible fossil fuels including hydrocarbons being used in industrial plants and ships. The effective methods of control of these emissions particularly SO_x and NO_x discharges from ship propulsion engines become the essence of discourse of this paper. The effective measures using seawater scrubbers had in fact been in used for sometimes in ships as quick responses and accomplishments by the parties concerned in the effort to reduce the releases of the detrimental gases and particulates to the atmosphere. The potential results from these continuous efforts of researches and initiatives undertaken by the research establishments and industries and adoption by ship operators or owners are discussed and presented herein. The method of the effective control using natural seawater scrubbers complement the costly refinery processes in producing engine fuels of low sulphur contents with less post-combustion product of NO_x.

Md. S. Kamil (✉) · M. Muslim · A. M. Saat
Universiti Kuala Lumpur, Malaysian Institute of Marine Engineering Technology, 32200 Lumut, Perak, Malaysia
e-mail: mdsalim@unikl.edu.my

M. Muslim
e-mail: mazlanm@unikl.edu.my

A. M. Saat
e-mail: asmalina@unikl.edu.my

C. L. Saw et al. (eds.), *Advancement in Emerging Technologies and Engineering Applications*, Lecture Notes in Mechanical Engineering,
https://doi.org/10.1007/978-981-15-0002-2_10

Keywords International maritime organisation (IMO) · Marine pollution · Propulsion engines · Sulphur oxides (SO_x) · Nitrogen oxides (NO_x) · Ozone depleting substance (ODP) · Volatile organic compound (VOC) · Marine environment protection committee (MEPC) · Seawater scrubbers

1 Introduction

The MARPOL Annex VI on limits of emissions from ships contained in Chap. 3 [1, 2] elaborates the Regulations on NO_x and SO_x emission from ships as follows:

1.1 Regulation 13

Nitrogen Oxides (NO_x). The regulation applies to marine diesel engine on a ship with output power of more than 130 kW and marine diesel engine of a ship of the same range of output that undergoes major conversion on or after 1 January 2000. It does not apply to marine diesel engine on ship solely for emergencies or engaged in voyages within waters subject to the sovereignty of the State of the flag the ship entitled to fly. The Administration may exclude from the regulation for marine diesel engine on ship or that undergoes major conversion before 19 May 2005 and is solely engaged in voyages to ports or offshore terminals subject to the sovereignty of the State of the flag the ship entitled to fly. The applicability on the limits of NO_x of a marine diesel engine on a ship under Regulation 13 is sub-divided into Tier I, Tier II and Tier III.

Tier I—Operation of a marine diesel engine installed on a ship constructed on or after 1 January 2000 and prior to 1 January 2011 is prohibited except when the calculated NO_x meets the following limits:

(i) 17.0 g/kWh when the engine speed revolutions per minute $n < 130$.
(ii) $45n^{-0.2}$ g/kWh when the engine speed revolutions per minute $n = 130$ or more but $n < 2000$.
(iii) 9.8 g/kWh when $n \geq 2000$.

Tier II—Operation of a marine diesel engine installed on a ship constructed on or after 1 January 2011 is prohibited except when the calculated NO_x meets the following limits:

(i) 14.4 g/kWh when $n < 130$.
(ii) $44n^{-0.23}$ g/kWh when $n = 130$ or more but $n < 2000$.
(iii) 7.7 g/kWh when $n \geq 2000$.

Tier III—Operation of a marine diesel engine installed on a ship constructed on or after 1 January 2016 is prohibited except when the calculated NO_x meets the following limits:

(i) 3.4 g/kWh when $n < 130$.
(ii) $9n^{-0.2}$ g/kWh when $n = 130$ or more but $n < 2000$.
(iii) 2.0 g/kWh when $n \geq 2000$.

1.2 Regulation 14

The Sulphur content of any fuel oil used on board ships shall not exceed the following limit:

(i) 4.5% m/m prior to January 2012.
(ii) 3.5% m/m on and after 1 January 2012.
(iii) 0.5% m/m on and after 1 January 2020 or 2025 subject to the outcome of IMO review by 2018.

Operation of a marine diesel engine with a power output of more than 5000 kW and a cylinder displacement at or above 90 l installed on a ship constructed on or after 1 January 1990 but prior to 1 January 2000 shall comply with the following limits:

(i) 17.0 g/kWh when $n < 130$.
(ii) $45n^{-0.2}$g/Wh when $n = 130$ or more but $n < 2000$.
(iii) 9.8 g/kWh when $n \geq 2000$.

2 Issues of Ship Emissions Regulations

It is well said that NO_x and SO_x emissions from ships affect the air quality of the environment and threatening human, vegetation and bacterial life. NO_x emissions depend on the engine design whereas SO_x emissions depend on the sulphur contents of the marine diesel oil (MDO).

Nitrogen oxides NO_x are reactive gases occur naturally in the atmosphere through oxidation process of nitric oxide NO releases from the environmental sources such as emissions from industrial plants and ships. NO_x made the biggest composition of the ozone and contributes to the formation of acid rain which give effects to vegetation in swampy and land areas, visibility deficiency, water acidification, eutrophication, toxicity to human and aquatic life. Nevertheless NO_x is an essential nutrient for plants and a variety of bacteria.

SO_x and other gases deposition into atmosphere could be in wet and dry forms. SO_x dissolve in the atmosphere and clouds resulting to rain drops and precipitations. Chemical processes give much smaller particles of sulphate originated from high temperature combustion than derived from physical processes. In the presence of ammonia, the formation of sulphuric acid is removed by precipitation or neutralised to form ammonia sulphate.

Table 1 The pros and cons of options for compliance with MARPOL Annex VI on SO_x

Options	Pros	Cons
• Use of Low Sulphur Fuel Oil (LSFO)	• LSFO suitable for conventional marine diesel engines • Produce acceptable reductions in So_x emissions, avoid using scrubbing equipment • No requirement for scrubbing equipment	• 50% much higher refining cost • Less availability of LSFO and increase in price • Increase of catalytic (cat) fines for engines that use LSFO
• Clean Exhaust Gas Before Release	• Possibility to continue to use HFSO • HSFO is currently being only two-thirds the price of LSFO • Plentiful supply of HSFO • Low risks of cat Fines in LSFO • Avoid cat monitoring and removal • Efficient exhaust gas cleaning	• Scrubbing equipment is required • US$10 million to install such equipment • Incurs maintenance cost
• Use LNG or Liquid Hydrogen	• No issue of LSFO or HSFO • Greener option than MDO • Reduce SO_x emissions by 90–95%	• High conversion cost of ship propulsion engines from diesel to LNG • Lack of storage facilities in ports • Loss of cargo space onboard

The MASH Report Q1 2018 [3] revealed that the emissions from ship diesel engines is one of the major contributors of SO_x discharge to the surrounding environments. The relevant regulation imposed on ship emissions pose a big challenge to ship owners and diesel fuel producers to be compliant with the regulations enforced. There are several options for complying with MARPOL Annex VI for SO_x giving the pros and cons as given in Table 1. In 2008, ship emissions study was performed by the National Technical University of Athens, Laboratory of Maritime Transport [4] which provides useful results for reference.

3 Studies on Ship Emissions

An environmental group comprising of AirClim, Seas At Risk, Bellona Foundation, North Sea Foundation, Transport & Environment and European Environmental Bureau in November 2011 published a forecast of NO_x and SO_x emissions from ships and comparing with land-based sources of the 27 European Union member states [5]. The results of the projected emissions were quantified and shown in bar charts in Figs. 1 and 2. It is expected that between the year 2000 and 2020, the emissions of NO_x and SO_2 from international shipping trade in the region will increase by 40–50%. However with the enforcement of the new regulations in Annex VI, a significant reduction is anticipated.

In July 2002, Entec [6] published a study Report of Chapter 3 on Market Survey of Marine Distillates with 0.2% Sulphur content for 10 different scenarios of 3% of growth rate per annum for the year 2000–2010 projection as shown in Table 2. Generally, the results showed an increasing and decreasing trend of the NO_x, SO_2, CO_2, HC and PM year by year. Nevertheless in 2010 the emissions were at the highest level.

Seawater scrubbing system or seawater scrubber is a method that could reduce NO_x and SO_x emissions from ships. This can be applied for exhaust emissions of NO_x and SO_x originated from MDO combustion by marine diesel engines which operate on HSFO or LSFO. The main objective of reducing the emissions is for marine diesel engine on HSFO, however it would be further enhanced for marine diesel engines on LSFC. This technological method is very practical and viable

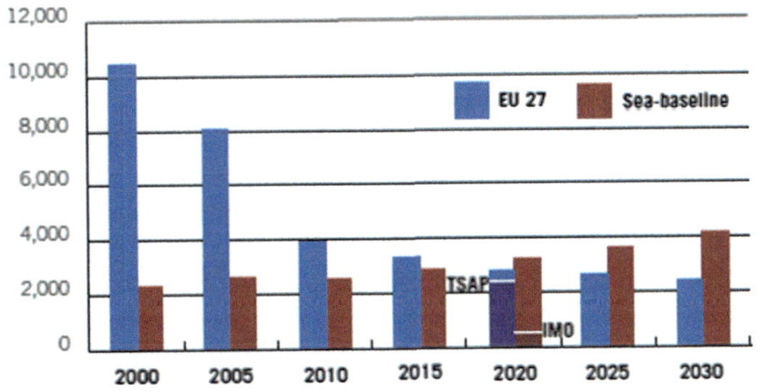

Fig. 1 Emissions of SO_2 2000–2030 (kilotonnes)

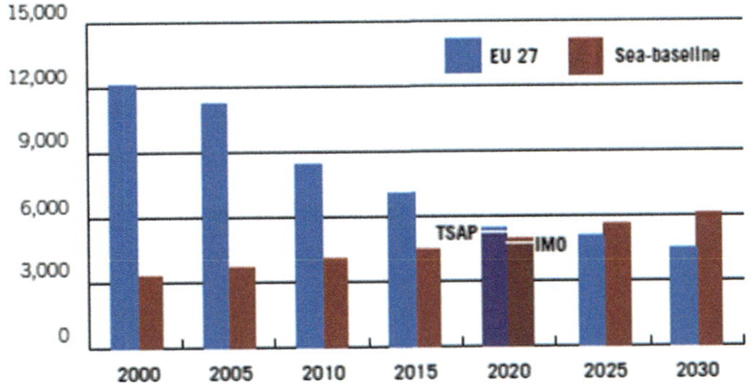

Fig. 2 Emissions of NO_x 2000–2030 (kilotonnes)

Table 2 Projected emissions with 3% assumed growth rate

Scenario	NO_x (Kte/annum)	SO_2(Kte/annum)	CO_2 (Kte/annum)	HC(Kte/annum)	PM (in port) (Kte/annum)
2000 BAU	3617	2578	157,298	134	21
1 2006 BAU	4186	2968	180,639	153	25
2	4186	2601	180,639	153	22
3	4183	2512	180,432	153	13
4	4179	2477	180,218	153	10
5 2006 BAU	4411	3127	190,085	162	23
6	4411	2739	190,085	162	24
7	4408	2642	189,865	162	14
8	4403	2602	189,640	162	11
9	4403	2248	189,640	162	11
10 2010 BAU	4649	3294	200,105	171	28

option [7] as there is no requirement to store lime mineral onboard ships, no by-product that requires removal on land and plentiful readily available of seawater at sea that contains substantial amount of sulphate and nitrate. The NO_x and SO_x removal performance success which has been proven is 90–95%. Seawater scrubber also removes of about 80% particulates and more than 20% hydrocarbons particularly carbon dioxides (CO_2). The particulates can be filtered and subject to the qualitative level may not be discharged outboard right away. It will be treated as necessary in compliance with the marine pollution (MARPOL) regulations enforced before it can be led overboard to the open sea or may be for disposal on land through ports or discharge barges.

In 2016, a research was conducted by the Department of Energy and Environment, Chalmers University of Technology, Gothenburg, Sweden [8] on process evaluation of a SO_x and NO_x exhaust gas cleaning concept for marine application. A selective catalytic reduction method was studied to reduce the amount of NO_x from the marine diesel engine exhaust gas by converting NO_x plus O_2 using NH_3 as the reductant into N_2 ad H_2O with the reduction success of 75–95%. The chemical reactions of the process are given below.

$$4NO + 4NH_3 + O_2 \rightarrow 4N_2 + 6H_2O \tag{1}$$

$$2NO_2 + 4NH_3 \rightarrow 3N_2 + 6H_2O \tag{2}$$

$$NO + NO_2 \rightarrow 2N_2 + 3H_2O \tag{3}$$

Table 3 Combined NO_x and SO_2 removal technologies in industrial plants processes

Technology	SO_x removal (%)	NO_x removal (%)
Electron beam flue gas treatment [16]	<98	<82
Wet scrubbing (using an aqueous $NaClO_2$) [17]	88–100	36–72
LoTOx™ (ozone injection and wet scrubber) [18]	>95	90–95
NeuStream®-MP (ozone oxidation) [18]	97	>90

Table 4 Global 0.5% Sulphur cap from 2020 for emissions from international shipping (kilotonnes)

	2020	2020	2025	2025
	(0.5% S)	(3.5% S)	(0.5% S)	(3.5% S)
SO_2	2500	10,900	2700	11,700
PM	760	1500	810	1600
NO_x	21,000		22,500	
CO_2	862,000		917,000	

Another technological method applied was to reduced emissions of NO_x and SO_2 simultaneously using seawater scrubbing system which make up of seawater intake system, water spray system, scrubbing tower, wash water system and gas monitoring system. The technology gave to promising results of 99% SO_x, 66% NO_x and 77% CO_2 reduction. Other prospective technologies in industrial plants processes studied gave the emissions reduction as shown in Table 3.

The global 0.5% Sulphur cap from 2020 to 2015 and the projected NO_2 and SO_x of the European Union 28 member states for land-based sources versus the international shipping [9] around Europe from 2010 to 2030 are shown in Table 4 and Fig. 3. More studies about emissions of exhaust gas from ships had been carried out continuously by researchers and industries of oil refineries and some could be found in references [10–14].

4 Technological Method of Effective Control

Figure 4 shows the schematic diagram of a proposed multi-stage seawater scrubbing system or installation for ship application. In the first stage the hot combustion exhaust gas out from the diesel engines is dipped into pool of water in a special dip tank for absorbing the NO_x and SO_x and capturing the particulates. The exhaust gas is further passed through a multi-layer high pressure seawater spraying system for the second stage emissions reduction process in which cleaner emissions are discharged out to the environments. The level of NO_x, SO_x and particulates are monitored at different points such as immediately at the exit of the exhaust manifold of the engines, in

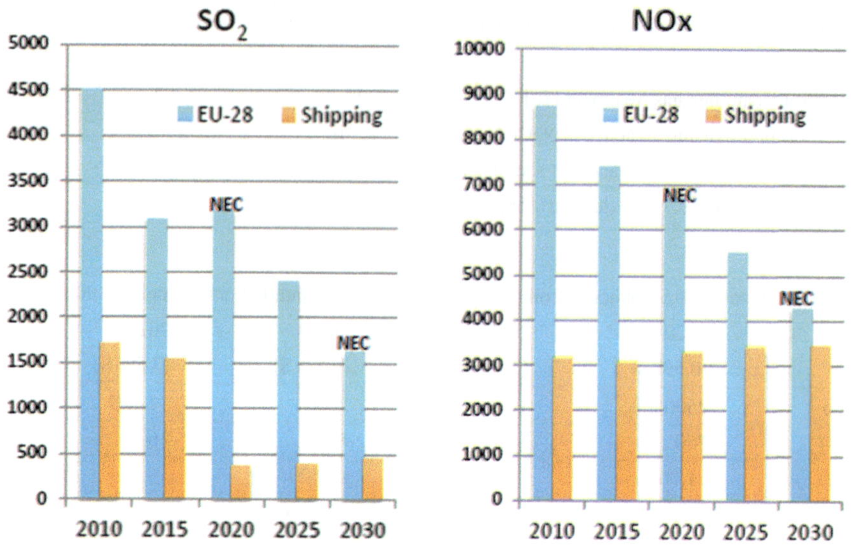

Fig. 3 EU28 emissions of land-based versus shipping around Europe 2020–2030 (kilotonnes)

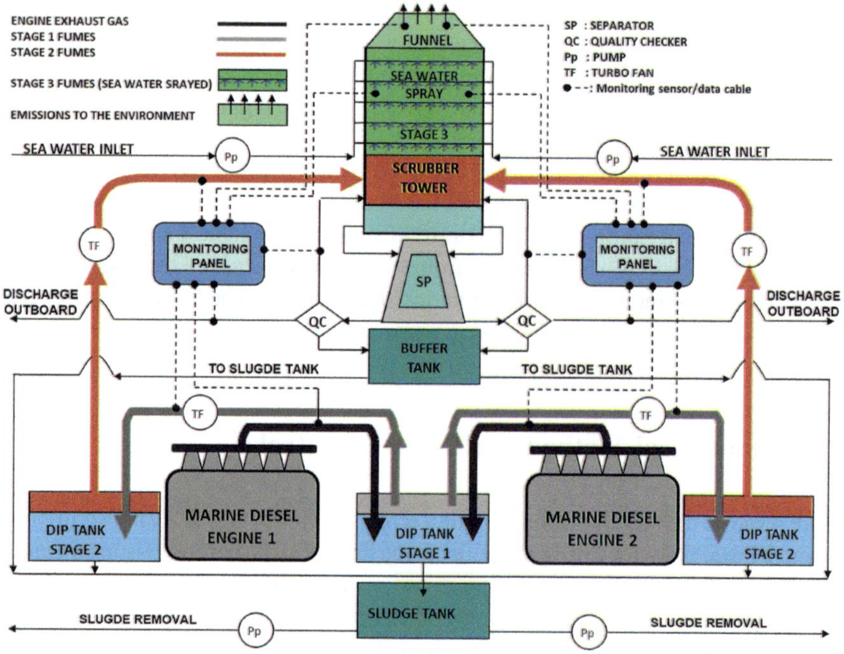

Fig. 4 Block diagram of a multi-stage open loop seawater scrubber

the exhaust gas dip tank and at different positions along the tower of the seawater scrubber upwards as necessary.

5 Conclusion

The Regulations 13 and 14 of the Annex VI of the IMO on ship emission limits of SO_x and NO_x have been continuously presented in many publications as important reminders to the ship owners, the fuel producers i.e. MDO refineries and other interested parties and stakeholders. The reduction of emissions particularly NO_x and SO_x could be realised efficiently using seawater scrubber which is a method of effective control of the emissions of these noxious gases to the surroundings. It is one of the viable options to achieve the compliance with the requirements of the regulations with the minimum cost possible. The relatively cheap seawater scrubbing system for ship installation provide the alternative to the costly refinery processes in producing MDO of low sulphur contents with less post-combustion product of NO_x from the engines. This method also avoids the options for the very expensive replacement or conversion of the existing ship propulsion engines of the ships it may be to be able to operate on low sulphur fuel oil. Over and above the cost of LSFO would also be expensive as the production cost at the refinery plants is high.

References

1. The Marine Environment Protection Committee (MEPC), Resolution MEPC 176 (58) Adopted on 10 October 2008 (Revised MARPOL Annex VI)
2. Frequently Asked Question, IMO Regulations to Reduce Air Pollution from Ships and the Review of Fuel Oil Availability, http://www.imo.org/en/MediaCentre/HotTopics/GHG/Documents/sulphur%20limits%20FAQ_20-09-2016.pdf. Last accessed 14 July 2018
3. Marsh-Emissions-Regulations-Concerns-for-the-Marine-Industry-2018, https://safety4sea.com/wp-content/uploads/2018/01/Marsh-Emissions-Regulations-Concerns-for-the-Marine-Industry-2018_01.pdf. Last accessed 14 July 2018
4. National Technical University of Athens, Laboratory for Maritime Transport, Ship Emissions Study for Hellenic Chamber of Shipping, May 2008
5. Air Pollution from Ships, Joint Pamphlet Publication of Environmental Organisations Air-Clim, Seas At Risk, Bellona Foundation, North Sea Foundation, Transport & Environment and European Environmental Bureau, Nov 2011
6. Entec Report, Chapter 3 Market survey of marine distillates with 0.2% sulphur content, pp 75–88, July 2002
7. Behrends B, Liebezeit G (2003) Reducing SO_2 and NO_x emissions from ships by a seawater scrubber. Research Centre Terramare, Wilhelmshaven, Germany
8. Ibrahim S (2016) Process evaluation of a SO_x and NO_x exhaust gas cleaning concept for marine application. Master thesis, Department of Energy and Environment, Chalmers University of Technology, Gothenburg, Sweden
9. Agren C. Air pollution from ships. https://www.unece.org/fileamin/DAM/env/docments/2016/AIR/EB/PPT/36thSessionPPT/Item5a/3Air_pollution_from_shipping.pdf. Last accessed 14 July 2018

10. Lamas MI, Rodriguez CG, Rodriguez JD, Telmo J (2016) Numerical model of SO_2 scrubbing with seawater applied to marine engines. Pol Marit Res 23:42–47
11. Svensson E. Regulation of the global SO_x emissions from ships. Thesis for the Degree of Licentiate of Philosophy. In: IMO proceedings 1988–2008. ISSN 1652-9189, Report No. 11:127, Chalmers Reproservice, Gothenburg, Sweden
12. Aksoyoglu S, Baltensperger U, Prévôt ASH. Contribution of ship emissions to the concentration and deposition of air pollutants in Europe. https://www.atmos-chem-phys.net/16/1895/2016/acp-16-1895-2016.pdf. Last accessed 14 July 2018
13. Luttenberger A, Rukavina B, Rak L. Challenges in regulating the air pollution from ships. University of Rijeka, Faculty of Maritime Studies. https://bib.irb.hr/datoteka/633941.Luttenberger_Rukavina_Rak_-_ICTS_05-10.pdf. Last accessed 14 July 2018
14. Blumberg KO, Walsh MP, Pera C. Low-sulphur gasoline & diesel, the key to lower emissions from vehicles. https://www.theicct.org/sites/default/files/publications/Low-Sulfur_ICCT_2003.pdf. Last accessed 14 July 2018

Study of the Stratification Effect at the Reactor Installation in the Smolenskaya Area

Thanh Binh Nguyen, Aleksey S. Shelegov and Igor A. Chusov

Abstract This paper studies the Stratification Effect (SE) in the cooling basin for the Smolensk Nuclear Power Plant (NPP) and its dependence on the climatic condition. We investigated how climatic condition is the cause of the SE and analyzed all variations of the SE via program complex ANSYS-CFX. The obtained results were then compared to each other to assess the consequences of the possible impact of SE on the economy of the power units in its normal operation mode. The results indicate that SE happens when there is no cooling evaporation on the surface of the reservoir, moreover, air velocity greatly impact the SE, if the wind flows faster, the SE is less likely to appear, as the result, NPP has the better net efficiency.

Keywords Stratification effect · Smolensk NPP · NPP net efficiency

1 Introduction

When operating a NPP from many of its units, both main and auxiliary, it is necessary to divert a large amount of heat into the environment. The heat transferred to the environment is mainly carried out by water. For the operation of NPP, technical water supply is very important, in many ways determining the reliability and economic efficiency of the plant [1]. After discharging hot water into the environment, the warmer liquid does not mix with the cooler in the water basin—that phenomenon is called the SE. The SE possibly violates the normal operation of the NPP. In addition, the SE affects the economy of NPP. In order to assess the SE, the following works were performed for the Smolensk NPP site, located on the bank of the Desna river:

- Developing a model for the propagation of the heat flow of liquid into the water-cooling pool, taking into account the climatic conditions of the region.

T. B. Nguyen (✉)
Duy Tan University, Da Nang 550000, Vietnam
e-mail: binh.nguyen.1406.93@gmail.com

A. S. Shelegov · I. A. Chusov
Obninsk Institute for Nuclear Power Engineering, National Research Nuclear University MEPhI, Obninsk 249020, Russia

C. L. Saw et al. (eds.), *Advancement in Emerging Technologies and Engineering Applications*, Lecture Notes in Mechanical Engineering,
https://doi.org/10.1007/978-981-15-0002-2_11

- Assessing the consequences of the possible impact of the SE in the normal operation mode of the power units on its economy.
- Executing a forecast of changes in water temperature in the river under different climatic conditions in the normal operation of power units.
- Calculating the impact of the SE on the profitability of NPP.

2 Climatic Condition and Hydrography of Smolensk Area

Smolensk region—the subject of the Russian Federation, is part of the Central Federal District. It borders with Moscow, Kaluga, Bryansk, Pskov and Tver regions of Russia, as well as with the Mogilyov and Vitebsk regions of Belarus. The climate of the area is temperate continental, in which:

- The average yearly temperature is in the range of [4.5, 4.8] °C.
- The wind speed in the warm period is in the range of [3, 4] m/s, in the cold period is [4, 5] m/s.
- The maximum temperature of the warmest month is July 34 °C, the minimum temperature of the coldest month is January −43 °C.
- The average annual relative humidity in Desnogorsk is 70%.

The purpose of water used in NPP is to take water resources from a river to cool the technological circuits of NPP, to compensate for irrecoverable losses and, accordingly, to generate electricity. The cooling systems of the technical water, designed to dispose unused heat for the production of electric power, represent a set of hydraulic structures, pumping stations, pipelines, and heat exchange equipment. The use of recycling water management system leads to an economical consumption of natural water.

The speed of the river flow varies depending on the season of the year. The greatest velocity during the spring high water and during the low water period, is up to 1 m/s on the rifts, and is about [0.3, 0.4] m/s on the reaches.

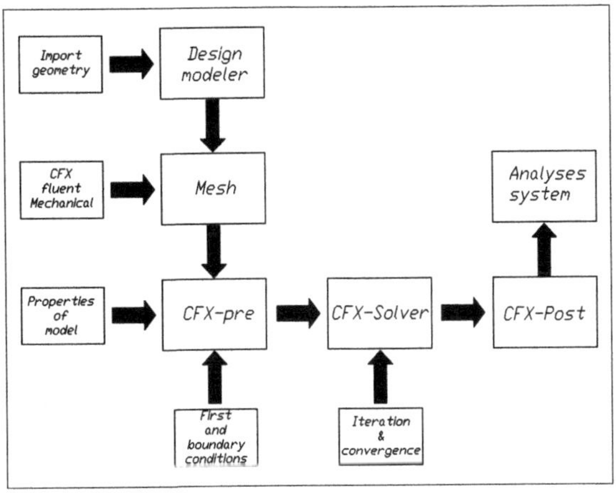

Fig. 1 The scheme for setting and solving the problem using program ANSYS-CFX

3 Creating the Forecasting Model of the Smolensk NPP Area

3.1 Methodology

In order to solve the problem, the program complex ANSYS-CFX[1] was used, which consists of 5 applications, between these applications there is an exchange of information flows that arise during the formulation and solution of hydrodynamic problems, the operating scheme of the complex is showed in Fig. 1.

3.2 Model of the Calculated Area

For the numerical modeling of the Smolensk NPP and the Desna River, the task of external flow around the buildings of the NPP site was designed to develop a solid vector model of the site, accurate to the main production buildings. In our case, the enclosing space in the 6-kilometer zone of NPP area was considered. The solid model of Smolensk NPP area is shown in Fig. 2 with a scale of 1:10 and consists of 2 domains:

- Aerial area with a height of 20 m;
- Water area with a depth of 1 m.

[1]The license of the ANSYS simulation belongs to Obninsk Institute for Nuclear Power Engineering of the National Research Nuclear University MEPhI & authors are permitted to utilize the method for solving the problem in this paper.

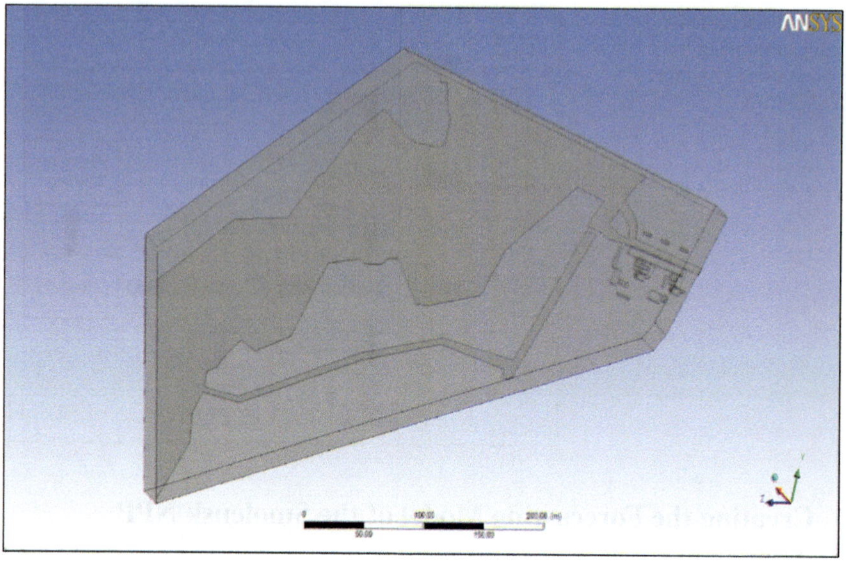

Fig. 2 The calculating model of the area

3.3 Boundary Conditions and First Conditions

Figure 3 indicates the boundary conditions of the problem, which illustrates the flow direction of the river, and the air. The picture also illustrates locations, where the hot water is disposed, and where the recycling water is inputed for the main condensers.

Next, we implied the first conditions based on the climate data, environment data and technical parameters of the Smolensk NPP:

- Air velocity $v_{air} = [0, 5]$ m/s, river flow velocity $v_{river} = 0.4$ m/s;

Fig. 3 Boundary conditions

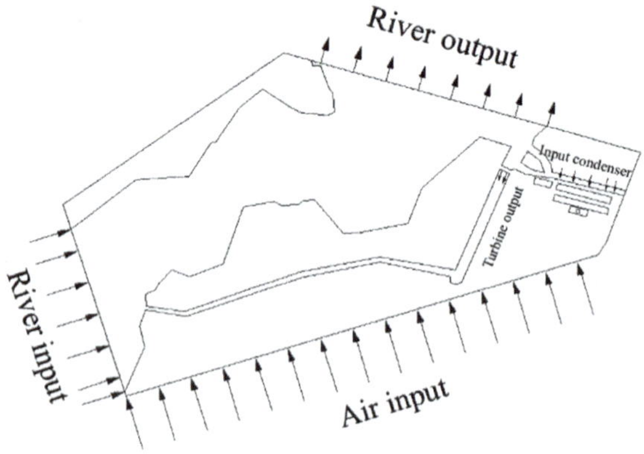

- Water temperature at the outlet of the turbine is T = 35 °C, the temperature inside the river T_{river} = 15 °C, air temperature: T_{air} = {10, 15} °C;
- Intensity of evaporation of the water in the basin is $I_{evapor.river} = 10^{-9}$ g/s.
- Cooling water flow rate for K-10120 type capacitor $G_{turbine}$ = 23,000 kg/s, Smolensk NPP has 3 working units, so we get G = 69,000 kg/s, the model with a scale of 1:10, thus G_{model} = 6900 kg/s.

4 Results and Discussion

In this section, we considered all of the possible variations of the SE by including various air velocity and two heat transfer methods—convection and cooling evaporation.

4.1 First Case—Heat Transfer by Convection

The first case provided variants, in which the velocity of the fluid flow was 0.4 m/s, the velocities of the air flow were {0.5, 1, 2, 3, 4, 5} m/s with convective heat transfer and evaporation was neglected. The obtained results of all variations in the first case are showed in Fig. 4.

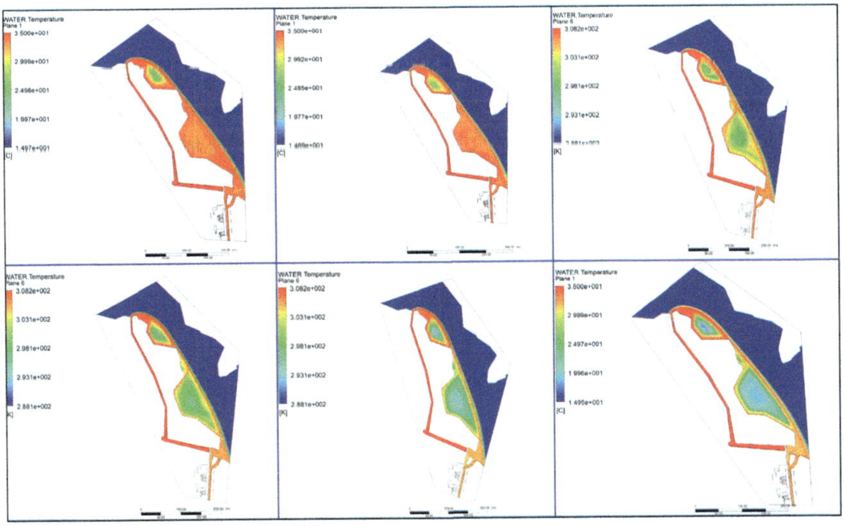

Fig. 4 The SE with the air velocity v_{air} was 0.5, 1, 2 m/s on the first row & 3, 4, 5 m/s on second row respectively

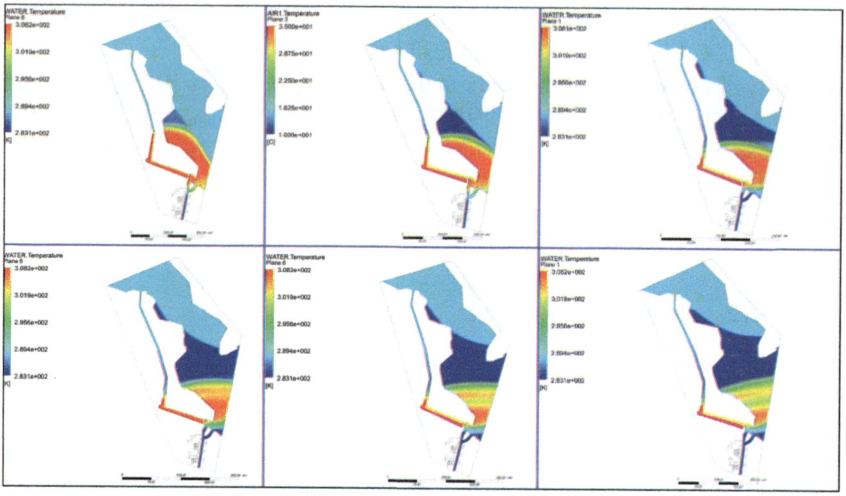

Fig. 5 The effect with the air velocity v_{air} was 0.5, 1, 2 m/s on the first row & 3, 4, 5 m/s on second row respectively

4.2 *Second Case—Heat Transfer by Cooling Evaporation*

Next, with the same procedure, the second case provided calculating variants, in which the velocity of the fluid flow was 0.4 m/s, the velocities of the air flow were {0.5, 1, 2, 3, 4, 5} m/s with heat transferred by cooling evaporation (intensity of evaporation to the surface of the river is 10^{-9} g/s). The obtained results are showed in Fig. 5.

Figure 4 shows how the air velocity effects the SE. Based on the color shown, there are changes in water temperature correspond with the changes of the air velocity, the faster the air flows the less likely SE to appear. Based on the color change shown in Fig. 5, it is worth noting that if cooling evaporation happens, the SE has yet to be appeared, on the other hand, if heat is only transferred by convection, the SE will occur, another words, the more evaporated water the better heat transferred from the reservoir. However, the process of evaporation heavily depends on the climatic conditions of the area such as the wind and the humidity. For instance, if the humidity is close to saturation (90–95%), the water barely evaporates, water contains a vast amount of heat, as the result, the SE takes place.

5 Processing the Calculated Results

Several points were set in the river and temperature at each point were measured, an average value of which was then computed. Comparison of the average temperatures of two cases correspond with different air velocities is shown in Fig. 6.

Fig. 6 Average water temperature in the river in two cases

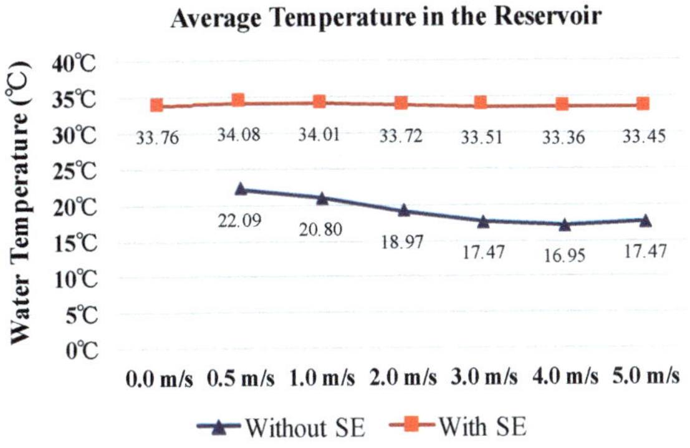

Next, the temperatures at the entrance of main condensers of the NPP complex was measured correspond with different air velocities, result of which is shown in Fig. 7.

The water used for condensing the post-turbine vapor plays an importance role in defining the efficiency of the power units. The cooler water input into the condenser, the better efficiency the NPP gets [2, 3]. In order to determine the impact of the SE on the economics of the NPP, we gathered a column chart, which shows the comparison of the net efficiency in normal condition and in the SE (shown in Fig. 8).

The graph shows that overall the NPP without the SE has better net efficiency than the NPP with the SE. Without SE, net efficiency barely depends on the air velocity, on the other hand, with the SE, the air velocity greatly impacts the NPP net efficiency.

Fig. 7 Water temperature at the entrance of the main condensers

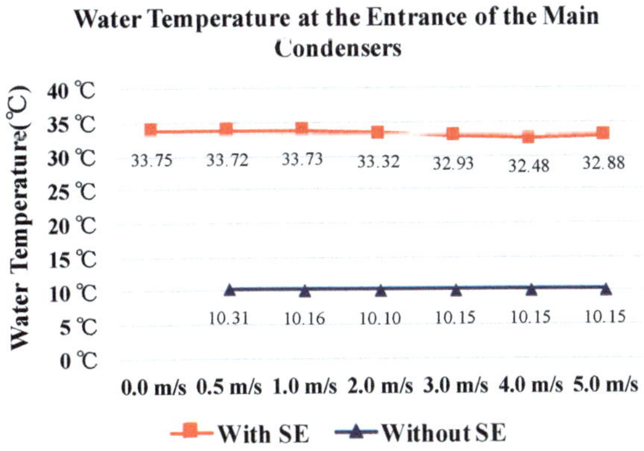

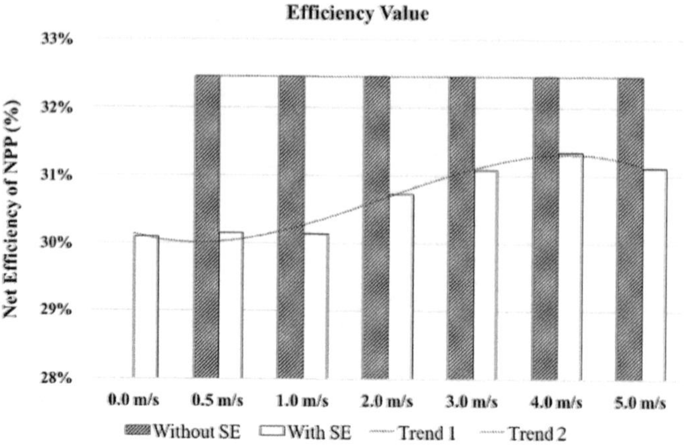

Fig. 8 The net efficiency comparison of the NPP in normal condition and in the SE

6 Conclusion

The fluid stratification is common effect that we come across on a daily basis such as SE in water tanks, pools, basin, sea etc. This study has revealed how the climate would be a factor of the SE,[2] and the impact of the SE on the profitability of NPP. When looking at the bigger scenario, different parts of the world with diverse climates, SE would occur in various way. The places for construction NPP should be researched carefully and have appropriate climatic conditions to reduce the impact of the SE, moreover, a desired profitability is achieved.

This study has unwrapped the preliminary version of the SE, however, it has created the foundation for further researches of the SE such as the SE between fluid layers of the cooling pools, the intensity of evaporation of the water surface of cooling basin, research the SE in various climatic conditions, etc.

References

1. Slobochuk VI, Shelegov AS, Leskin ST (2010) Basic systems of power units of nuclear power plants
2. Rivkin SL, Aleksandrov AA (1980) Thermophysical properties of water and vapor. "Energy", Moscow
3. Avdeev AA, Zudin Yu B, Ryasny SI (2004) Analysis and evaluation of thermal and hydraulic processes in the branch pipe of water of steam generator PGV-1000. Heat Power Eng 1

[2] In this study, the dependence of the SE on the wind was considered.

Piezoelectric P(VDF-TrFE) Thick Film Based Micro-power Generator Using Flexible Substrate for Wearable Applications

Khoon-Keat Chow, T. K. Woo, Swee Leong Kok, Kok-Tee Lau and Ali Mohammed Abdal Kadhim

Abstract Energy harvesting, non-toxic and flexible sensing capability have attracted enthusiasm for the potential variety of devices in Internet of things (IoT), wearable and biomedical applications. In this paper, usage of non-toxic lead-free piezoelectric material, P(VDF-TrFE) as a polymer piezoelectric material for developing thick-film based flexible piezoelectric micro-power generator. The P(VDF-TrFE) thick film was prepared by solution casting in between of top and bottom electrodes on top of PET substrate. Fabrication and characterization of the P(VDF-TrFE) thick film were examined by EDX, FESEM, XRD and ferroelectric tester. In order to prove the flexible P(VDF-TrFE) piezoelectric micro-power generator applications, a sample of the fabricated P(VDF-TrFE) film was mounted in such way that it directly hit by the blades of a rotating D.C fan. It generated a maximum open-circuit output voltage of 14 V when the D.C fan operated at rotational speed of 20 cps.

Keywords Energy harvesting · P(VDF-TrFE) · Thick film · Wearable applications

K.-K. Chow (✉)
Department of Mechanical Engineering, Politeknik Ungku Omar, Jalan Raja Musa Mahadi, 31400 Ipoh, Perak, Malaysia
e-mail: markchowmy@yahoo.co.uk

Electrical Technology Program, Unit for Research and Innovation, Kolej Komuniti Sungai Siput, Kampung Sungai Sejuk, 390, 31100 Sungai Siput (U), Perak, Malaysia

K.-K. Chow · S. L. Kok · A. M. A. Kadhim
Advanced Sensors and Embedded Control System Research Group, Center for Telecommunication Research & Innovation, Fakulti Kejuruteraan Elektronik dan Komputer, Universiti Teknikal Malaysia Melaka, Hang Tuah Jaya, 76100 Durian Tunggal, Malacca, Malaysia

T. K. Woo
Department of Electrical Engineering, Politeknik Ungku Omar, Jalan Raja Musa Mahadi, 31400 Ipoh, Perak, Malaysia

K.-T. Lau
Carbon Research Technology Research Group, Fakulti Teknologi Kejuruteraan Mekanikal dan Pembuatan, Advance Manufacturing Center, Universiti Teknikal Malaysia Melaka, Hang Tuah Jaya, 76100 Durian Tunggal, Malacca, Malaysia

C. L. Saw et al. (eds.), *Advancement in Emerging Technologies and Engineering Applications*, Lecture Notes in Mechanical Engineering,
https://doi.org/10.1007/978-981-15-0002-2_12

1 Introduction

Nowadays, wearable technology have captured a large market share in latest trending products. Consumers become increasing conscious of their daily activities such as monitoring their health and exercise using wearable devices with Internet of thing capability [1–3]. These flexible electronics consists of sensors, display, and smart system. The consumers also look for wearable devices which are flexible, environmental friendly, comfortable and safe to be worn. Wearable device allows the device to conform to our bodies by seamlessly integration into any attires that we wear. The flexible and wearable devices are stretchable, thus do not constrain and fix to any rigid surface. Currently, most of the wearable devices are powered by bulky batteries with limited lifespan. This problem is overcome by using micro-power generator based on P(VDF-TrFE) polymer piezoelectric thick film.

Poly(vinylidene fluoride trifluoroethylene) [P(VDF-TrFE)] is a high performance polymer piezoelectric materials used in the energy harvesting applications. This P(VDF-TrFE) could be used as sensors, wearable devices, flexible electronic devices, actuators and energy harvester. The polymer piezoelectric materials incur low cost in manufacturing, has mechanical flexibility and low processing temperature. It is an environmentally friendly materials as compared with PZT piezoelectric materials [4–6]. The PVDF material exists in crystalline form of alpha, α-phase and is nonpolar. In order to increase the degree of crystallinity of P(VDF-TrFE), it required hot plate annealing and poling are applied [7, 8].

In this paper, thick-film based P(VDF-TrFE) piezoelectric using solution casting is demonstrated. Then, the P(VDF-TrFE) thick film were examined by EDX, FESEM, XRD and ferroelectric tester. Finally, mechanical tapping on the P(VDF-TrFE) was conducted to prove the piezoelectric response to mechanical energy and the conversion of electrical energy in application of micro-power generator.

2 Experiment Setup

The P(VDF-TrFE) solution was prepared by dissolving P(VDF-TrFE) powders (Kureha, 12.5 wt.%, 75/25 mol%) in dimethylformaide liquid medium (DMF, Sigma Aldrich) by mechanical stirring at room temperature for 12 h. At this point, the solution was drenched in ultrasonic bath for 30 min so that the solution was fully dissolved. The well-mixed P(VDF-TrFE) solution were casted on a flexible substrate, PET which pre-coated bottom electrode and then slow evaporated under infra-red light at 80 °C for 30 min. After that, the samples were cured in oven at 60 °C for 4 h in order to remove unwanted residual solvent. Finally, the P(VDF-TrFE) thick film was sandwiched by top electrode layer (see Fig. 1).

Subsequently, the materials processing and fabrication, the fabricated P(VDF-TrFE) thick film micro-power generator was demonstrated by using finger tapping method (see Fig. 2) for output voltage in open circuit [9].

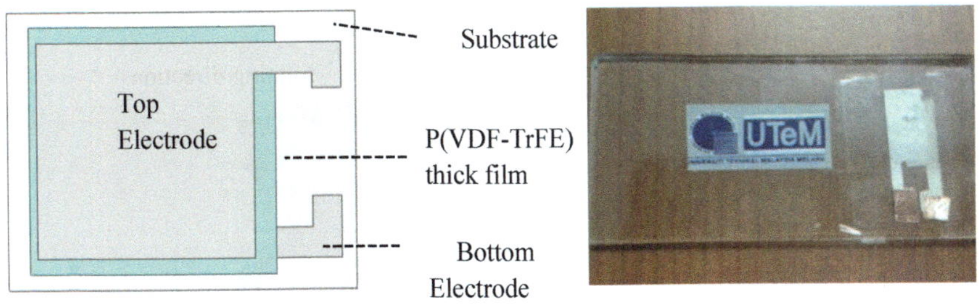

Fig. 1 **a** Schematic diagram. **b** Fabricated device

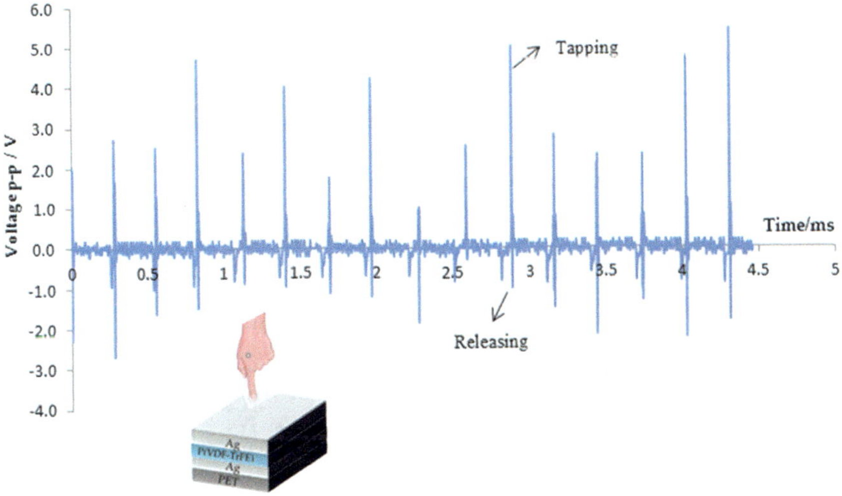

Fig. 2 Peak-to–peak voltage of P(VDF-TrFE) harvester versus tapping time

Due to the simple experiment setup using finger tapping and releasing where the output voltage will be varied from time to time because variable such as force of finger must be controlled so that the output voltage is increasing linearly to the force. To improve the experiment setup, the P(VDF-TrFE) device as attached to the stand of direct current (D.C) fan motor [366B, Mini D.C Fan—MRDIY] (see Fig. 3). The blades of the van act as force variable to tap the edge of the P(VDF-TrFE) device by controlling the speed of the blades using direct current power supply.

P(VDF-TrFE) harvester was attached in the cantilever configuration to a stand located next to a DC electric fan motor The motor position was adjusted until two of the fan blades were located at same distance from the top and bottom surface of the free end of the P(VDF-TrFE) cantilever. After P(VDF-TrFE) harvester was connected to an oscilloscope (see Fig. 4), the DC electric motor was switched on to rotate the fan blade. Periodical tapping on the P(VDF-TrFE) cantilever were performed by the fan blades every time the blades hitting the cantilever end. The tapping frequency

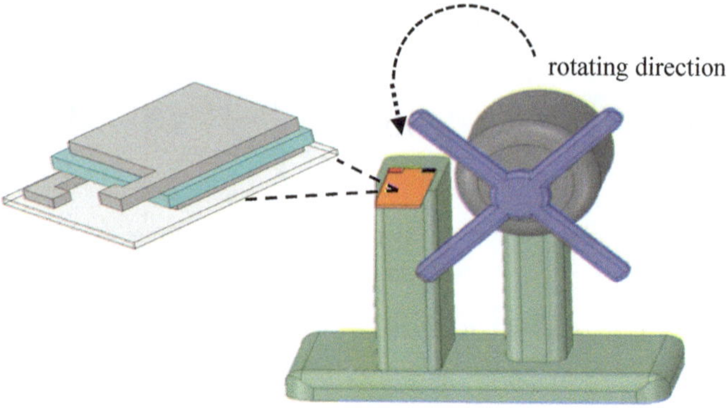

Fig. 3 Experimental setup of force tapping by rotating fan blade

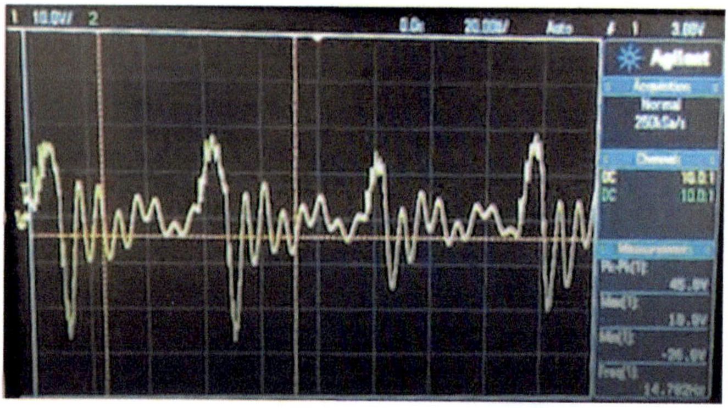

Fig. 4 Voltage (P-P) profile generated by the experimental setup

was controlled by adjusting the motor speed through its power supply [PM3006-2, Digimess]. Speed of the blades were measured by portable digital vibrometer, (PDF-100, Polytec). The current innovation improves the control of tapping force and increases the accuracy of the output voltage measurement.

3 Results and Discussion

Their cross-section were captured using field emission scanning electron microscopy (FESEM, 15 kV accelerating voltage, scanning electron emission mode, Evo.50, Carl Zeis) (see Fig. 5a) where the thickness, 40 μm of P(VDF-TrFE) layer, top and bottom electrodes of thickness, 8 μm on the PET substrate, 175 μm. The distribution of P(VDF-TrFE) harvester elements were determined using Energy-Dispersive Microanalysis (EDS, Oxford Instrument) integrated to Field-Emission Scanning Electron

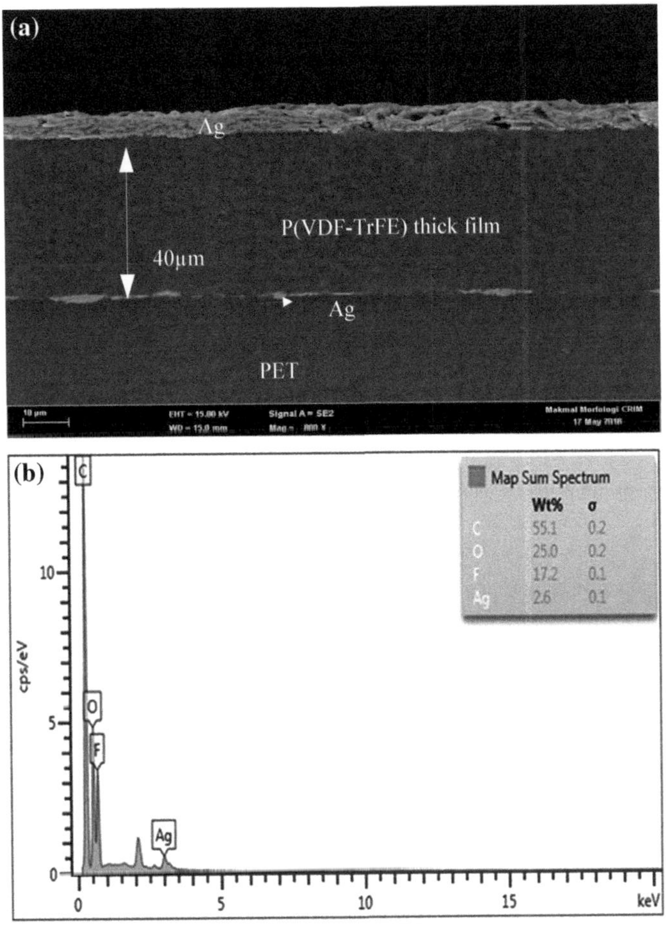

Fig. 5 **a** FESEM micrograph of the electrical harvester structure's cross-section with thickness of 40 μm P(VDF-TrFE) thick film. **b** EDS spectrum elements, P(VDF-TrFE) harvester

Microscope (FESEM, 3 kV accelerating voltage, secondary electron emission mode, Merlin compact-60-25, Carl Zeis) (see Fig. 5b). The elemental composition F, C, Ag, and O confirmed the presence of P(VDF-TrFE) and Ag polymer electrode on the PET substrate [10].

The XRD peak of fabricated P(VDF-TrFE) thick film (see Fig. 6) appeared at $2\theta = 21.03°$ indicates the presence of β-phase at (110/200) reflection plane [11, 12]. The process temperature for annealing and poling was reported in the previous work [13].

The hysteresis loop of ferroelectric of the P(VDF-TrFE) thick film with thickness of 40 μm and electrode area of 2.99 cm^2. This measurement was done using Radiant Precision LC II Ferroelectric tester and the result shows the remnant polarization is about 0.093 mC/m^2 during external voltage of 330 V (see Fig. 7).

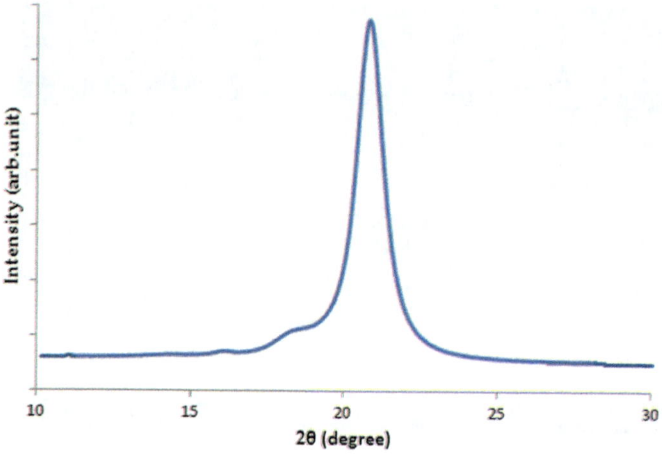

Fig. 6 XRD peaks of β- PVDF phase in P(VDF-TrFE) thick film harvester

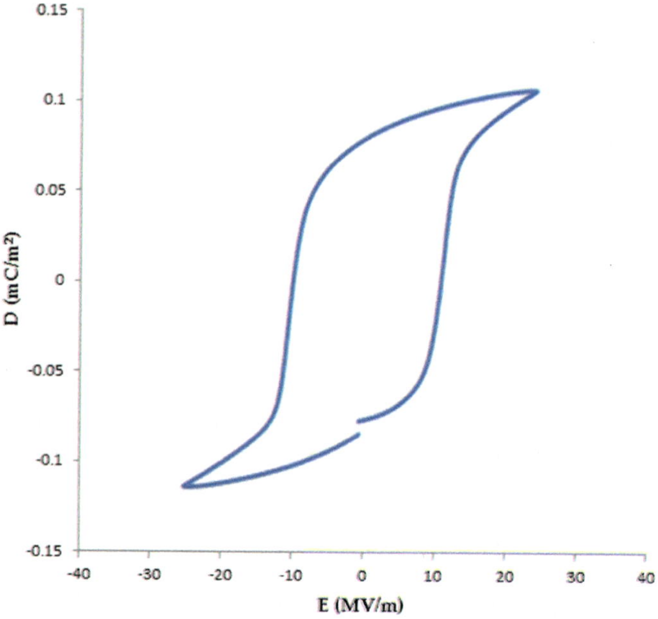

Fig. 7 P-V Hysteresis loop of P(VDF-TrFE) harvester

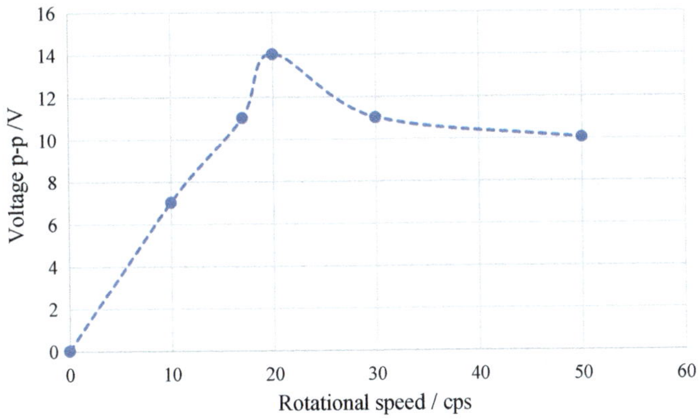

Fig. 8 Voltage output of P(VDF-TrFE) harvester

Maximum output voltage peak-to-peak achieved 7 V when mechanical energy by finger tapping and releasing finger of the hand at frequency of 40 Hz. The output voltage of P(VDF-TrFE) device as function of speed of the rotating blades measured by portable digital vibrometer, (PDV-100, Polytec) and digital oscilloscope. The maximum output voltage peak-to peak was generated about 14 V at rotational speed of 20 cps. The decrease of output voltage reduced slowly after 30 cps (see Fig. 8). The experiment result shows the P(VDF-TrFE) piezoelectric material is compatible with PET substrate, thus the generator is biocompatible for self-powered wearable applications. The P(VDF-TrFE) flexible power generator has mechanical flexibility, capable at low frequency, and has good sensitivity piezoelectric response [14–16].

4 Conclusion

P(VDF-TrFE) thick film was casted on PET substrate to produce piezoelectric micro-power generator device based on mechanical energy. For the first time of the experiment setup, the piezoelectric voltage output was recorded about 7 V (voltage peak-to-peak). A consistent mechanical tapping generated output voltage up to 14 V was obtained under external speed of 20 cps. The research work is clearly show that the P(VDF-TrFE) device micro-power generator have huge potential as self-powered micro-power generator to harvest electric energy from mechanical energy.

Acknowledgements The authors would like to express sincere appreciation to Low Dimensional Materials Research Centre, Department of Physics, Faculty of Science, University of Malaya, Kuala Lumpur, Malaysia for the usage of ferroelectric tester as well as the Ministry of Education of Malaysia for the scholarship and research grant of PRGS/1/2016/TK10/FKEKK-CETRI/02/T00016 and UTeM-Industry Matching GLUAR/IMPRESSIVE/2017/FKEKK-CETRI/I00024.

References

1. Lam Po Tang S (2007) Recent developments in flexible wearable electronics for monitoring applications. Trans Inst Meas Control 29(3–4):283–300
2. Park S, Chung K, Jayaraman S (2014) Wearables: fundamentals, advancements, and a roadmap for the future. Wearable sensors. Academic Press, Oxford, pp 1–23
3. DeRossi D, Lymberis A (2005) New generation of smart wearable health systems and applications. IEEE Trans Inf Technol Biomed 9(3):293–294
4. Zhang WL, Yu YC, Luo WB, Shuai Y, Pan XQ, Wu QQ, Wu CG (2016) Lead free KNN/P(VDF-TrFE) 0–3 pyroelectric composite films and its infrared sensor. Infrared Phys Technol 80:100–404
5. Ismael MR, Gan WC, Majid WH (2014) Abd.: hot plate annealing at a low temperature of a thin ferroelectric P(VDF-TrFE) film with an improved crystalline structure for sensors and actuators. Sensors 14–10:19115–19127
6. Ma S, Ye Tao, Zhang T, Wang Z, Li K, Chen M, Zhang Jing, Wang Z, Ramakrishna S, Wei L (2018) Highly oriented electrospun P(VDF-TrFE) fibers via mechanical stretching for wearable motion sensing. Adv Mater Technol 1800033:1–7
7. Li L, Zhang M, Rong M, Ruan W (2014) Studies on the transformation process of PVDF from α to β phase by stretching. RSC Adv 4:3938–3943
8. Sencadas V, Gregorio R, Lanceros-Méndez S (2009) α to β phase transformation and microestructural changes of PVDF films induced by uniaxial stretch. J Macromol Sci B 48(3):514–525
9. Singh D, Choudhary A, Garg A (2018) Flexible and Robust piezoelectric polymer nanocomposites based energy harvesters. ACS Appl Mat Interfaces 10(3):2793–2800
10. Chow K-K, Kok S-L, Lau K-T (2018) Study of screen printed polymer and ceramic based electrode on P(VDF-TrFE) flexible film. J Telecommun Electron Comput Eng 10(2–8):93–96
11. Pi Z, Zhanga J, Wena C, Zhangb Z, Wu D (2014) Flexible piezoelectric nanogenerator made of poly(vinylidenefluoride-co-trifluoroethylene) (PVDFTrFE) thin film. Nano Ener 7:33–41
12. Won SS, Sheldon M, Mostovych N, Kwak J, Chang B-S, Won Ahn C, Kingon AI, Kim WI, Kim S-H ()Piezoelectric poly(vinylidene fluoride trifluoroethylene) thin film-based power generators using paper substrates for wearable device applications. Appl Phys Lett 107:202901, (2015)
13. Chow K-K, Kok SL, Lau K-T (2018) Design and characterization of piezoelectric P(VDF-TrFE) thick film on flexible substrate for energy harvesting. J Telecommun Electron Comput Eng 10(1):63–68
14. Foster FS, Harasiewicz KA, Sherar MD (2000) A history of medical and biological imaging with polyvinylidene fluoride (PVDF) transducers. IEEE Trans Ultrasonics Ferroelectrics Frequency Control 47(6):1363–1371
15. Fan FR, Tang W, Wang ZL (2016) Flexible nanogenerators for energy harvesting and self-powered electronics. Adv Mater 28:4283–4305
16. Khan S, Tinku S, Lorenzelli L, Dahiya RS (2015) Flexible tactile sensors using screen-printed P(VDF-TrFE) and MWCNT/PDMS composites. IEEE Sens J 15(6):3146–3155

Thermal Degradation of Unsaturated Polyester and Composite Fiberglass Embedded with Aluminium Phosphate

Asmalina Mohamed Saat, Asmawi Abd Malik, Md. Salim Kamil, Mazlan Muslim, Fatin Zawani Binti Zainal Azaim, Adila Azmi, Mohamad Fadzil Abd Latif, Nur Ermadiana Ramlee, Muhammad Amiruddin Ahmad and Mohd Rafie Johan

Abstract The thermal degradation of unsaturated polyester resin and composite fiberglass containing aluminum phosphate as additive flame retardant were investigated. A new phosphate type fire retardant is developed using aluminum phosphate embedded in unsaturated polyester resin with varies weight ratio (0–50%). The modified unsaturated polyester then combined with seven layer of fiberglass reinforcement to produce composite fiberglass boat panels. The effects of aluminum phosphate on structural and thermal properties were evaluated on both unsaturated polyester resin and composite fiberglass using Fourier Transform Infrared (FTIR) and thermogravimetric analysis (TGA). Addition of aluminum phosphate in modified unsaturated polyester observed increases in weight residue due to formation char layer. Thermal properties and structural bonding observed in both modified unsaturated polyester and composite fiberglass show that oxygen has a great influenced in the interaction between phosphate group, aluminum and silica in both systems.

Keywords Composite fibreglass · Phosphate · Thermal degradation

1 Introduction

Thermal degradation is a behavior of a material in a reaction with heat or at elevated temperature which causes a loss of its chemical, physical and mechanical properties. Common polymeric materials and its composites experienced a decomposition process due to chemical and physical reaction. The chemical reaction of polymer produces volatiles flammable products, meanwhile physical reaction produces melting and charring that change the decomposition and burning characteristic of materials.

A. M. Saat (✉) · A. A. Malik · Md. Salim Kamil · M. Muslim · F. Z. B. Z. Azaim · A. Azmi · M. F. A. Latif · N. E. Ramlee · M. A. Ahmad
Malaysian Institute of Marine Engineering Technology, Universiti Kuala Lumpur, Lumut, Perak, Malaysia
e-mail: asmalina@unikl.edu.my

M. R. Johan
Nanotechnology and Catalyst Research, Universiti Malaya, Kuala Lumpur, Malaysia

C. L. Saw et al. (eds.), *Advancement in Emerging Technologies and Engineering Applications*, Lecture Notes in Mechanical Engineering,
https://doi.org/10.1007/978-981-15-0002-2_13

The decomposition of composite polymer starts from break down of molecules to smaller species with various different equilibrium. Lighter molecules will vaporize and heavier molecules will remain in the condensed phase either in solid or liquid. At higher temperature, the condense phase will further decomposed to lighter molecules and end as vapor. Some other composite polymer that does not break to lighter molecules will end as solid residues in the form of carbonaceous, inorganic or combination of both [1]. Char is a good thermal insulator, it significantly inhibit the flow of heat from the gaseous combustion zone back to the condensed phase and thus slowing down the decomposition process or decrease the flammability of a polymer through additive or reactive flame. Composite fiberglass boats are constructed from unsaturated polyester and fiberglass. High flammability of unsaturated polyester polymer is due to oxygen component which lead to high oxidation process. Flame retardants are chemicals which are added to flammable materials to improve their resistant from ignition from small heat source such as cigarette, candle or electrical faults. It delayed the spread of flame thus giving more time to save people and prevent severe losses. Flame retardant inhibits combustion process in a number of stages such as during heating, decomposition, ignition or flame spread. Phosphates type fire retardant is used to interrupt combustion process by formation of char layer and protect the heat flow on the surface of materials. The change in solid state decomposition and char layer are formed due incorporation of flame retardant that act by interfering with radical flame reaction. Phosphate additive reacts in the presence of heat source, releasing phosphoric acid that causes the material to char and form a thick glassy layer of carbon on the surface of specimen hence prevents the release of toxic gases and slow down the spreading of combustion [2]. Thermal stability and flame retardant properties of unsaturated polyester embedded with three different phosphorus types flame retardant additive observed that ammonium polyphosphate (APP) shows lower decomposition compared to melamine pyrophosphate (MPP) and silane-coated APP (S-APP) due to formation char layer [3]. APP also observed higher thermal stability, lower peak of heat release rate (PHRR) and lower heat combustion. Combination of APP with mineral additive, aluminum trihydroxide (ATH) in unsaturated polyester reduce the weight ratio of ATH, improved the thermal stability and flame properties due to occurrence of P–Al–O bond [4]. Aluminum phosphate is a new approach in unsaturated polyester and composite fiberglass system. Aluminum phosphate with 5 wt.% weight ratio shows good interaction with unsaturated polyester and composite fiberglass [5]. However the effect of aluminum phosphate on the thermal properties of unsaturated polyester and composite fiberglass were needed in order to have better understanding of its fire performance. Thus, this research is focused on the thermal degradation properties of unsaturated polyester and composite fiberglass embedded with aluminum phosphate.

2 Methodology

Unsaturated polyester resin Norsodyne 3338 W (Polynt Composite Malaysia Sdn. Bhd.) with 30–45% styrene monomer was modified with aluminum phosphate with various ratios of 0, 5, 10, 15, 20, 25 and 50 wt.%. Seven layers of composites fiberglass were prepared using the mixture of unsaturated polyester resin and aluminum phosphate on fiberglass mat from chopped strand mat and woven roving. Methyl Ethyl ketone Peroxide (MEKP) supplied by Pt Kawaguchi Kimia Indonesia was used as catalyst. Thermal analysis (TGA) of modified polyester and composite has been conducted in air atmosphere with heating rate of 10 °C/min and heated 0–1000 °C for both using Mettler Telode TGA/SDTA851 respectively. The sample used was 10–12 mg and the flow rate are 100 ml/min. Simultaneous differential thermal analysis (SDTA) provides the similar information as DSC/TGA. Sample preparation for modified unsaturated polyester and composite fiberglass are summarize in Table 1.

3 Results and Discussion

Figure 1 shows the FTIR spectra modified unsaturated polyester resin with various weight ratios of aluminum phosphate (0, 5, 10, 15, 20, 25 and 50 wt.%). Introduction of aluminium phosphate shows a significant improvement with occurrence and change spectra of 2928, 1718, 1254, 1119, 1065, 743, 701, 521 and 453 cm^{-1} peak that represent various band assignments C–H, C–O–Al, O–P–O–Al and O–Al respectively. These peaks become intense with increasing ratio of aluminum phosphate until 15 wt.% and then slowly reduced its intensity as observed broaden peak in sample with 50 wt.% aluminum phosphate. The amount of water also varies in the sample due to interaction with aluminum phosphate. All peak in the FTIR spectra shows good interaction between aluminium phosphate and unsaturated polyester resin that contribute to phosphate-carbonaceous char formation [6]. The FTIR spectra for unsaturated polyester composite fibreglass observed at 5 wt.% of aluminium phosphate ratio produce an optimum interaction between composited and aluminium phosphate have been reported earlier [5]. The oxygen element in unsaturated polymer and composite have interaction with phosphate and create bonding that reduced the oxygen available for a fire to ignite or propagate thus improve the fire retardant performance.

Figure 2 shows the TGA and SDTA pattern for modified unsaturated polyester with various aluminum phosphate ratios 0, 5, 10, 15, 20, 25 and 50 wt.%. The TGA pattern for both pure unsaturated polyester and modified polyester occur has simultaneously confirmed aluminum phosphate is an effective flame retardant and have good interaction with unsaturated polyester. Thermal degradation of modified unsaturated polyester produces a condensed phase mechanism which involves dehydration, crosslinking and char formation. The initial degradation observed below 300 °C was

Table 1 Thermal and flame properties of unsaturated polyester and composite fiberglass embedded with various aluminum phosphate ratios

Thermal properties	Fire retardant									
Fire retardant ratio (%)	0	5	10	15	20	25	50	0	5	10
Fiberglass layer (%)	–	–	–	–	–	–	–	7	7	7
TGA weight loss T(5%) (°C)	290	300	270	290	300	300	290	270	270	260
TGA weight loss T(50%) (°C)	406	415	407	413	422	423	420	400	408	400
Weight residue (%)	0	5	7	12	15	18	32	33	32	36

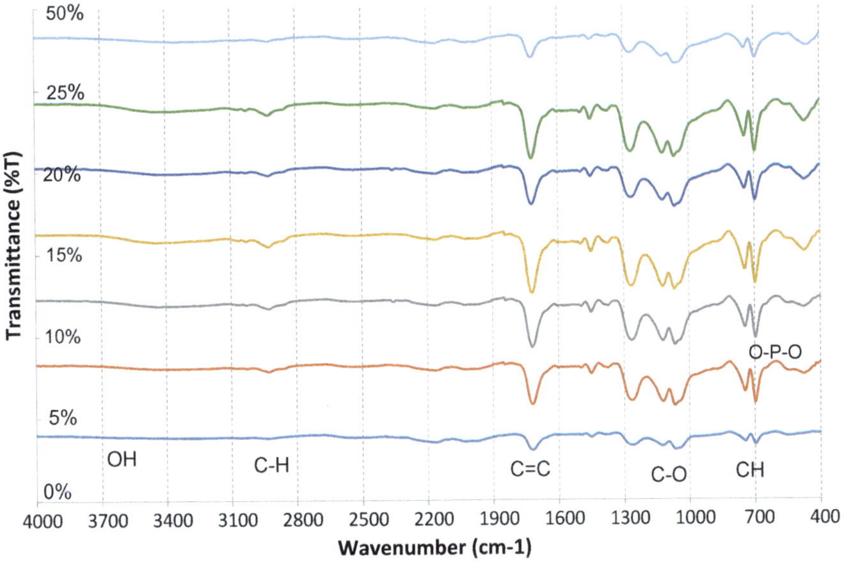

Fig. 1 FTIR spectra modified unsaturated polyester resin with various weight ratios of aluminum phosphate (0, 5, 10, 15, 20, 25 and 50 wt.%)

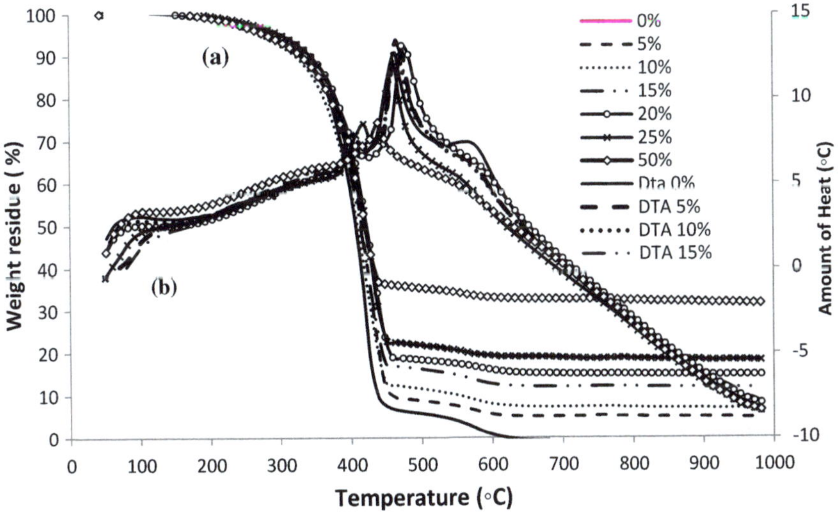

Fig. 2 Thermal analysis of modified unsaturated polyester with various aluminum phosphate ratios (0, 5, 10, 15, 20, 25 and 50 wt.%). **a** TGA; **b** DTA

due to elimination of water, volatile materials and acetate groups as well as crosslinking and formation of double bond. The second degradation stage at temperature 300–550 °C contributes to 90 wt.% occurs due to spontaneous pyrolysis decomposition of pure unsaturated polyester backbones. The degradation of pure unsaturated polyester reported earlier started at temperature 320 °C was within the same range [7]. The third degradation stage and weight losses observed at range of temperature 550–625 °C due to oxidation process and formation of char layer. The remaining unoxidized residue remains as char layer at temperature higher than 700 °C. The weight residue of modified unsaturated polyester increases as the aluminum phosphate ratio increases. The weight residue of modified unsaturated polyester decreases to 32 wt.% as aluminum phosphate reaches to 50 wt.%. During thermal process, the dissociated of water increased the potential of re-associated phosphate group in unsaturated polyester. At lower temperature phosphate creates bond and transform to di-phosphate and tri-phosphate at higher temperature. Decomposition of pure unsaturated polyester breaks the complexes with aluminum phosphate. The decomposition produces phosphoric acid and polyphosphoric acid that catalyze dehydration and crosslinking reaction which contributed to the high char formation and weight residue [8]. DTA shows the exothermic peak shifted to lower temperature as aluminum phosphate increases. However for sample 5, 10 and 20 wt.% of aluminum phosphate ratio observed contras finding with exothermic peak shift slightly to higher temperature compared to virgin unsaturated polyester. The initial weight loss is observed below 150 °C, due to the evaporation of moisture. The first stage of degradation of virgin unsaturated polyester occurs between 150 and 450 °C due to elimination of volatile materials and acetate groups as well as crosslinking and formation of double bond. The maximum peak of exothermic has been observed for all samples at temperature range between 300 to 550 °C due to the formation of carbonaceous char by decomposition of carbon monoxide and carbon dioxide [9]. The modified unsaturated polyester observed less volatile product and higher char retention consequence of good interaction between unsaturated polyester with aluminum phosphate. Degradation of pure and modified unsaturated polyester occurs simultaneously within the same temperature range which confirmed that aluminum phosphate is an effective flame retardant [1].

Figure 3 shows the thermal properties of unsaturated polyester composite fiberglass embedded with various weight ratios of aluminum phosphate (0.5 and 10 wt.%). The TGA pattern for all samples shows similar degradation stage. The weigh residue increases as the aluminum phosphate increase. However at lower weight ratio of aluminum phosphate the weight residue for sample wit 5 wt.% shows lower value than sample with 0 wt.%. Exothermic peak for sample with 5 wt.% shift to higher temperature compare to both sample with 0 and 10 wt.% due to increase in crosslinking in the system [10]. However the sample with 5 wt.% observed less moisture with the occurrence of broaden exothermic at range below than 200 °C due interaction of composite with aluminum phosphate produce more Si–O–Al bond [11]. The decomposition of unsaturated polyester composite is affected with the present of oxygen due to breakage of bond with Si, Al and P, at higher temperature. However sample with 5 wt.% of aluminum phosphate needed higher temperature in order to break

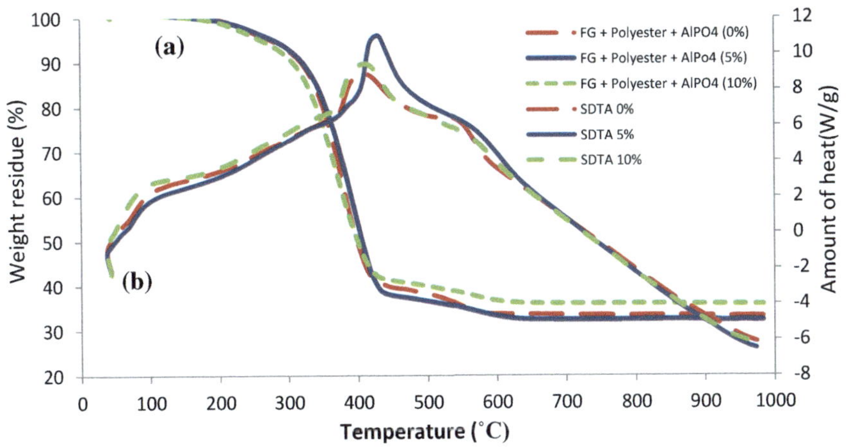

Fig. 3 Thermal analysis of composite fiberglass embedded with various ratio of aluminum phosphate (0, 5 and 10 wt.%). **a** TGA; **b** DTA

the complexes into lighter molecules before further degradation. At increasing temperature the complexes of phosphate, oxygen and Si found in fiberglass break to smaller component and decomposed. The oxygen content varies during the breakage and affected the performance of thermal degradation and formation of char layer. The addition of aluminum phosphate in 10 wt.% composite fiberglass encourage the ability of char layer to form in the materials during combustion [8]. The thermal decomposition of polymer may proceed by oxidative processes or simply by the action of heat. The TGA and DTA of modified polyester resin and composite fiberglass showed that the oxygen and water content plays and important roles to create bonding in polyester and composite fiberglass. Table 1 summarizes the thermal and flame retardant performance for both unsaturated polyester and composite fiberglass.

4 Conclusion

Thermal degradation for both unsaturated polyester and composite fiberglass embedded with aluminum phosphate show simultaneous degradation stage due to good interaction. Weight residue increases as the weight ratio of aluminum phosphate increases in unsaturated polyester due to the formation of more char layer. However, in composite fiberglass the residue varies due to the interaction with silica which is interesting to explore further. Thermal properties and structural bonding in both systems are influenced by the oxygen interaction with phosphate, aluminum and silica in fiberglass.

Acknowledgements This work was financially supported by UniKL Short Term Grant (UniKL/Cori/STRG/16035 & UniKL/Cori/STRG/15126). Author also appreciates of the collaboration with Nanotechnology and Catalyst Research, Universiti Malaya, Kuala Lumpur, Malaysia. The author gratefully acknowledge on both of support.

References

1. Beyler CL Hirschler MM (2005) Thermal decomposition of polymers. In: SFPE handbook of fire protection engineering, 3rd edn, pp 110–131
2. Leopoldshafen E, Dresden P, Str H (2012) Phosphorus polyesters as halogen-free polymeric flame retardants in poly(butylene terephthalate)—influence of the chemical structure
3. Ricciardi MR, Antonucci V, Giordano M, Zarrelli M (2012) Thermal decomposition and fire behavior of glass fiber-reinforced polyester resin composites containing phosphate-based fire-retardant additives. J Fire Sci 30(4):318–330
4. Hapuarachchi TD, Peijs T (2009) Aluminium trihydroxide in combination with ammonium polyphosphate as flame retardants for unsaturated polyester resin. Express Polym. Lett. 3(11):743–751
5. Saat AM, Malik AA, Azmi A, Latif MFA, Ramlee NE, Johan MR (2017) Effect of aluminum phosphate on structural and flame retardant properties of composites fibreglass. ARPN J Eng Appl Sci 12(4):1315–1318
6. Saat AM, Johan MR (2017) Enhanced thermal and structural properties of partially phosphorylated polyvinyl alcohol—aluminum phosphate (PPVA-AlPO4) nanocomposites with aluminium nitrate source. AIP Conf Proc 1901(130011):1–8
7. Bastiurea M, Rodeanu MS, Dima D, Murarescu M, Andrei G (2015) Thermal and mechanical properties of polyester composites with graphene oxide and graphite. Dig J Nanomater Biostructures 10(2):521–533
8. Pan LL, Li GY, Su YC, Lian JS (2012) Fire retardant mechanism analysis between ammonium polyphosphate and triphenyl phosphate in unsaturated polyester resin. Polym Degrad Stab 97(9):1801–1806
9. Al Bayaty SA, Farhan AJ (2015) Polyester and unsaturated polyester reinforcement by Toner Carbon Nano Powder (TCNP) composites. Int J Appl Innov Eng Mange 4(3):139–146
10. Siddharthan KS, Mohan S, Elayaperumal A (2014) Mechanical and thermal properties of glass/polyester composite with glycerol as additive. Int J Eng Trends Technol 7(2):61–64
11. Saat AM, Johan MR (2016) Effect of aluminum source on the structural and thermal properties of partially phosphorylated polyvinyl alcohol composite (PPVA)—aluminum phosphate (PPVA-AlPO4) nanocomposite. Key Eng Mater 701(3):291–294

Effect of Thermal Aging on the Mechanical Properties of Hybrid Multilayer Composite with Various Core Structure

Asmalina Mohamed Saat, Syarmela Alaauldin, Hanisah Johor, Hazwani Mohd Radzi, Fatin Zawani Zainal Azaim, Mohd Saidi Mohd Saad, Harith Johari and Mohamad Taufiq Hamidi

Abstract Steels well known for their sturdiness and long-term life expectancy; however, there are drawbacks such as it is expensive, corroded as exposed to air and difficult to be formed into shape. Hybrid composite fiberglass consists of the combination of matrix, reinforcement and core structure that are well known as cheap, non-corrosive and easy to fabricate with superior characteristic that attract in many applications including utility usage as replacing steel materials. Hybrid composite panels with two different cores (aluminum and plywood) are prepared by hand lay-up method and treated at various temperature 60, 90, 120, and 150 °C. Hybrid composite fiberglass with Aluminum core observed good mechanical properties compared to plywood in the properties of flexural strength, impact and moisture resistant. Flexural strength for both Plywood and Aluminum show significant p values, meanwhile p values for strain in Aluminum observed more significant compared to plywood. Both plywood and aluminum observed significant p values in moisture test and impact strength as temperature increases. However, tensile and hardness properties show lower significant p values, R and R-square value at increasing temperature.

Keywords Thermal aging · Composite · Core structure · Mechanical properties

1 Introduction

Composite is a heterogeneous material consisting of two or more solid phase materials that increase the properties of the material rather than the individual component properties. Composite additionally offers various preferences over conventional material such as resistance to chemicals, thermal and electrical insulation properties. The demand of structural composite as an alternative to replace metals in automotive, aircraft, marine and building industries is due to its easy fabrication, high

A. M. Saat (✉) · S. Alaauldin · H. Johor · H. M. Radzi · F. Z. Z. Azaim · M. S. M. Saad · H. Johari · M. T. Hamidi
Universiti Kuala Lumpur, Malaysian Institute of Marine Engineering Technology, Lumut Perak, Malaysia
e-mail: asmalina@unikl.edu.my

C. L. Saw et al. (eds.), *Advancement in Emerging Technologies and Engineering Applications*, Lecture Notes in Mechanical Engineering,
https://doi.org/10.1007/978-981-15-0002-2_14

strength to weight ratio, lightweight and corrosion resistant [1]. Structural composite is expected to withstand various operational condition related to humidity and temperature. The combination effect of long term loading and environment surrounding can provide harmful effect toward hybrid composite mechanical strength and properties. In general, exposure to lower and elevated temperature lead to brittle behaviour and degradation of mechanical properties (cracking, chalking and flaking) of polymer respectively [2]. It is in need to design composite polymer structure that can sustain the great environment diversity such as large variation of temperature and moisture. Combination of unsaturated polyester and fibreglass composite have a strength that is dependent to the mechanism of load transfer at the interphase and the adhesion strength between the fibre and the polymer matrix [3]. However the properties also can be varied due to the change in matrix, loading condition, microstructure and environment condition (temperature, moisture and corrosive environment) [4]. Buck et al. [5] reported that the combination of moisture and sustained load at elevated temperatures causes a significant decrease in the ultimate tensile strength of E-glass/vinyl-ester composite materials. Thermal aging is a slow and irreversible alteration of a material chemical or physical structure. This alteration has normally a detrimental effect on the material properties. It leads to gradual loss of the design function and ultimate failure or unacceptable loss of efficiency. So basically, thermal aging is the process of irreversible change in the properties of polymers under the action of heat. The key factor in this aging process is thermal which lead to the alterations of the properties of the polymer itself. Aging can occur during storage or treatment of polymers. Aging is caused by chemical transformations of micro molecules, which lead to the degradation of the macromolecules and the formation of branch or structure. The rate of aging depends on the sensitivity of polymer itself. The composite fibreglass in four layers system reported obtained maximum tensile strength (250 MPa) and impact (160 kJ/mm^2) at temperature 90 °C [6]. Laoubi et al. [7] also observed in four layers of composite fibreglass the tensile and flexural strength increased at temperature below 100 °C and dramatically dropped once treated above 200 °C. Impact strength of composite fibreglass reported to have increased in impact strength until temperature 90 °C and then decreased as the temperature increased [6]. Similar finding reported by Elahi et al. 2014 with optimum value for tensile, impact and hardness at 90 °C with 200 MPa, 150 kJ/mm^2 and 55 BHN respectively [8]. Belaid et al. [9] reported that thermal aging is strongly affecting the mechanical properties of polyester fibreglass composite. The Young modulus and stress recorded decreased due to thermal aging effect treated at 80 °C in various periods of time. Similar observation reported for polyester composite fibreglass untreated and treated for 1 h at range of temperature 60–150 °C [10]. Visco, 2011 reported that polyester that is commonly used in marine field immersed in seawater and conducted accelerated aging test at 60 °C. They observed that polyester resin has high degradation compare to other vinyl ester due to different network organization [11]. The water absorption test was also reported increase due to penetration and diffusion of water molecule within polyester composite fibreglass [10]. Moisture absorption was also reported influenced by temperature and time of experimentation. Higher temperature induces greater diffusion and results in higher percentage of moisture absorbed and achieve saturation stage [12]. Meng and Wang [13], concluded that during operation, machining, storage and application the thermal aging of composite materials are affected to the climate environment such as oxygen, moisture and sunlight. Regression analysis is conducted for exploration of interaction and relationship between variables. The main benefit using regression is the capability to determine relative influence of one or more predicted variables on the dependent variables. It is also the diagnose tools to identify the abnormalities in a set of experimental results. The *p*-value less than 0.05 proved that the relationship between variables is highly significant. The R square is a coefficient to determine the percentage of the data fit to the model. Meanwhile, coefficient correlation shows how strong the relationship is positive value means positive relationship, 0 means no relationship and negative value means negative relationship among variables. In this present study the hybrid composite fibreglass with various core have been developed and subjected to various thermal aging and moisture test to analyse the effect of core on the mechanical properties of hybrid composite unsaturated composite fibreglass through experiment and regression analysis.

2 Methodology

Composite fiberglass panels consist of two different core aluminum and plywood were prepared using hand lay-up method with a thickness less than 9 mm. Matrix of unsaturated polyester with trade name Norsodyne 3338 W was used with 2% catalyst of Methyl Ethyl Ketone Peroxide (MEKP). Composite fibreglass was prepared by hand lay-up method in five-design layers using two different core materials plywood and aluminium. The laminated panels were then cured for 72 h and cut into respective sample size according ASTM for mechanical testing such as tensile (D3039), flexural (D790), impact (D6110), hardness (D2240) and moisture test (D5229). Samples were then treated at various temperature 60, 90 and 120 °C in a close oven for 2 h. Tensile and flexural tests were conducted using universal testing machine Instron 1195 with crosshead 10 mm/min and 2 mm/min respectively. Meanwhile impact test conducted by Charpy method used Charpy Izod Impact Tester 300 J (LS-22006-300 J) with a V notch. Hardness test was conducted using Brinnel method with 187.5 kg(f), ball indenter size 2.5 mm and indented at five different points to obtained average. Moisture test was conducted for 30 days and percentage weight change was recorded every 5 day. Regression method of statistical analysis was conducted for all samples in correlation to the change in temperature.

3 Results and Discussion

Figure 1 shows the flexural and impact strength of untreated (30 °C) and treated (60, 90, 150 °C) composite fiberglass with aluminum and plywood core. Meanwhile Fig. 2 shows the moisture test of composite fiberglass with aluminum and plywood core.

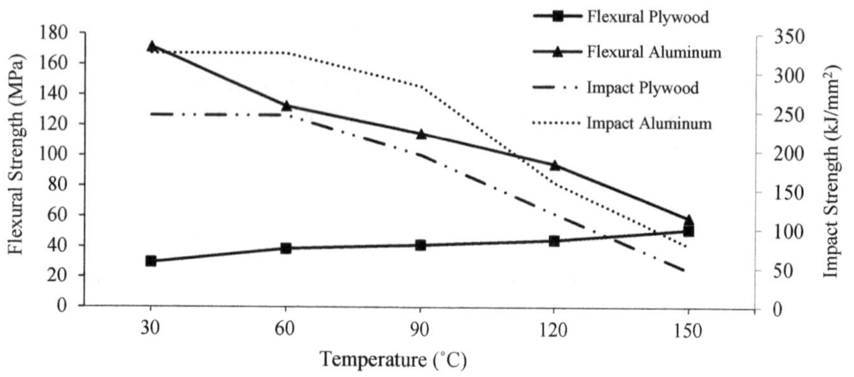

Fig. 1 Flexural and impact strength of untreated (30 °C) and treated (60 °C, 90 °C, 150 °C) composite fiberglass with aluminum and plywood core

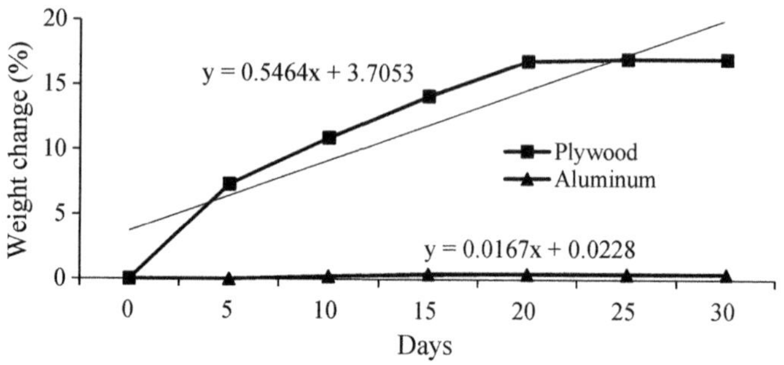

Fig. 2 Moisture test of composite fiberglass with aluminum and plywood core

Table 1 shows the summary of mechanical testing and moisture test for untreated and treated composite fibreglass with aluminum and plywood core. Tensile stress for both core aluminum increase as temperature increase up to 90 °C and it tend to decrease exceeded this temperature. Similar trend observed in four layer of fibreglass conducted by Elahi, et al. [8]. The decreasing of tensile stress is due to the decrease interaction between layer of fibreglass and aluminum core as the temperature exceeds 90 °C. Tensile strain for both core shows increasing in strain, however plywood experience higher strain value and optimum at temperature 120 °C. Meanwhile aluminum observed small increased up to 60 °C then slowly decrease in strain as temperature increased. Flexural stress observed significant increased in plywood core meanwhile aluminum shows a slight decreased as temperature increased. Flexural strain shows decreased pattern as temperature increased and in plywood strain increased optimum up to 90 °C, then decreased as temperature increased to higher temperature. The impact strength for aluminum and plywood recorded as 325.3 and 245.8 kJ/mm^2 respectively which are higher than 140 kJ/mm^2 observed in four layers of fibreglass composite without core reported earlier [8]. Impact strength observed

Table 1 Summary of mechanical testing and moisture test for untreated (30 °C) and treated (60 °C, 90 °C, 150 °C composite fibreglass with aluminum and plywood core

Properties		Aluminum				Plywood			
		p-value	CC,R	R-square (%)	Linear equation	*p*-value	CC,R	R-square (%)	Linear equation
Flexural	Stress	0.0017	−0.9907	98	y = −0.7207x + 174.9052	0.0012	0.9899	98	y = 0.1366x + 29.5179
	Strain	0.0243	0.9252	86	y = 0.000032x + 0.010067	0.9408	0.0465	0	y = 4.67E−06x + 0.0165
Tensile	Stress	0.8579	−0.1119	1	y = −0.0108x + 57.8191	0.7768	0.1762	3	y = 0.0305x + 26.3586
	Strain	0.9785	0.0169	0	y = -9.5465E-07x + 0.0306	0.1610	0.7304	53	y = 0.0002x + 0.0577
Impact	Impact strength	0.0159	−0.9437	89	y = −2.193x + 431.92	0.0097	−0.9596	92	y = −1.74x + 327.54
Hardness	BHN	0.1817	−0.7071	50	y = −0.112x + 183.94	0.7821	0.1720	3	y = 0.0793x + 168.02
Moisture	Moisture content	0.0069	0.8922	80	y = 0.0167x + 0.0228	0.0032	0.9217	85	y = 0.5464x + 3.7053

a decreased pattern as treated with increasing temperature. At temperature 150 °C all samples observed to have less than 100 kJ/mm^2 impact strength value. Hardness value for both aluminum and plywood core observed the same value 187.3 BHN, due to similar fibreglass outer layer in the untreated samples. In the treated samples, both samples observed a decreased in hardness value, however as the temperature increased to 150 °C the value increases slightly in plywood core. Aluminum core observed very small decreased in hardness value as temperature increased. Moisture absorption test observed increased in weight percentage as number of days increased in both aluminum and plywood core. Ounies, 2018 reported that the moisture concentration is higher on the surface and reduce towards the core material which explain only small increase in hybrid composite fibre especially in aluminum core. In hybrid composite fibre a dissymmetric absorption occur, produced lower diffusion due to it need to travel twice distance compared to symmetric absorption. The moisture absorption was also affected by the temperature and time of exposure that contributed to the higher moisture absorbed [12]. The regression analysis shows flexural strength, for both aluminum and plywood core produce significant *p* value square R value for Flexural strength observed significant values 0.0017 and 0.0012 respectively. However in the flexural strain value aluminum observed significant *p* value compare to plywood at increasing temperature. Meanwhile tensile stress and strain for both samples give non significant p values. Increasing temperature affected the hybrid composite fibreglass to sustain better in flexural load compare to tension load. The hybrid fibreglass also produced significant p values for both aluminum and plywood core with 0.0159 and 0.097 values respectively in the impact test. Similar observation recorded in moisture test for both core produce significant *p* value 0.0069 and 0.0032 for aluminum and plywood respectively. All the R-square of significant P values show high percentage more than 80% of the value fit into the linear model from regression analysis. The *p*-value less than 0.05 proves that the relationship between types and temperature is highly significant. The R-square near to 100% shows that the data fit to the linear relations between two variables properties and thermal aging. Coefficient correlation near to one show a strong relationship also supported similar significant finding. It can be concluded that both hybrid composite fibreglass core have a linear relation that produce significant value in flexural stress, impact and moisture test. The different in tensile and flexural stress value for aluminum and plywood core due to different corresponding microscopic failure mechanism. The embrittlement of unsaturated polyester in treated samples produce more growth of microcracks even at lower load even a single microcrack affected tensile strength of hybrid composite. In compression the biaxial/triaxial load leads to larger array of microcrack that leads to an overall toughening in the hybrid composite specimen [14]. The non-significant values in tensile stress, tensile strain, hardness testing were due to thermal aging did not produce linear relation. All samples observed a fluctuating pattern decrease value at lower temperature and reach optimum value then increases up to temperature 150 °C. Increased of hardness value was also reported earlier due to compressive triaxial during hardness which also have similar effect as in flexural test [14]. Table 1 summarize all the mechanical properties and moisture test for untreated (30 °C) and treated (60, 90, 150 °C) composite fibreglass with aluminum and plywood core.

4 Conclusion

Hybrid composite fiberglass with Aluminum core observed good mechanical properties compared to plywood in the area of flexural strength, impact and moisture resistant. Flexural strength for both Plywood and Aluminum shows significant p values, meanwhile p values for strain in aluminum observed more significant compared to plywood. Both plywood and aluminum observed significant p values in moisture test and impact strength as temperature increase. The results showed that composite fibre with Aluminum core suitable in the application related to flexural, impact and better moisture resistant compared to plywood. Thus concluded that hybrid composite fibre both core aluminum and plywood suitable for structural application at various thermal aging.

Acknowledgements This work was financially supported by UniKL Short Term Grant (UniKL/Cori/STRG/15126 and UniKL/Cori/STRG/16135).

References

1. Campbell F (2010) Introduction to composite materials. Struct Compos Mater 1–29
2. Sanders B (ed) (2004) Characterization and failure analysis of plastics. ASM Int Mater 125–136
3. Jesson DA, Watts JF (2012) The interface and interphase in polymer matrix composites: effect on mechanical properties and methods for identification. Polym Rev 52(3–4):321–354
4. Wang J, GangaRao H, Liang R, Liu W (2016) Durability and prediction models of fiber-reinforced polymer composites under various environmental conditions: a critical review. J Reinf Plast Compos 35(3):179–211
5. Buck SE, Lischer DW, Nemat-Nasser S (1998) The durability of E-glass/vinyl ester composite materials subjected to environmental conditioning and sustained loading. J Compos Mater 32(9):874–892
6. Ganeshan P, Raja K (2016) Study on the mechanical properties of glass fiber reinforced polyester composites. Int J Adv Eng Technol VII(II):261–264
7. Laoubi K, Hamadi Z, Ahmed Benyahia A, Serier A, Azari Z (2014) Thermal behavior of E-glass fiber-reinforced unsaturated polyester composites. Compos Part B Eng 56:520–526
8. Elahi F, Hossain MM, Afrin S, Khan MA (2014) Study on the mechanical properties of glass fiber reinforced polyester composites. In: International Conference on Mechanical Industrial and Energy Engineering, vol ICMIEE-PI, pp 1–5
9. Belaid S, Chabira SF, Balland P, Sebaa M, Belhouideg S (2015) Thermal aging effect on the mechanical properties of polyester fiberglass composites. J Mater Environ Sci 6(10):2795–2803
10. Hossain MM, Elahi AHMF, Afrin S, Mahmud MI, Cho HM, Khan MA (2017) Thermal aging of unsaturated polyester composite reinforced with e-glass nonwoven mat. Autex Res. J. 17(4):313–318
11. Visco AM, Brancato V, Campo N (2011) Degradation effects in polyester and vinyl ester resins induced by accelerated aging in seawater. J Compos Mater 46:2025–2040

12. Ounaies M, Harchay M, Dammak F, Ben Daly H (2018) Prediction of hygrothermal behavior of polyester/glass fiber composite in dissymmetric absorption. J Compos Mater
13. Meng J, Wang Y (2016) A review on artificial aging behaviors of fiber reinforced polymer-matrix composites. MATEC Web Conf 67:6041
14. Odegard GM, Bandyopadhyay A (2011) Physical aging of epoxy polymers and their composites. J Polym Sci Part B Polym Phys 49(24):1695–1716

Design and Fabrication of a Parallel Mechanism for Foot/Ankle Rehabilitation Therapy

Subramaniam Krishnan, Ananda Krishna, Amer Nordin, Victor Amirtham, A. M. A. Rani and T. V. V. L. N. Rao

Abstract This paper presents the development of a low-cost and mobile parallel manipulator for treatment of ankle sprain rehabilitation. The manipulator is called Ankle Rehab Robot (ARR) which has bio-inspired design devised after careful studies of improvement opportunities in the existing ankle rehabilitation devices. Ankle Rehab Robot employs lightweight but powerful pneumatic muscle actuators (PMA) which imitates skeletal muscle in actuation. Some of the key existing sprained ankle rehab robots/prototypes (Intermediate Complexity Devices) was carefully studied and reviewed in order to find the limitations or improvement opportunities. The research approach has been conversed in brief and the ARR design has been described and analysed. Actuator and prototype testing were conducted to obtain specification data of the proposed Ankle Rehab Robot (ARR). For evaluation of mobility and economics, some of the data collected were used for comparison between some of the key (reviewed) existing ankle rehabilitation devices and the ARR Manipulator. From the results attained, target of the research was successfully met. Future works and refinements have been discoursed to improvise the existing Ankle Rehab Robot (ARR).

Keywords Ankle rehabilitation robot (ARR) · Intermediate complexity devices (ICD) · Pneumatic muscle actuator (PMA)

S. Krishnan (✉) · A. Krishna · A. Nordin
Faculty of Mechanical Engineering, Manipal International University (MIU), 71800 Nilai, Negeri Sembilan, Malaysia
e-mail: krishnan.subramaniam@miu.edu.my

V. Amirtham
Department of Mechanical Engineering, Politeknik Ungku Omar, 31400 Ipoh, Perak, Malaysia

A. M. A. Rani
Department of Mechanical Engineering, Universiti Teknologi PETRONAS (UTP), 32610 Seri Iskandar, Perak, Malaysia

T. V. V. L. N. Rao
Department of Mechanical Engineering, SRM Institute of Science and Technology, Kattankulathur 603203, India

C. L. Saw et al. (eds.), *Advancement in Emerging Technologies and Engineering Applications*, Lecture Notes in Mechanical Engineering,
https://doi.org/10.1007/978-981-15-0002-2_15

1 Introduction

Foot is one of the most important limbs in human body which supports the human body. Foot movements are controlled by ankle joint which is one of the most important joints in the human body. However due to several reasons such as vigorous physical exercise or walking on an uneven surface, forces the foot and ankle into an unnatural position [1] which leads to ankle injuries such resulting in the functional loss of the ankle and this affects our daily life. Moreover among the types of ankle injuries, ankle sprains are the most common injuries [2] where literature reveals that the number of people who got affected by ankle sprain in Malaysia is estimated to be 73,507 which is 0.0024% of the people in Malaysia per annum. So in order to regain full functionality of ankle/foot, rehabilitation therapy was introduced [3] and various methods were developed ever since in order to fulfil the ankle rehab therapy. One of the methods of approaching ankle rehabilitation therapy was usage of automation (robots) which displayed enhanced performance compared to other methods. Currently there are various types of existing ankle rehab robot or prototypes in this rehabilitation field which can be mainly classified into serial robots or parallel robots. So the focus of this study will be on proposing an Ankle Rehabilitation Parallel Robots/Mechanism for sprained ankle treatment by improving weakness of previous or existing designs. Various robot-assisted sprained ankle rehabilitation devices (Intermediate Complexity Devices) have been developed by using parallel mechanism or shortly known as parallel robots [4, 5].

2 Parallel Mechanism

There are two types of mechanism or in particular, which is serial mechanism and parallel mechanism which has been used in the system of bodies (Linkages) designed to convert motions of, and forces on, one or several bodies into constrained motions of, and forces on, other bodies. Parallel mechanism as shown in Fig. 1 is defined as

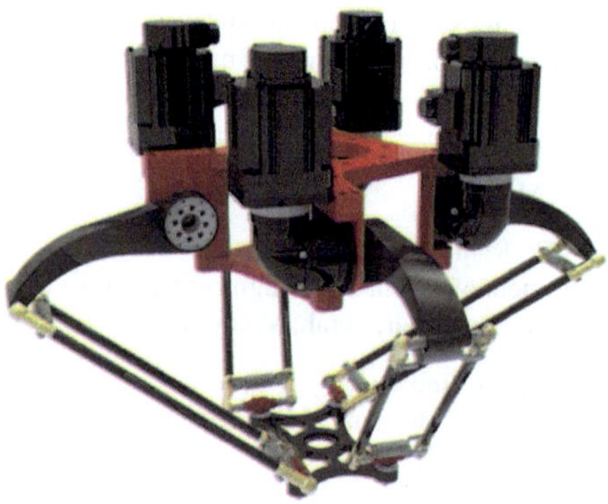

Fig. 1 6 DOF DELTA robot [6]

a closed-loop mechanism in which the end-effector is connected to the base by at least two independent kinematic chains in which the links will be functioning in the parallel manner [6, 7]. From the above definition, parallel robot can be described as a developed version of serial robot, which is formed when the serial chains (At least 2 serial robots) are attached together forming a loop working mechanism. In most cases, only the first joints are actuated and the rest are passive which is only meant for motion transfer. Over the years the continued development of pneumatic drives has moved from the utilization of individual components to the implementation of complex automation system [8]. Industrial robots have used three primary actuator types; electric motors (AC or DC), hydraulic cylinders and pneumatic cylinder. Their performance is characterized by parameter such as power/weight ratio, strength response rate, physical size and cost. Therefore, pneumatic actuators are finding increased use in robotic system due to low cost, quickness of response and high power/weight and power/volume ratios [9–12].

3 Intermediate Complexity Device

ICD or also known as robotic platforms manipulate the user's foot using their end effectors and are generally developed to facilitate the treatment of ankle sprains. Robotic platforms are usually designed to be used while patients are sitting [13, 2]. Most of these electromechanical systems allow patients to move and stretch the muscles and tendons gently; usually their movements are similar to the basic ankle movements as shown in Fig. 2 [3]. These machines focus more in ROM restoration and improving the flexibility of ankle muscles [13, 14].

4 Ankle Rehabilitation

Rehabilitation is a process by which, through physiotherapy it is possible to recover physical abilities lost due to a traumatic incident. The objective of rehabilitation is to regain full function without limitations. If this is not possible, basic ability to perform daily activities are tried to be regained. The general rehabilitation therapy for ankle sprains could be divided into three phases in which duration of each phase depends on the individual healing process [1, 2]. Initial phase is focused on reducing effusion and swelling at the affected. The second phase is rehabilitation phase which involves ROM and muscle strengthening exercises and proprioceptive exercise [15]. The final phase is functional phase which involves preparation for return to full activity and includes jumping and running as well as isokinetic exercises. If full practice is tolerated without any pain in the injured part, the patient may return to their daily activity or sports for athletes [15].

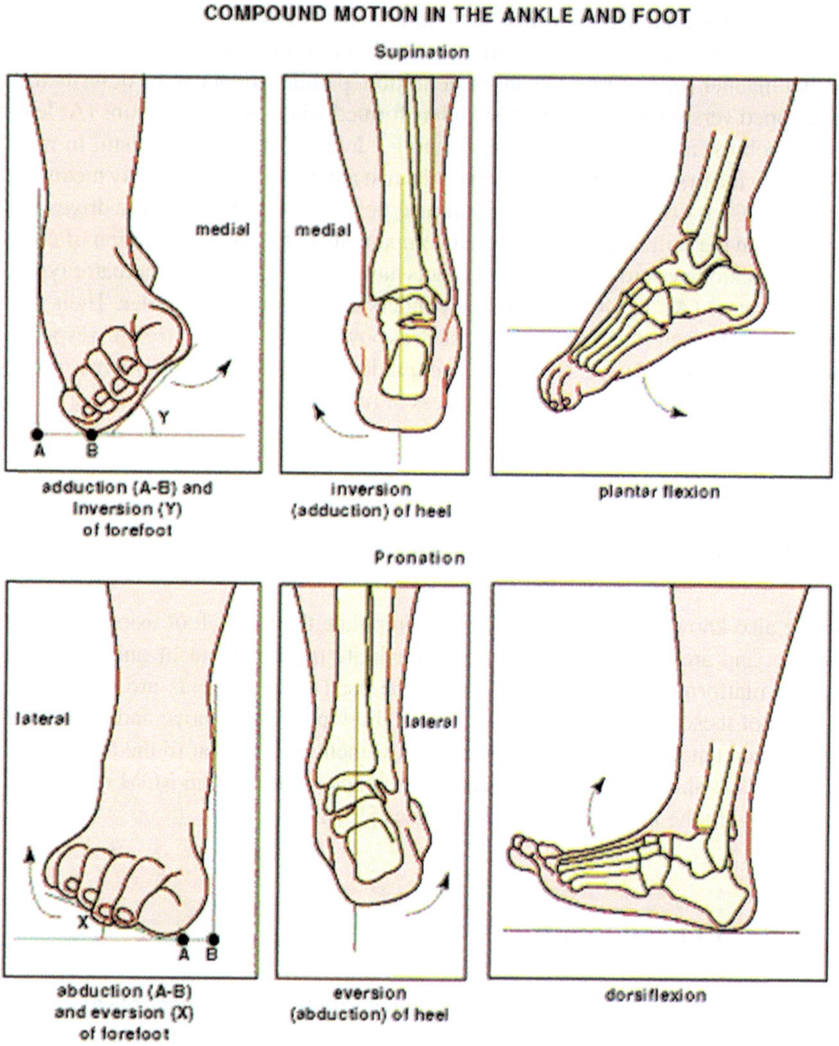

Fig. 2 Compound motion in the ankle/foot [3]

5 Methods and Materials

5.1 Ankle Rehab Robot Design Phase

The designing process of Ankle Rehab Robot (ARR) involves 4 main stages which is shown in Fig. 3. The first stage is Design Specification in which the design constraints or target are set to be followed throughout the designing course. Then was proceeded to Conceptual Design, where several concepts were generated through

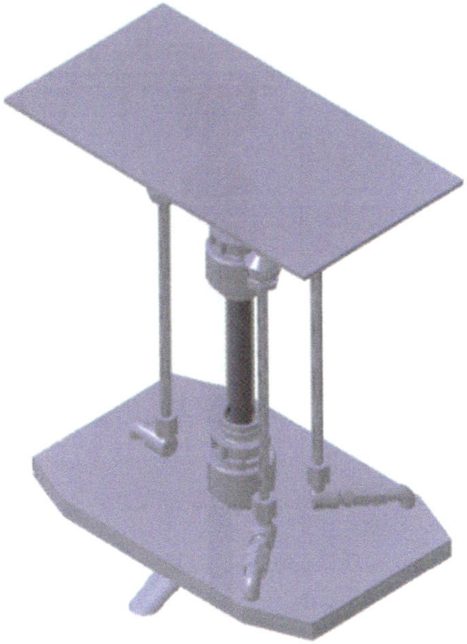

Fig. 3 3D CAD model of ankle rehab robot (ARR)

systematic methods and the best feasible concept was chosen. The following stage is Embodiment Design which is a process of improvising the final concept design to come up with a logical and functionable design/product and finally ends with detailed design which is compilation of scale drawing for each and every assembly and the whole robot with required dimensions.

The final design of Ankle Rehab Robot (ARR) which was developed in CAD software CATIA V5 is shown in Fig. 3. The manipulator design consists of 4 major assemblies. These assemblies are top assembly, central strut assembly, bottom assembly and actuator assembly as shown in Fig. 4. For top assembly it consist of 2 parts, which are foot platform which acts as the base which patients place their foot/ankle and top platform which is the supporting sub-assembly to the foot platform. Followed by central strut assembly which is the main support to the whole structure which is located in between Top and Bot assembly and it is the strongest assembly which will be experiencing most of the force from both top and bottom assembly. It also allows movement of the top assembly through specific joint system. Central strut together with the top and bot assembly forms "I" shaped structure which is very efficient form for carrying both bending and shear loads as proved by Beam theory. For bottom assembly it consists of 2 parts which are the bottom platform and extenders. Bottom platform is the weight- bearing sub-assembly which is located beneath central strut where compression load will be applied on.

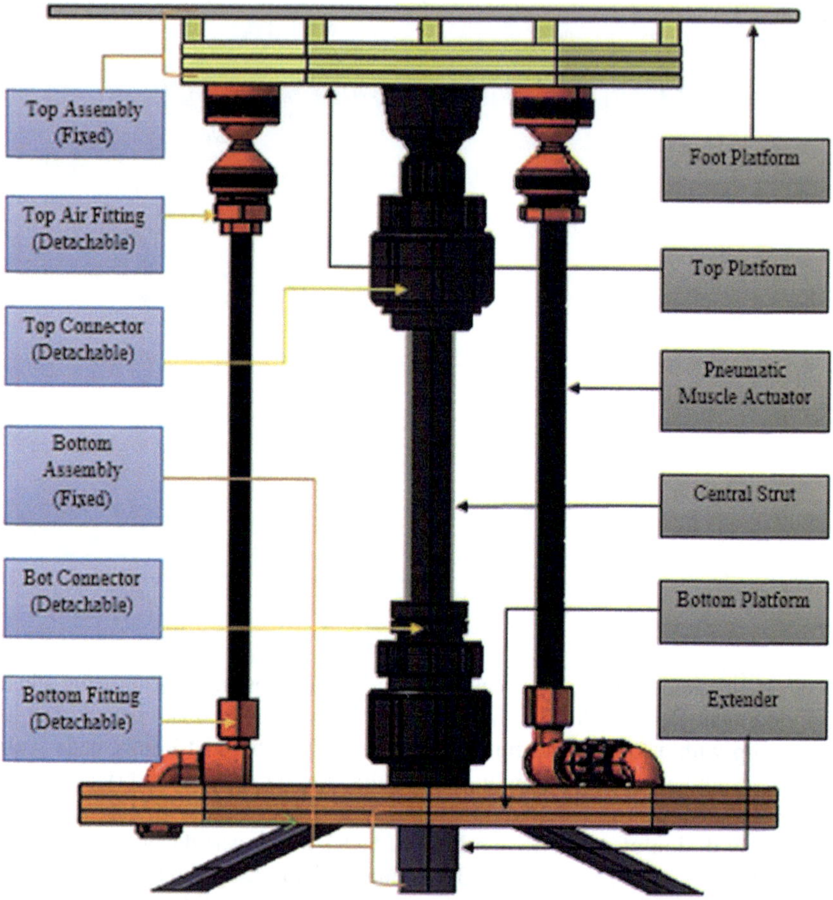

Fig. 4 Configuration of ankle rehab robot (ARR)

5.2 *Ankle Rehab Robot Fabrication Phase*

The fabrication of Ankle Rehab Robot (ARR) was done in 3 stages which are as shown in Fig. 3. The first stage involves actuator construction in which the PMA's to be utilized in the prototype is fabricated based on guides from several literatures [16, 17]. Most of the fabrication process of the actuator is similar to the stated literatures except for few modifications done to end fittings of the actuators to suit the design of the prototype. The next part of fabrication involves the frame construction in which the structure or body of the manipulator is built. Each assembly are built systematically by using appropriate engineering techniques. The final stage of fabrication is complete prototype construction, in which all the main assemblies are brought together as design plan to complete the Ankle Rehab Robot (ARR). The fabrication of ARR was carefully done based on design plan, and the overall specifications of the ARR prototype will be presented in the results and discussion section.

5.3 Ankle Rehab Robot Testing Phase

The testing process of Ankle Rehab Robot (ARR) was done in 2 stages which are preliminary testing and prototype testing. For the preliminary testing [18–20], these tests are generally done to evaluate the fabricated air muscle (PMA) which will be deployed as actuators in the prototype that is being built. Various crucial specification of the fabricated PMA's such as the maximum force produced and maximum contraction percentage are obtained. As for this report, the data obtained from testing the fabricated PMA's will not be presented in detail but general data will be shown in the results and discussion part. For the prototype testing, it is done on the whole assembly of prototype to obtain some of the technical specification of the Ankle Rehab Robot. The testing involves range of motion (ROM) testing and also simple weight testing. All the data obtained from the testing will be presented in the results section and discussed accordingly.

6 Results and Discussion

The pressure, force and contraction percentage is shown in Figs. 5 and 6.

All the readings obtained are average value for all three muscles from three trials. Some of the important data's collected are the actuating pressure for all the PMA's starts approximately at 15 PSI. The maximum force that could be produced from a single PMA is 7.25 N and the maximum contraction that a single PMA can achieve is up to 18%. Currently the maximum pressure that each muscle can withstand is up to 45 PSI after consideration of safety factors. Form the results obtained, the force

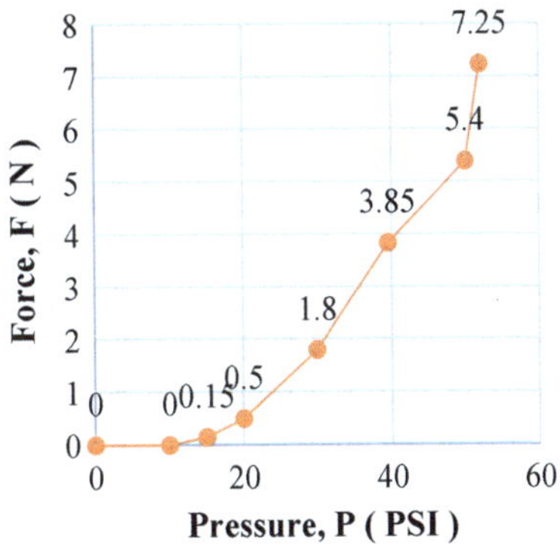

Fig. 5 Pressure versus force relationship of ankle rehab robot actuator

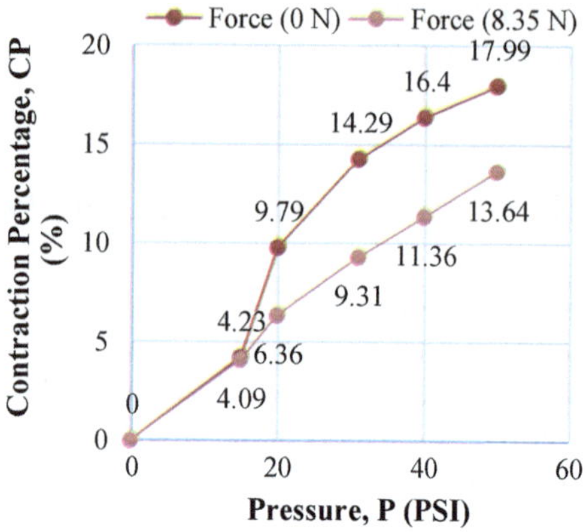

Fig. 6 Pressure versus contraction percentage relationship of ankle rehab robot actuator

produced and contraction percentage are not as expected which has been analyzed and will be improved in the second prototype. Overall the patterns produced from the preliminary testing's are almost similar to the literature [16–19], result which proves that the muscle has a stable performance and can be used.

7 Conclusions

In this study, a highly mobile and low-cost Ankle Rehab Robot (ARR) has been successfully developed. The design of ARR is bio-inspired and was done via specific methods that comprises of design target, conceptual design, embodiment design and finally detailed design. Various design factors were considered such as part count, sufficient kinematic links, material usage, assembly configuration and joinery methods (assembly and disassembly) to achieve the target of the research. The fabrication process of the ARR Manipulator involves systematic engineering methods which are necessary and minds the target of the research at the same time. The developed Ankle Rehab Robot (ARR) Manipulator has "I" beam structure with 4 major parts which are the top assembly, bottom assembly, central strut assembly and actuator assembly. The 4 main assemblies are attached together by threaded components (Modular Design) to enable easy and fast assembly and disassembly process which facilitates high mobility. The manipulator consist of 3 kinematic links with FPS joint configurations to enable 2 DOF motions which are (+20.50) for dorsiflexion, (−23.00) for plantar flexion, (+19.50) for inversion and (−17.50) for eversion. The ARR manipulator uses pneumatic muscle actuators (PMA) as source of actuation, with total of 3 sets of actuators which are 1.2 cm in diameter, 18 cm in length and can be pressurized up to 45 PSI. Currently the ARR prototype was able to fit average foot size up to 11. From the analysis and comparison made, the fabricated ARR prototype has been effectively improved in sense of mobility and cost compared to other key existing ankle rehab devices.

References

1. Alcocer Rosado W, Vela Valdéz LG, Blanco Ortega A, Gonzalez Rubio JL, Oliver Salazar MA (2012) Major trends in the development of ankle rehabilitation devices. DYNA 79(176):45–55
2. Jamwal PK, Xie SQ, Hussain S, Parsons JG (2014) An adaptive wearable parallel robot for the treatment of ankle injuries. IEEE/ASME Trans Mechatron 19:64–75
3. Ben-Dov D, Salcudean SE (1995) A force-controlled pneumatic actuator. IEEE Trans Robot Autom 11:906–911
4. Krishnan S, Nagarajan T, Rani A, Ambaraj W, Ramiah R (2006) Rehabilitation for foot/ankle-continuous passive motion (CPM) using shape memory alloy (SMA) actuated stewart platform
5. Abdul-Rani AM, Krishnan S, Nagarajan T, Rao T, Ramiah R, Ambaraj W (2017) Early mobilization using heat treated shape memory alloy (SMA) for rehabilitation phases. Key Eng Mater 112–116
6. Girone M, Burdea G, Bouzit M (1999) The Rutgers ankle orthopedic rehabilitation interface. Proc ASME Dyn Syst Control Div 67:305–312
7. Abdul-Rani AM, Krishnan S, Nagarajan T, Rao T (2017) Inverse kinematics-stewart platform actuated by shape memory alloy for immobilization of ankle-foot rehabilitation. Key Eng Mater 105–111
8. Deaconescu T (2007) Studies regarding the performance of pneumatic muscles. In: Proceedings of international conference on economic engineering and manufacturing systems, pp 271–275
9. Nagarajan T, Krishnan S, Amirtham V, Abdul-Raniand AM, Rao T (2013) Experimental investigation-natural fiber braided sleeve for pneumatic artificial muscles actuation. Asian J Sci Res 6:596
10. Krishnan S, Nagarajan T, Rani AMA, Rao T (2012) Silk pneumatic artificial muscle (SPAM) construction for bio-medical engineering application. In: Business engineering and industrial applications colloquium *(BEIAC)* IEEE, pp 302–306
11. Krishna S, Nagarajan T, Rani A (2011) Review of current development of pneumatic artificial muscle. J Appl Sci 11:1749–1755
12. Schulte H (1961) The characteristics of the McKibben artificial muscle. In: The application of external power in prosthetics and orthotics. National Academies of Science, National Research Council, Washington, DC
13. Krishnan S, Abdul-Rani A, Nagarajan T, Darus AN, Ab Kadir M, Nanimina A (2017) Experimental investigation of shape memory alloy (Sma)—Force and angle analysis
14. Nagarajan T, Krishnan S, Rani A, Amirtham V, Rao T, Ambaraj W et al (2015) A novel approach of using SMA wires actuated Stewart platform for continuous passive motion (CPM) of ankle rehabilitation therapy. Adv Mater Res 494–498
15. Saglia JA, Tsagarakis NG, Dai JS, Caldwell DG (2009) A high-performance redundantly actuated parallel mechanism for ankle rehabilitation. Int J Robot Res 28:1216–1227
16. Obiajulu SC, Roche ET, Pigula FA, Walsh CJ (2013) Soft pneumatic artificial muscles with low threshold pressures for a cardiac compression device. In: ASME 2013 International Design Engineering Technical Conferences and Computers and Information in Engineering Conference
17. Chou C-P, Hannaford B (1996) Measurement and modeling of McKibben pneumatic artificial muscles. IEEE Trans Robot Autom 12:90–102
18. Tondu B (2012) Modelling of the McKibben artificial muscle: a review. J Intell Mater Syst Struct 23:225–253
19. Belforte G, Eula G, Ivanov A, Sirolli S (2014) Soft pneumatic actuators for rehabilitation. Actuators 84–106
20. Al-Ibadi A, Nefti-Meziani S, Davis S (2017) Efficient structure-based models for the McKibben contraction pneumatic muscle actuator: the full description of the behaviour of the contraction PMA. Actuators 32

Mechanical Sieve Grading of Silica Sand from Tin Mining for Metal Casting Mould

A. Azhar, D. Kamaruzzaman and M. Y. Khairul Azuan

Abstract Malaysia has an estimated resource of 640 Mt of silica sand, of which 492 Mt from tin mine tailing sand and 148 Mt are natural silica. Tin mine tailing sand was one of residue produced during the extraction of tin, which the silica content is 94–99.5%. States of Perak, Selangor, Pahang and Johor of the Malaysia Peninsular have been mined for alluvial tin deposits over the past 100 years. The tin mining industry was once a major contributor to the Malaysian economy. By the end of the 19th century, it was the largest tin producer which was supplying about 55% of the world's tin, but since the collapse of the world tin market in 1985, the tin mining industry in Malaysia has been declined. The objectives of this study are to determine the grain size distribution through the cumulative curve and average grain size by conducting mechanical sieve grading of silica sand from tin mining. Five samples were taken from states of Perak, Selangor, Johor and Pahang and grain size above 425 μm was eliminated in order to comply the foundry sand size. The samples were graded using mechanical sieve shaker accordingly to prescription by Foseco Ferrous Foundryman's Handbook and analysed for the grain size distribution and average grain size. These samples were compared with typical grading of foundry sand as stated in Foseco Ferrous Handbook and sand from RCS Manufacturing Company, the company who supply foundry sand for Proton Casting Company. This study found that the grain size distribution and average grain size of silica sand from identified tin mines are suitable for making metal casting mould. The majority of the silica sand for all samples distributed within the acceptable range which is 75–95% of the total weight samples tested. The average grain size of the samples found within the acceptable size which is ranging 214–281 μm. As a conclusion this research discovered that in terms of mechanical sieve grading, silica sand from identified tin mining in these three states are within the requirement underlines by Foseco. It is

A. Azhar · D. Kamaruzzaman (✉) · M. Y. Khairul Azuan
Mechanical Engineering Department, Politeknik Ungku Omar, Ipoh, Perak, Malaysia
e-mail: mandaud@puo.edu.my

A. Azhar
e-mail: azhar@puo.edu.my

M. Y. Khairul Azuan
e-mail: azuan@puo.edu.my

C. L. Saw et al. (eds.), *Advancement in Emerging Technologies and Engineering Applications*, Lecture Notes in Mechanical Engineering,
https://doi.org/10.1007/978-981-15-0002-2_16

suggested to investigate other properties such as chemical, physical and mechanical properties to identify the suitability of silica sand from tin mining as foundry sand for metal casting mould.

Keywords Silica sand · Tailing sand · Mechanical sieve grading · Grain distribution · Grain size

1 Introduction

SILICA sand is the most popular sand in metal casting mould, other are zircon, olivine and chromite. Silica sand can be broadly grouped into two types, depending essentially upon their chemical composition, which are clay-free sand (synthetic sand) or washed sand and clay-bearing sand (naturally-bonded sand). Malaysia has an estimated resource of 640 Mt of silica sand, of which 492 Mt from tin mine tailing sand and 148 Mt are natural silica [1]. States of Perak, Selangor, Pahang and Negeri Sembilan of the Malaysia Peninsular have been mined for alluvial tin deposits over the past 100 years. The tin mining industry was once a major contributor to the Malaysian economy. By the end of the 19th century, it was the largest tin producer which was supplying about 55% of the world's tin [2], but since the collapsed of the world tin market in 1985, the tin mining industry in Malaysia has been declined. Sand called as tailing sand is the most common by—product of the tin mining operation and benefited for the construction industry for many years due to abundant supply [3]. Silica sand from tin mine produced during the extraction of tin, soil was washed out by using gravel pump where produced tremendous amount of tailings sand, which the silica content is 94–99.5% [2]. Previous research on the silica sand from Perak River found that it's physically and chemically suitable for making greensand casting mould [4]. The ideal purity for silica sand as a foundry sand is that the material should have a silica content of 95–96% although 98–99% is often preferred [5].

Grading by sieving using mechanical shaker is preliminary test to identify the grain size distribution, cumulative curve and average grain size of the silica sand from identified tin mining areas. These properties have an important influence upon mechanical properties such as permeability, green compression strength [6]. It is determined by grading on the standard sieves to separate selected fractions for weighing. Comparison can be made by plotting a histogram and a curve graph of the cumulative percentage against each sieve number [7]. Grain size distribution and cumulative curve show whether the sand is well sorted or not while the average grain size determine the size of the sand is compiled with the requirement from Foseco Foundryman's Handbook or common practices in foundry industry. Table 1 shows the requirement

Table 1 Foseco's requirement and typical UK and German foundry sands [8]

Sieve size		Sand type					
		Foseco	UK sands		German sands		
μm	BSS no	Typical grading of sand suitable for greensand system	Chelford 60	Chelford 50	H32	H33	F32
1000	16	Nil	Trace	Nil			
700	22	0.14	0.7	0.4			
500	30	1.42	4.5	2.3	1.0	0.5	1.0
355	44	5.44	19.8	10.0	15.0	7.5	7.0
250	60	22.88	44.6	25.7	44.0	30.0	30.0
210	72	19.64	21.6	23.8	39.0	60.0	60.0
150	100	26.74	8.2	28.7			
100	150	9.7	2.6	7.6			
75	200	2.84	Nil	1.3	10	2.0	2.0
−75	−200	1.1	Nil	0.2	Nil	Nil	Nil
Average grain size, mm		0.219	0.275	0.23	0.27	0.23	0.23

Note Haltern 32, 33 and Frechen 32 are commonly used, high quality German sands. German sieve grading are based on ISO sieves [8]

of grain size distribution and average grain size according to Foseco, typical UK and German foundry sands.

Generally, average grain size of foundry sands range between 150–400 μm and the most commonly used is 220–250 μm. Majority of the grain size in the UK falls in the range of 150–300 μm, where for ferrous casting within 210–300 μm and 170–270 μm for non-ferrous work [9]. For ferrous casting such as steel is a size of AFS 60-40 and for gray iron is a size of AFS 60-50 [3]. Majority of the sand is preferred with a grain size distribution spread over four to five consecutive sieves [6]. Finer sand is associated with better surface finish but resulting in poor permeability compared to coarser grains which is better in permeability. Another disadvantage of finer grains is it appears to be more easily fused than coarser ones. The spread of the grain size distribution is 95% of the sand retained on four or five screens, or more than 10% on each sieve [8]. The percentage sand retained on 425 μm shall be 9–14%, on 300 μm is 20–30%, on 212 μm is 30–40% and 150 is 15–25% [10]. Professor Boswell, the eminent geologist who suggested that moulding sands should be classified by sieving into grades named 'gravel' (greater than 2 mm), 'very coarse' (2–1 mm), 'coarse' (1–0.5 mm), 'medium' (0.5–0.25 mm), 'fine' (0.25–0.1 mm), 'coarse silt' (0.1–0.05 mm), 'fine silt' (0.05–0.01 mm) and 'clay grade' (less than 0.01 mm) [7]. According to American standard sand, all the grains pass through a 420 μm sieve,

95% pass the 297 μm sieve, and remain on the 210 μm. Any grain passing the 210 μm sieve must remain on the 149 μm sieve [11].

The grain size distribution is determined by the cumulative curve plotted from the data obtained after the mechanical sieve grading. The quality of the castings is affected by the grain size distribution of the sand. Poor surface finish of the casting can be occurred because of coarse-grained sands allow metal penetration into moulds and core. Better surface finish can be obtained with fine-grained sands but need higher binder content and may cause gas defects in castings due to low permeability [8]. Figure 1 shows the 'S' curves of foundry sands from various locations in the UK while Fig. 2 shows the cumulative 'S' curves for well graded and poorly graded sand. A wide distribution of the grains, with a small percentage of the grains on each of the sieves, produce curve that begins in the bottom left-hand corner and slopes at an angle of 45° to the upper right-hand corner [12].

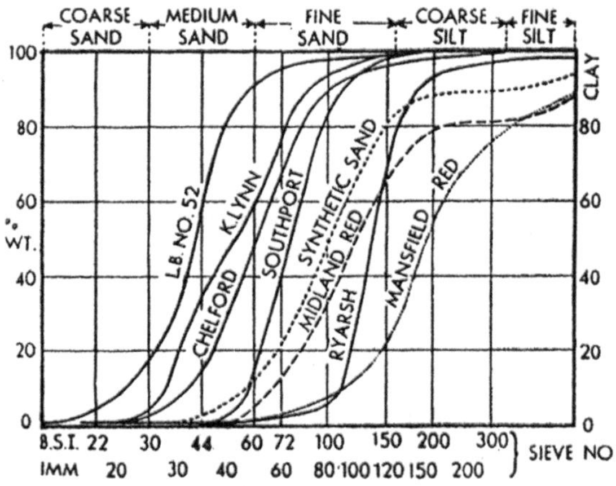

Fig. 1 Cumulative curves for various foundry sands in the UK [12]

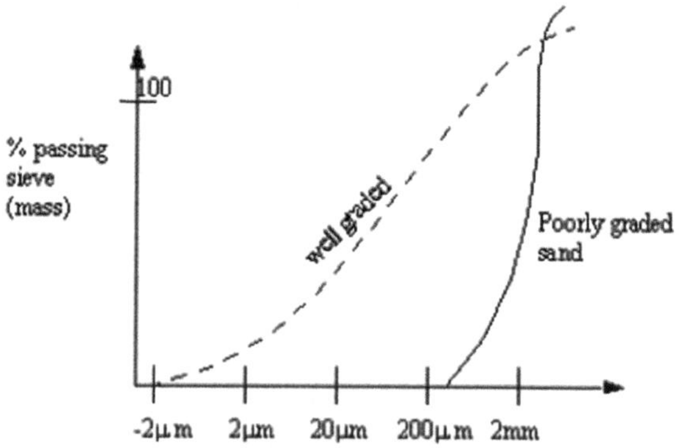

Fig. 2 Cumulative curve of grain size distribution [13]

The objectives of this study are to determine the grain size distribution through the cumulative curve and average grain size by conducting mechanical sieve grading of silica sand from identified tin mining in Malaysia. Five samples were taken; two from Perak which are Tronoh and Tanjung Tualang; Bestari Jaya in Selangor; Gambang in Pahang and Jemaluang in Johor. Samples size over 425 μm were eliminated in order to comply the foundry sand size. The samples were tested using mechanical sieve shaker accordingly to foundry sand testing standard of procedures from Foseco Ferrous Handbook for greensand casting mould and analysed for the grain size distribution and average grain size. These samples were compared with the typical foundry sand from Foseco and foundry sand from RCS Manufacturing Company who supply sand for Proton Casting Company which is a company that producing car parts such as engine block and cylinder head for Proton cars.

2 Material and Method

A total of five samples representing deposit from tin mining locations in Perak (Tanjung Tualang and Tronoh), Selangor (Bestari Jaya), Pahang (Gambang) and Johor (Jemaluang), weighing approximately 5 kg were obtained for mechanical sieve grading.

2.1 Grain Sand Distribution

The specimen sand is grading by using Matest Auto Sieve Shaker machine with aperture size 710, 500, 355, 250, 212, 150, 90, 63 μm and pan. The sieve grading as follows:

(a) Sample of 100 g dry sand is weighed.
(b) The sample is placed into the top sieve of a nest of ISO sieves on a vibrator and vibrate for 15 min.
(c) The sieves is removed, beginning with the top sieve, weight the quantity of sand remaining on each sieve.
(d) The percentage of the sample weight retained on each sieve is calculated, and arrange in a column as shown in Table 2 [5].

The percentage of the sand retained in each sieve was then calculated using Eq. 1:

$$\%\text{wt. of sand in different sieves} = \frac{\text{Weight of sand in a particular sieve}}{\text{Total sand weight}} \times 100 \quad (1)$$

Table 2 The calculation of average grain size for typical sand suitable for iron and steel casting from FOSECO [8]

Aperture size (μm)	Sand retained (%)	Multiplier	Product
710	0.14	1180	165.2
500	1.42	600	852.0
355	5.44	425	2312.0
250	22.88	300	6864.0
212	19.64	212	4163.7
150	26.74	150	4011.0
90	9.70	106	1028.2
63	2.84	75	213.0
Pan	1.10	38	41.8
Total	89.9		19650.9

Average grain size = 19,650.9/89.9 = 219 μm

2.2 Average Grain Size

The adoption of the ISO metric sieves means that the old AFS grain fineness number can no longer be calculated [8]. Instead, the average grain size, expresses as micrometres (μm) is now used. Table 2 shows the calculation of the average grain size for typical sand suitable for iron and steel casting from FOSECO [8] and Table 3 shows the calculation of foundry sand obtained from RCS Manufacturing Company, which supply foundry sand for Proton Casting Plant.

Table 3 The calculation of average grain size for sand from RCF manufacturing company

Aperture size (μm)	Sand retained (%)	Multiplier	Product
710	0.010	1180	11.8
500	0.313	600	188.0
355	18.850	425	8011.3
250	34.980	300	10,494.0
212	20.530	212	4352.4
150	20.480	150	3072.0
90	4.660	106	494.0
63	0.187	75	14.0
Pan	0.033	38	1.3
Total	100		26,638.6

Average grain size = 26,638.6/100 = 266 μm

3 Result and Discussion

3.1 Grain Sand Distribution

Figures 3, 4, 5 and 6 show the cumulative 'S' curve of the grain size distribution for all samples. The figures display the cumulative 'S' curve for sample from Tronoh, Bestari Jaya and Gambang demonstrate that the samples have cumulative curve in-between the cumulative curve of RCS and Foseco. The 'S' curve for all samples indicate that sand retained on 250 μm is slightly higher. Since the size of 250 μm is within recommended size, it's acceptable. Mostly all samples have been discovered well sorted grain where the 'S' curve has 45° slope and the curves are not far away as compared to cumulative 'S' curve Foseco and RCS as shown in Fig. 7.

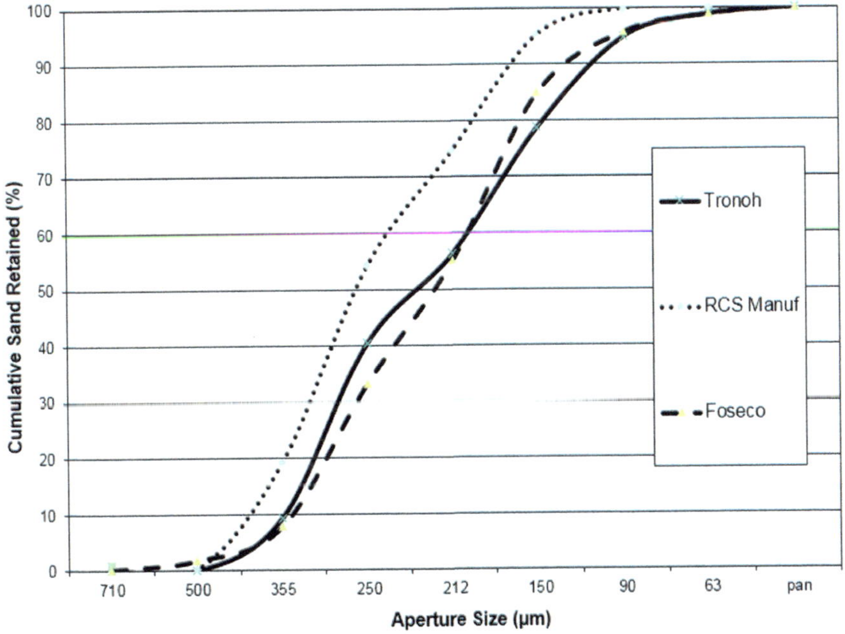

Fig. 3 Cumulative sieve grading for Tronoh

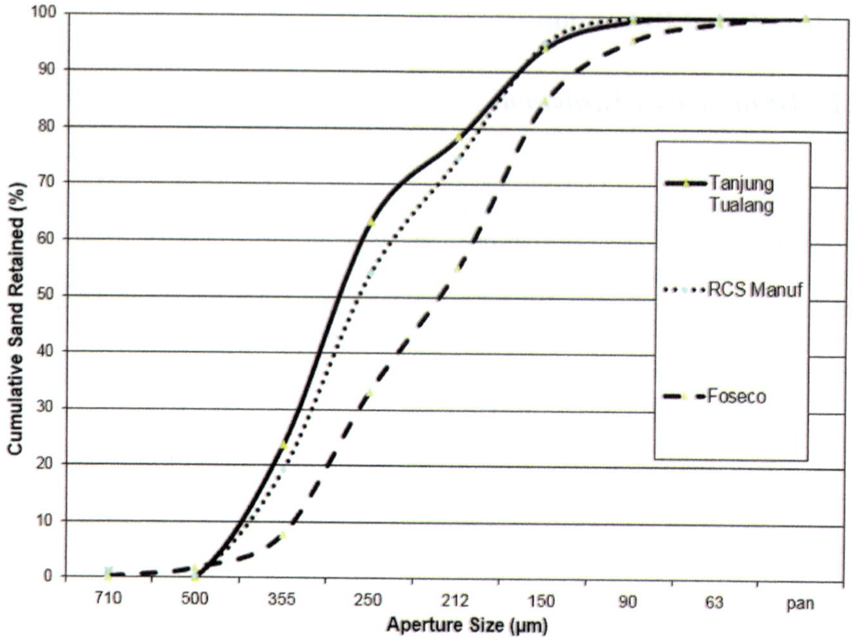

Fig. 4 Cumulative sieve grading for Tanjung Tualang

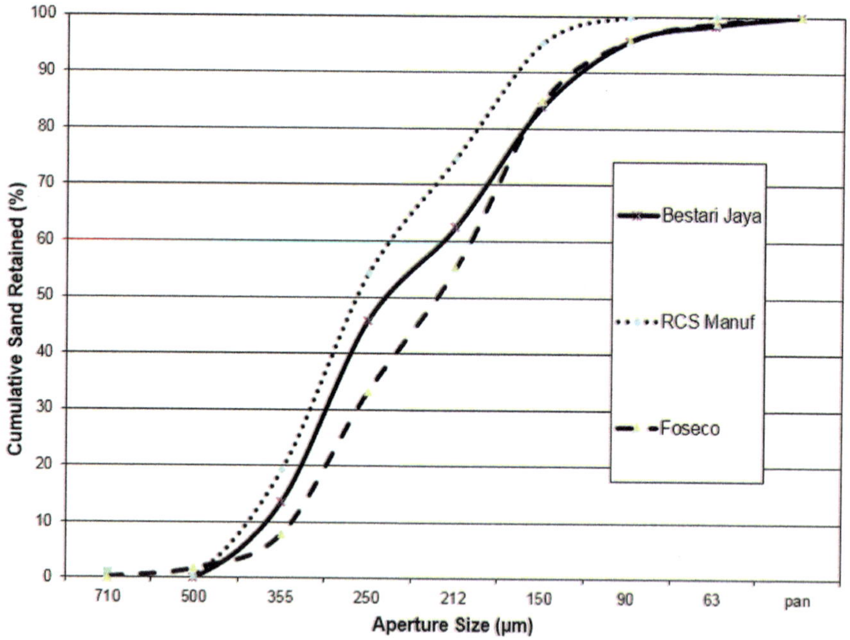

Fig. 5 Cumulative sieve grading for Bestari Jaya

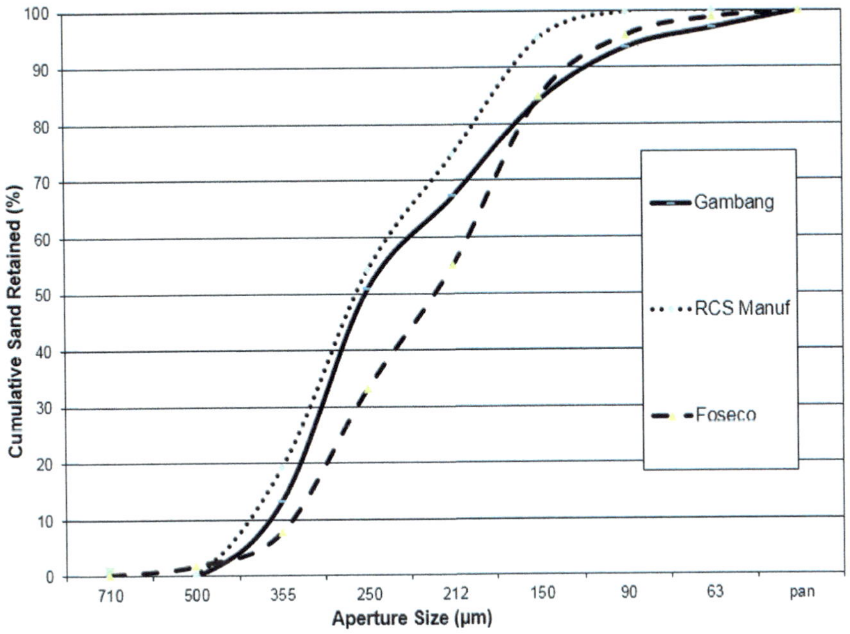

Fig. 6 Cumulative sieve grading for Gambang

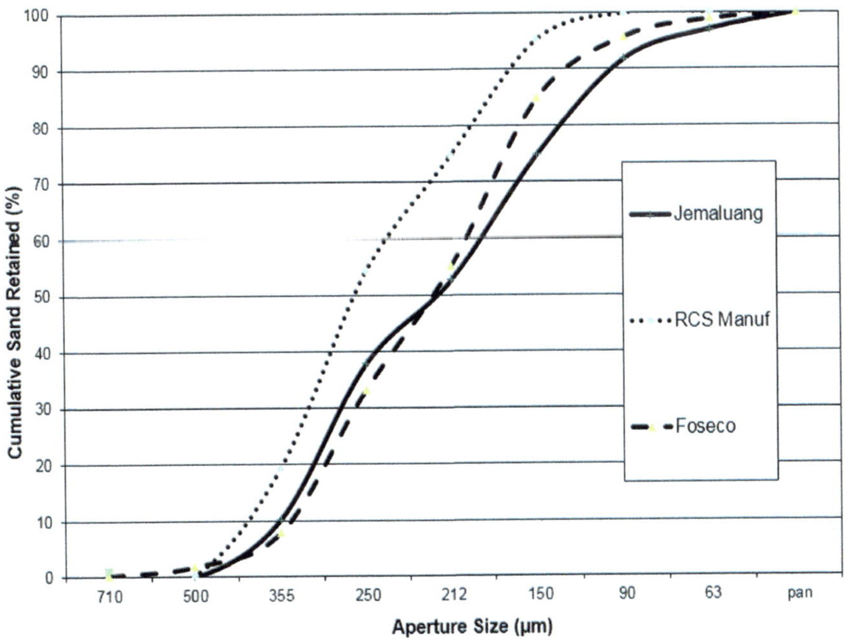

Fig. 7 Cumulative sieve grading for Jemaluang

3.2 *Average Grain Size*

Tables 4, 5, 6, 7 and 8 show the calculation of average grain size for all samples. Table 4 for Tronoh shows that 79% sand retained on these four adjacent sieve seizes with the average grain size is 221 μm, Table 5 for Tanjung Tualang is 95% with 281 μm, Table 6 for Bestari Jaya is 84% with 237 μm, Table 7 for Gambang is 84% with 243 μm and Table 8 for Jemaluang is 75% with 214 μm. These results show that majority of the sand samples spread over four consecutive recommended sieves (150, 212, 250 and 335 μm) and the average grain size is suitable for making

Table 4 The calculation of average grain size for Tronoh

Aperture size (μm)	Sand retained (%)	Multiplier	Product
710	0.0	1180	0.0
500	0.0	600	0.0
355	9.3	425	3962.4
250	31.2	300	9348.0
212	15.8	212	3351.7
150	22.4	150	3357.0
90	16.1	106	1702.0
63	4.0	75	300.3
Pan	1.2	38	45.2
Total	100		22,066.6

Average grain size = 22,066.6/100 = 221 μm

Table 5 The calculation of average grain size for Tanjung Tualang

Aperture size (μm)	Sand retained (%)	Multiplier	Product
710	0.0	1180	0.0
500	0.0	600	0.0
355	23.7	425	10,065.4
250	39.8	300	11,942.0
212	15.0	212	3177.9
150	16.1	150	2410.0
90	4.8	106	507.7
63	0.7	75	51.0
Pan	0.1	38	3.7
Total	100.1		28,157.7

Average grain size = 28,157.7/100.1 = 281 μm

Table 6 The calculation of average grain size for Bestari Jaya

Aperture size (μm)	Sand retained (%)	Multiplier	Product
710	0.0	1180	0.0
500	0.0	600	6.0
355	13.5	425	5741.8
250	32.3	300	9701.0
212	16.6	212	3519.2
150	21.4	150	3217.0
90	11.6	106	1227.8
63	2.9	75	216.0
Pan	1.6	38	59.9
Total	99.9		23,688.7

Average grain size = 23,688.7/99.9 = 237 μm

Table 7 The calculation of average grain size for Gambang

Aperture size (μm)	Sand retained (%)	Multiplier	Product
710	0.0	1180	0.0
500	0.0	600	0.0
355	13.2	425	5593.0
250	37.5	300	11,262.0
212	16.4	212	3474.0
150	17.0	150	2543.5
90	9.3	106	984.0
63	3.4	75	257.3
Pan	3.1	38	117.0
Total	99.8		24,230.8

Average grain size = 24,230.8/99.8 = 243 μm

ferrous casting mould where the recommended size is 210–300 μm. Except Tanjung Tualang, all samples are also suitable for making non-ferrous casting mould where the recommended size is 170–270 μm.

Table 8 The calculation of average grain size for Jemaluang

Aperture size (μm)	Sand retained (%)	Multiplier	Product
710	0.0	1180	0.0
500	0.0	600	14.0
355	10.2	425	4336.4
250	27.7	300	8298.0
212	14.8	212	3129.1
150	22.2	150	3336.5
90	16.8	106	1784.0
63	5.3	75	397.0
Pan	3.0	38	114.9
Total	100.0		21,409.9

Average grain size = 21,409.9/100.0 = 214 μm

4 Conclusion

This study found that all silica sand samples from identified former tin mining locations in Malaysia have well sorted grain distribution and the average grain size within the requirement for metal casting industry. Cumulative 'S' curve for all sample show that all sample are well sorted where the majority of the sand retained on four adjacent sieve sizes. The calculation of average grain size shows that all samples have the size within the recommended size for making ferrous and non-ferrous casting mould, except Tanjung Tualang only meet for ferrous casting. The finding from this research found that silica sand from former tin mining in Malaysia is suitable for making metal casting mould in terms of mechanical sieve grading analysis.

Acknowledgements The authors greatly acknowledge the support of Department of Mechanical Engineering, Politeknik Ungku Omar and Department of Polytechnic and Community College Education, Ministry of Education Malaysia.

References

1. Ali A (2002) The silica-based industry in Malaysia. Minerals and Geoscience Department Malaysia, Kuala Lumpur
2. Mackay IMC, Schnellmann (2000) Final report of mineral processing consultancy silica sand. International Mining Consultants Limited, Kuala Lumpur
3. Min YK (2006) Gravel pump tin mining in Malaysia. Jurutera, Kuala Lumpur

4. Hussain MAB, Abdullah AB, Abdullah RB (2018) Physical and chemical properties of Perak river sand for greensand casting molds. Engineering applications for new materials and technologies. Springer, pp 13–24
5. Bolger R (1996) Foundry minerals: recycling: the new name of the game. Ind Miner 29–39
6. Beeley P (2001) The moulding material: properties, preparation and testing. In: Foundry technology, 2nd edn. Butterworth-Heinemann, Oxford, pp 178–238
7. Salmon WH, Simons EN (1966) Foundry practice. Sir Isaac Pitman & Sons Ltd., London
8. Brown JR, John RB (1999) Sands and green sand, Foseco non-ferrous Foundryman's Handbook, Eleventh edn. Butterworth-Heinemann, Oxford, pp 149–163
9. Beadle JD (1971) Castings. The Macmillan Press Ltd., Hampshire
10. Jain PL (2008) Principles of foundry technology, 8th edn. Tata McGraw-Hill, New Delhi
11. Parkes WB (1971) Clay-Bonded foundry sand. Applied Science Publishers Ltd., London
12. Howard ED (1958) Modern foundry practice, 3rd edn. Odhams Press Limited, London
13. Heine RW, Carl R, Loper J, Rosenthal PC (1967) Principles of metal casting, 2nd edn. McGraw-Hill Inc., San Francisco

Design of Full-Bridge Rectifier for Underwater Remotely Operated Power Supply Unit

W. M. Dahalan, Norhafizah Othman, Noorazlina Mohamid Salih and Mohd Harkikie Fitri Bin Abdul Karim

Abstract Rectifiers or AC/DC converters are widely used in industrial applications. Input AC voltage is rectified into DC voltage and smoothen using the filtering circuit which consists of electrolytic capacitors. The 220VAC, 50 Hz input voltage source is increased and converted to get 400VDC by using a step-up transformer and full-wave bridge. The 620 μF filtering capacitor is chosen based on the calculation and simulation. This paper described the step to design rectifier and proposes the component configurations by which voltage drop and ripple voltage can be reduced. The results of respective simulations are shown through Proteus software program and their parameters are calculated.

Keywords Rectifier · Full-bridge · ROV · AC/DC converter

1 Introduction

Rapid developments in the marine operations field and activities have demanded the growth of underwater vehicle called Remotely Operated Vehicles (ROV). The automation of such vehicle requires sufficient power sources capacity to make all components build inside the ROV are working as expected. The power transmission is completed through tether cable along with communicating systems. The longer the cable is needed to transfer the power to the ROV to the deeper area of the ocean.

The AC/DC voltage converter for the ROV power supply unit (PSU) is designed using a conventional full bridge rectifier with a filter capacitor. In the first stage, the single-phase power source, 220 VAC with frequency 50 Hz is increased into 400 VAC using a step-up transformer. Later the 400 VAC is then converted into 400 VDC using full-bridge rectifier and a smoothing capacitor. Then the supply is transmitted through the 100 m long umbilical cable. The design of the rectifier is based on the power requirements from the thrusters and other electronics part in the

W. M. Dahalan · N. Othman (✉) · N. M. Salih · M. H. F. B. A. Karim
Malaysian Institute of Marine Engineering Technology (UNIKL-MIMET), Universiti Kuala Lumpur, 32200 Lumut, Perak, Malaysia
e-mail: wardiah@unikl.edu.my

C. L. Saw et al. (eds.), *Advancement in Emerging Technologies and Engineering Applications*, Lecture Notes in Mechanical Engineering,
https://doi.org/10.1007/978-981-15-0002-2_17

ROV. The primary constraint with regards to the power system is the space available for DC to DC power conversion on board. The required voltage at the output of the rectifier must compensate the transmission losses in addition to the voltage required by the ROV's load.

The smoothing capacitor value is achieved by calculation and simulation iteration. The simulation of the rectifier is done by using Proteus software. The suitable capacitor value is chosen based on the comparison with ripple voltage.

2 Design of Single-Phase Full Bridge Rectifier

The primary function of the PSU was to convert 220 VAC mains into 400 VDC and supply it at a rate of 5A to the ROV. Based on Roger et al., the most safest and reliable ways to provide an isolated or ungrounded DC supply is by using toroidal transformer with double windings to step-up the mains supply voltage before rectifying it [1]. This is because the toroidal transfer would provide complete isolation between the inputs and outputs. Besides, the double windings would provide greater reliability than single wounded-auto-transformers.

The circuit topology for the conventional full bridge rectifier with filter capacitor is shown in Fig. 1.

Some equations and calculations are done in order to finalize the design [2, 3].

i. Input voltage to the step-up transformer
   Input voltage the using single-phase 220 VAC mains with 50 Hz frequency

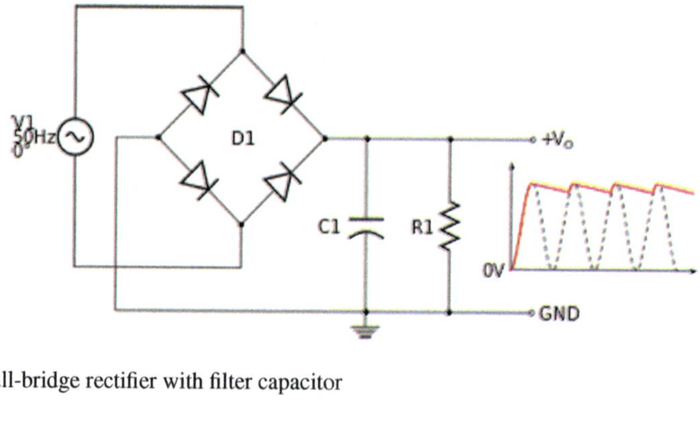

Fig. 1 Full-bridge rectifier with filter capacitor

ii. Transformer primary and secondary windings

$$\frac{V_s}{V_p} = \frac{N_s}{N_p} \tag{1}$$

where

V_s secondary voltage
V_p primary voltage
N_s secondary winding
N_p primary winding

$$N_s = \frac{400\ \text{V}}{220\ \text{V}} \times 1 = 1.82 \approx \mathbf{2}$$

iii. Output voltage of the rectifier

$$\text{Output voltage} = \sqrt{2} \times \text{Input voltage}$$
$$\text{Output voltage} = \sqrt{2} \times 400\text{V} = \mathbf{565.69\ V} \tag{2}$$

iv. Peak diode current

$$\text{I}_\text{m} = \frac{\text{Vm}}{\text{R}} \tag{3}$$

$$\text{I}_\text{m} = \frac{400\ \text{V}}{100} = \mathbf{4\ A} \tag{4}$$

v. Load resistor = **100 Ω**

2.1 *Simulation of Rectifier*

The simulation was carried out using Proteus software as shown in Fig. 2 in order to find the suitable capacitor value. The result of simulations is shown in figures below. The smoothing capacitor converts the full-wave rippled output of the rectifier into a smoother DC output voltage.

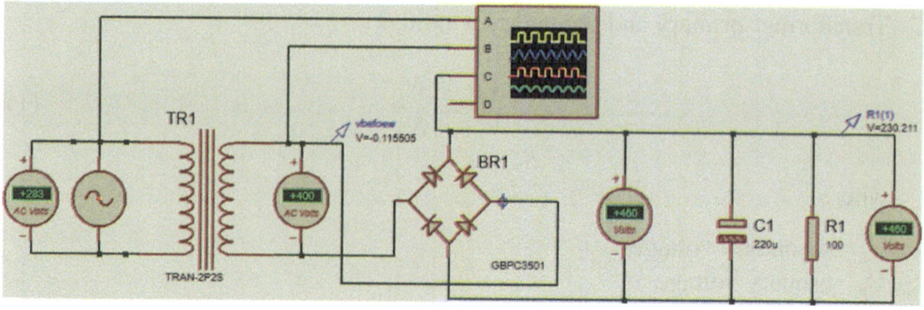

Fig. 2 Circuit configuration on Proteus software

2.2 *Filtering or Smoothing Capacitor*

Figure 3 shows the waveform of the result of using a 220 μF across the rectifier output. This results in the capacitor discharging down to about 283.1 V – 208.6 V = 74.5 V. The capacitor maintains the voltage across the load resistor until the capacitor recharges once again on the next positive slope of the DC pulse. Thus, the DC voltage applied to the load resistor in higher voltage drops. This can be improved by increasing the value of the smoothing capacitor [4–7]. The ripple voltage is done by using Eq. 5 [4]:

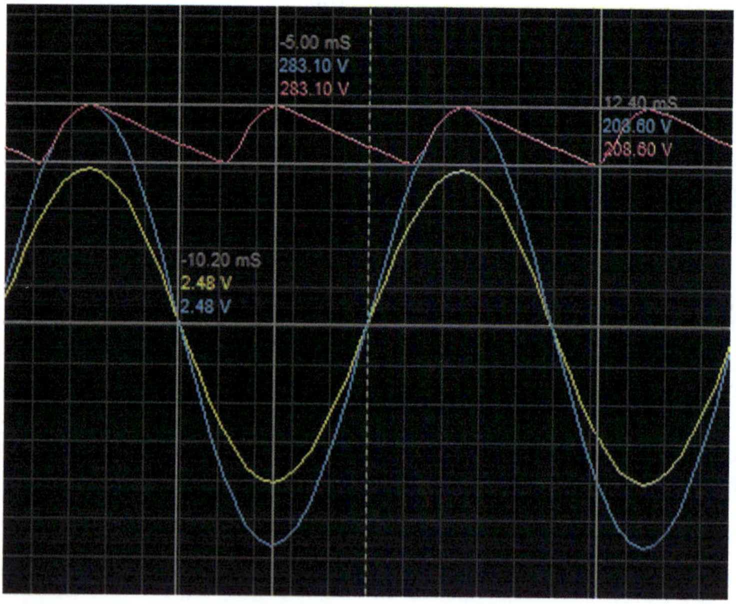

Fig. 3 Waveform of rectifier by using 220 μF

$$V_{ripple} = \frac{I_{load}}{f \times C} \quad (5)$$

where f is twice that of the AC supply frequency.

Simulations are conducted in iteration in order to get the suitable capacitor value. Table 1 shows the result of ripple voltage with different capacitor value.

It is seen from the above table that the value of voltage drops and ripple voltage for the rectifier is decreasing with the increasing value of capacitor. Changing the capacitor is not the deciding factor of a good rectifier, so it can be concluded that by using 620 μF with reduce the voltage drop and ripple voltage. The voltage waveform is shown in Fig. 4.

Table 1 Smoothing capacitor simulation and calculation

Capacitor value (μF)	Voltage drop	Ripple voltage (V)
220	283.1 V – 208.6 V = 74.5 V	227.27
320	283.1 V – 225.9 V = 57.2 V	156.25
420	283.1 V – 238.4 V = 44.7 V	119.05
520	283.1 V – 245.9 V = 37.2 V	96.15
620	283.1 V – 258.3 V = 24.8 V	80.65

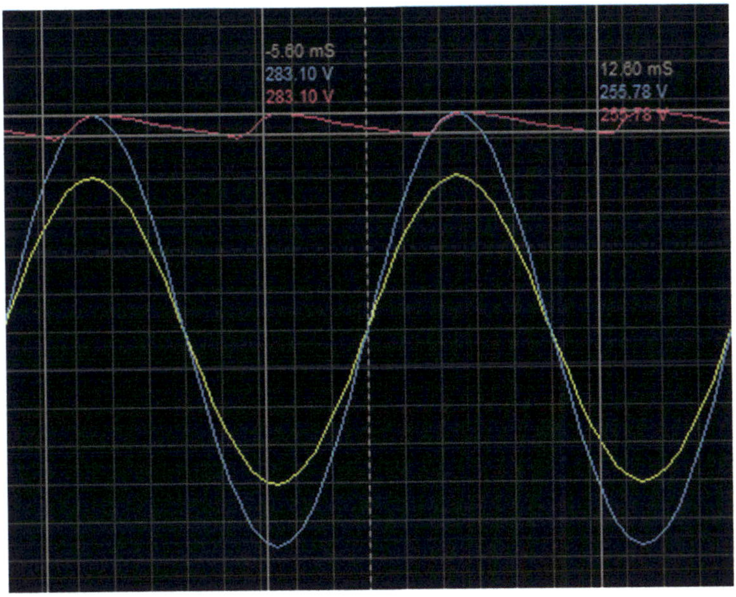

Fig. 4 Waveform of rectifier by using 620 μF

3 Conclusions

From the above design and simulation, step-up transformer is used to increase the input AC voltage source. Then, the voltage source is converter into DC voltage source by using full-bridge rectifier. The 620 μF smoothing capacitor is chosen to reduce the ripple voltage. The circuit topologies can be used for low and high voltage applications.

References

1. De Smidt R (2014) Development of the electronics pod for an underwater remotely operated vehicle, pp 1–6
2. Sadasivuni H, Thayumanavan P, An S (2014) Design of a 5KW power converter for powering thruster of a remotely operated submersible. 3(1):229–231
3. Batard C, Poitiers F, Millet C, Ginot N (2012) Simulation of power converters using Matlab-Simulink. MATLAB—A Fundam Tool Sci Comput Eng Appl 43–68
4. Hambley AR, Rizzoni G (2015) 3. Diode, rectifiers, and power supplies. No. i 1–14
5. Rahm M (2010) Ocean wave energy: underwater substation system for wave energy converters
6. Santos M, Salcedo F, Ben Haim D, Mendia JL, Ricci P, Villate JL, Khan J, Leon D, Arabi S, Moshref A (2013) Integrating wave and tidal current power: case studies through modelling and simulation
7. Soltani MN, Sichani MT, Mirzaei M (2014) Model predictive control of buoy type wave energy converter

Optimization Structure Design of Offshore Oscillating Water Column (OWC) Wave Energy Converter

Khairul Anuar bin Mat Saad and Ahmad Khairil bin Azman

Abstract Oscillating water column (OWC) is one of optional renewable energy device that been used for converting kinetic energy from waves energy into electrical energy. Suitable structural design must be measured and determine to make sure structure is capable to resist wave load. The scope of the research is to design of closed structure of OWC. A model has been set up in Ansys AQWA and Ansys Static Structural which has been undergone through several steps. The model has been run with selected significant wave frequency which is from article finding of the research. Hydrodynamic diffraction analysis has been carried out to identify maximum wave pressure acting on model. Maximum wave pressure be one of component to determine maximum equivalent stress (von Mises) and total deformation of OWC closed structure. There are 4 detail design structure models in this research. Furthermore, there are also 4 different position of maximum wave pressure acting on structure model. Parameter that involve in research including type of material, type of stiffener, number of stiffener and number of frame. Optimization process also carried out by selected suitable detail design structure model to minimize maximum equivalent stress (von Mises) and total deformation. Result shows the optimization process de-creased on structure up to 20–70% at different position for maximum equivalent stress (von Mises). Total deformation also decreased on structure up to 25–80% at different position.

Keywords Oscillating water column (OWC) · Finite element analysis (FEA) · Structural analysis

---

K. A. bin Mat Saad · A. K. bin Azman (✉)
Malaysian Institute of Marine Engineering Technology, Universiti Kuala Lumpur, Kuala Lumpur, Malaysia
e-mail: akhairil.azman@s.unikl.edu.my

K. A. bin Mat Saad
e-mail: khairulanuar@unikl.edu.my

C. L. Saw et al. (eds.), *Advancement in Emerging Technologies and Engineering Applications*, Lecture Notes in Mechanical Engineering,
https://doi.org/10.1007/978-981-15-0002-2_18

1 Introduction

Green energy development is one of the current issues to reduce the pollution during the operation such as alternative to coal operation. OWC is one of green energy development in industry. The principle of OWC is motion of the wave causes the captured water column to move up and down like a piston thus pressurizing and depressurizing the air though an opening connected to a turbine. Wave impact on structure can cause the device fail to operate in sea. Besides that, to design renewable wave energy converter, suitable structural design must be measured to make sure the structure can operate without fail to resist wave impact load. The structural design use optimization approach and meet the requirement of classification rules. Objective of study research is to analyze and determine structural strength of OWC wave energy converter. It is including optimizing on structural design of OWC wave energy converter. This investigation will give information of concept of OWC. The research also will clarify the difference of several types of OWC based on principle. This investigation also will test and analyze the selected design of OWC on FEA software. It will help the optimization of structural design of OWC. Several parameters of OWC will be change to find best result of design analysis. This research will focus for closed structural design of OWC. The entire test will perform by using FEA software that including equivalent stress (Von-Mises) and total deformation analysis only [10–12].

This project will use existing design of OWC in world. Benchmark model use as reference in FEA to make some comparison between optimization models. Wave data will used for simulate in FEA software. Benchmark model will be design using CAD software. Benchmark model will be design based on limited data from journal or article and guide by classification rules that can be found to validate the structure. In this research, wind and currents will be neglected because wave is the only parameter that will be concentrated.

2 Literature Review

This study presents a comprehensive review of OWC technologies and air turbines [9]. This is followed by a survey of theoretical, numerical and experimental modelling techniques of OWC converters. Its explain types of OWC, location and principle for each type. Study about structure design of BBDB type of OWC [8]. Detail structure on design also identify in this article. In this article, conceptual design for an Oscillating Water Column Wave Energy Converter (WEC) device appropriate for the modeled reference resource site was identified, and a detailed backward bent duct buoy (BBDB) device design was developed using a combination of numerical modeling tools and scaled physical models.

Study about hydrodynamic loads acting on OWC structure [4]. Wave data that used for analysis in this research obtained from this article. This article provides historical development of the OWC device, an explanation of the determination of the hydrodynamic characteristics, a full explanation of a complete motion response formulation, choices regarding device modelling and the presentation of hydrodynamic loads and the resulting transfer functions.

Study about structural analysis on OWC wave energy converter [7]. Measurement of OWC reference model is obtained from this article before detail structure is design. In this article, the dynamic loads determined for the SEA-OWC-Clam wave energy device, treated as a floating offshore structure with six degrees-of-freedom with partial internal sealed-off channels, are applied to assess its structural integrity. This task necessitates the matching of the boundary element determined dynamic pressures to the corresponding Finite Element Analysis (FEA) to determine the effective von Mises stress levels.

Study about understanding of fluid-structure coupled analysis [11]. This research focusing on wave load acting on structure that understanding of fluid- structure coupled analysis is reviewed to achieve the objective. In this study, three design variables (draft, diameter, ballast weight) with 3 levels in the substructure were selected and 9 corresponding models were produced based on design of experiment (DOE). By performing fluid-structure coupled analysis under severe wave conditions with normal current and hydrostatic pressure, they evaluated the dynamic motions (maximum resultant displacement) and maximum von Mises stress and analyzed the influence of those three design variables on the dynamic motion and stress distribution of spar type floating the substructures of spar type floating wind turbines.

3 Methodology

First process of research is to design the reference model by using combination of AutoCAD and Solidworks. AutoCAD will design the model and integrate into Solidworks to check the detail dimension and change the format of analysis software. Then, maximum wave pressure is analyzed and determine by using Ansys AQWA [1, 2]. All the wave data is from journal finding of reference model. This research using coupled analysis to get valid pressure for structural analysis. Maximum wave pressure will be used at analysis of detail structure in Ansys Static Structural. There are 4 design detail structure to identify maximum equivalent stress (von Mises) and total deformation of closed structure. Comparison between various models can identify best model (less equivalent stress and total deformation). Optimization process is continuing by redesign the best model and reanalysis the model to get the final result.

4 Data Collection

OWC design is choose based on the existing OWC that is currently under development. All the information of the model and their wave characteristic is gain from article of [3, 7] (Fig. 1) (Tables 1, 2 and 3).

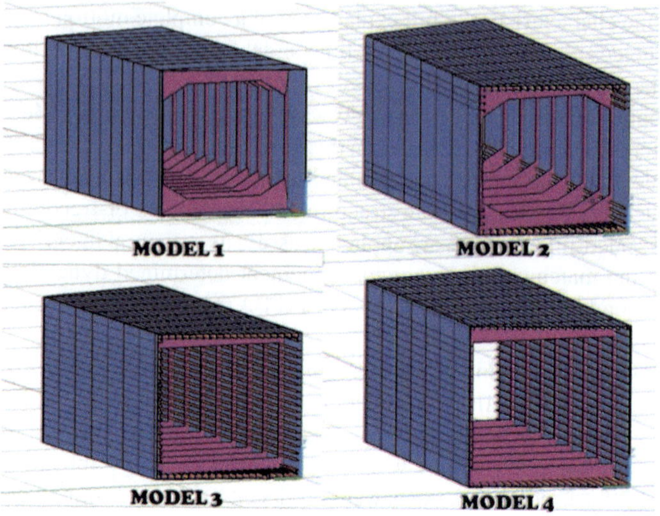

Fig. 1 3D AutoCAD of 4 OWC detail structure design

Table 1 OWC particular

Particular	Full-scale (m)
Length	77.3
Draught	6
Depth	8
Shape	Dodecagon
Type	Multi-device OWC

4.1 Wave Load Configuration

Maximum wave load must be identified as pressure component in FEA analysis [5]. The loads that a floating structure is subjected to are divided in two categories which due to the function on the structure (Functional loads) and due to the environment (Environmental loads). This research only focusing on environmental load which is wave load acting on structure. Other environment load such as wind load or current load not include in this research. The setup of model in this configuration is full body model. This is because full body model can give overall impact of wave into structure. Several mesh element size is compared to generate maximum wave load. The model is modified in order to smoothly be imported into ANSYS Workbench. In AutoCAD, the following modifications to the hull are made:

(i) The body of model is lowered in the coordinate system so the z-axis cuts through the hull on draught level (6 m). The hull is then separated into two

Table 2 Wave data [4]

No.	Frequency (rad/s)	Wavelength (m)	No.	Frequency (rad/s)	Wavelength (m)
1	0.1	1379.72	11	0.93	71.25
2	0.3	428.23	12	1	61.63
3	0.465	244.52	13	1.035	57.54
4	0.5	219.81	14	1.055	55.38
5	0.522	205.81	15	1.068	54.05
6	0.62	154.89	16	1.085	52.36
7	0.669	135.11	17	1.125	48.7
8	0.75	108.9	18	1.24	40.09
9	0.798	96.51	19	1.375	32.61
10	0.87	81.36	20	1.5	27.39

Table 3 Comparison between various models of OWC

	Stiffener			
	Type	No.	No. of frame	Weight (kg)
Model 1	FLAT BARS	62	11	96,385
Model 2	NAB	54	8	1,06,980
Model 3	NAB	80	8	1,18,490
Model 4	FLAT BARS	84	11	94,603

surfaces with the z-axis separating the two. This is also necessary in order to allow Ansys AQWA to easier analyse the body.

(ii) All the body surface is combined together using the UNION command in AutoCAD software to ease the import process of the hull body in Ansys AQWA.

(iii) The axis is located in the centre of body.

The model is then imported into ANSYS AQWA. On the model, a point mass is added, containing all the necessary mass data.

4.2 *Structure Analysis Configuration*

Structural analysis is the determination of the effects of loads on physical structures and their components [6]. Maximum stress and deformation are identified for each detail structure model in this analysis. Model that used for this research is single body that consist of detail structure. There are several setups in the outline of the analysis such as direction of pressure, fixed support, mesh and their solution.

Mesh element size is obtained from analysis of maximum wave pressure. For configuration, there are several steps that carried which are setup the mesh sizing, face meshing and method of the mesh. Method of the mesh is Multizone that constructs Hexa mapped mesh type for each model.

5 Results and Discussions

The discussion separated into 2 discussions. First discussion is result obtained from Ansys AQWA that focus on identify the maximum pressure as shown in Figs. 2, 3 and 4. Second result is result obtained from Ansys Static Structural which identifies maximum equivalent stress (von Mises) and total deformation for each model. Furthermore, selection model from comparison study also through optimization process to get better result and achieve objective of research (Fig. 5).

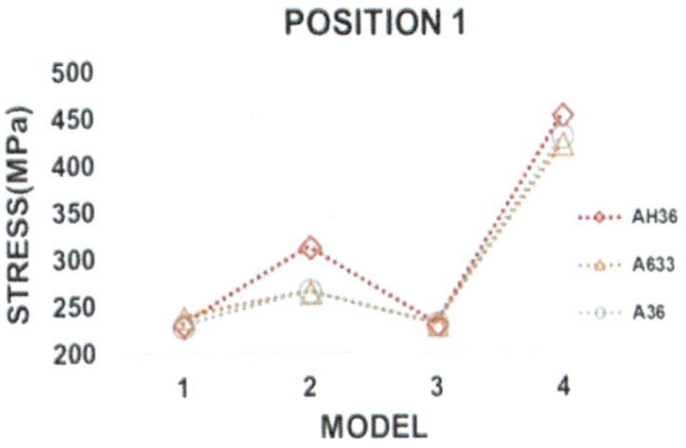

Fig. 2 Line chart of maximum equivalent stress between materials of position 1

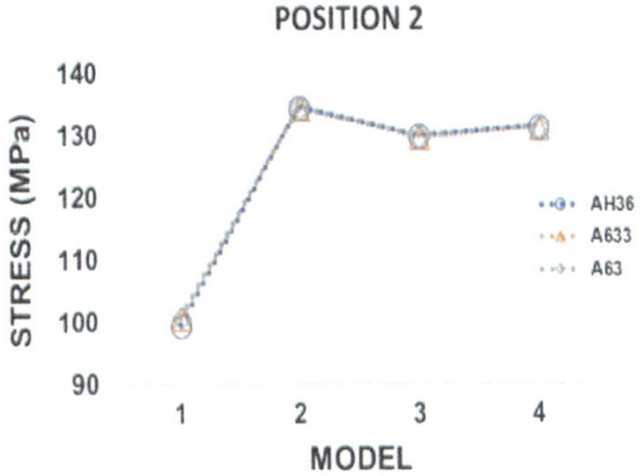

Fig. 3 Line chart of maximum equivalent stress between materials of position 2

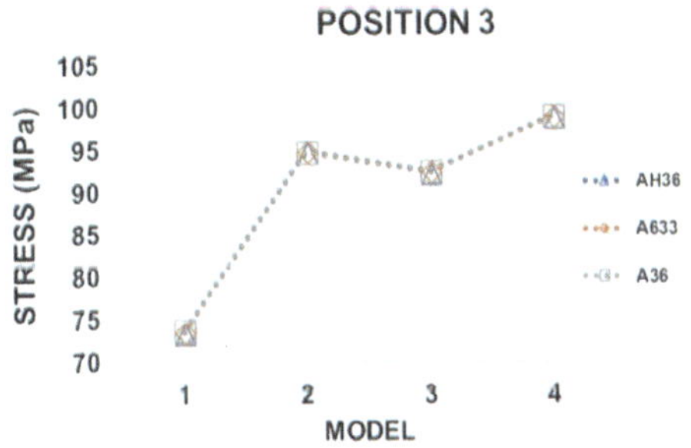

Fig. 4 Line chart of maximum equivalent stress between materials of position 3

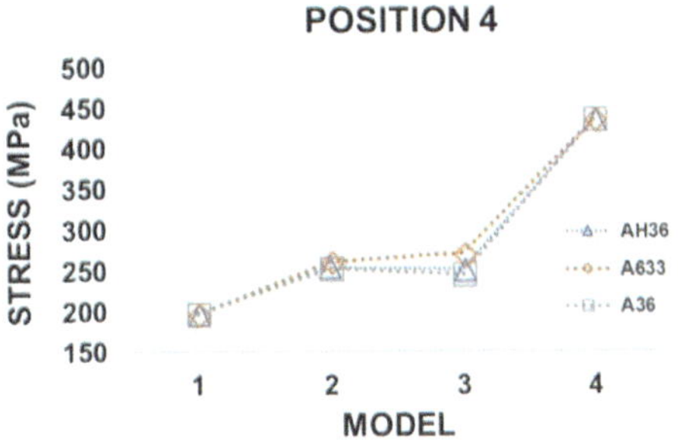

Fig. 5 Line chart of maximum equivalent stress between materials of position 4

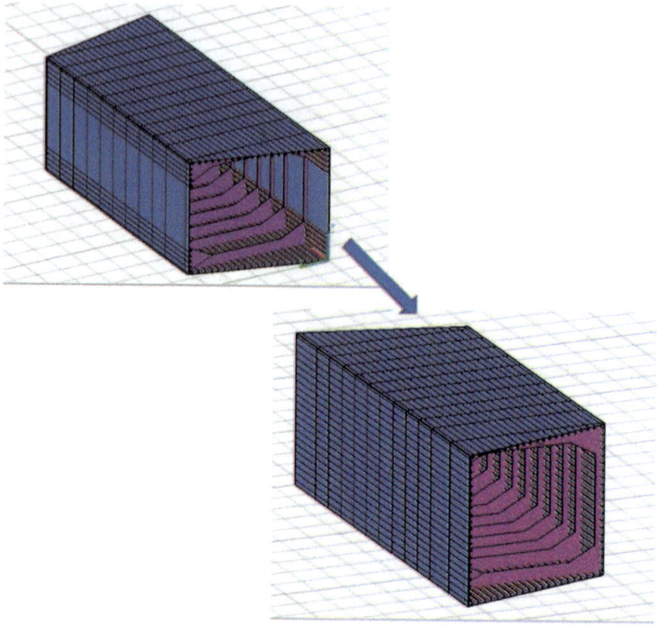

Fig. 6 Optimization of model 1 detail structure

5.1 Optimization of Selection Model

From observation, Model 1 with material of A633 is selected for optimization process. Model 1, referred to Fig. 6, has shown positive result for both maximum equivalent stress and total deformation but for material of A633, there are opposite view that total deformation shows A36 has low total deformation compared to other material. Material A633 can be selected based on result at all Model 1 that use material A633. It shows low result of deformation for Position 1, 2 and 4 as shown in Figs. 7, 8, 9 and 10.

Optimization process is started with redesign the detail structure. Maximum equivalent stress and total deformation can be reduced by optimization process. Stiffener is added to side structure from 62 stiffeners to 84 stiffeners. After stiffener is added, the weight of structure is increase to over 6 tones but the added weight is not exceeded compare to weight of Model 3.

Optimization process has decreased the structure up to 20–70% of Maximum Equivalent Stress. Others, total deformation also decreased up to 25% into 80%.

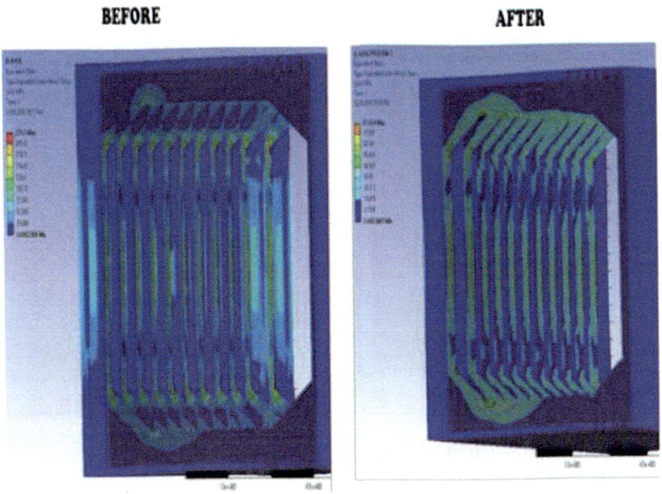

Fig. 7 Before and after optimization for position 1 of model 1

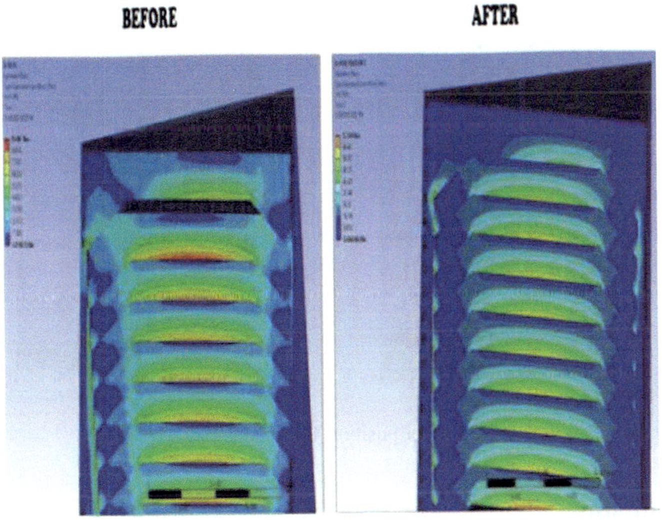

Fig. 8 Before and after optimization for position 2 of model 1

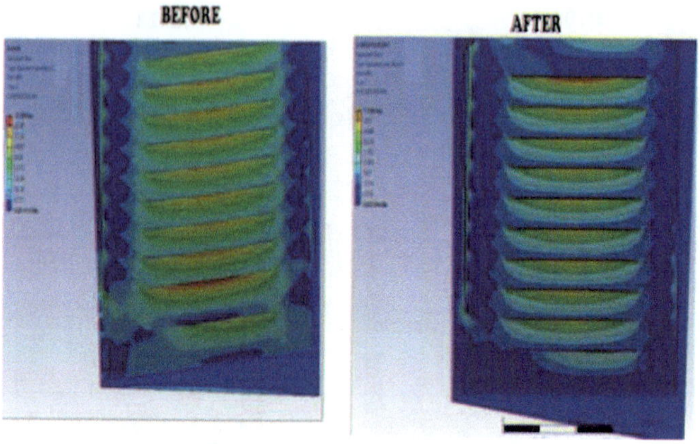

Fig. 9 Before and after optimization for position 3 of model 1

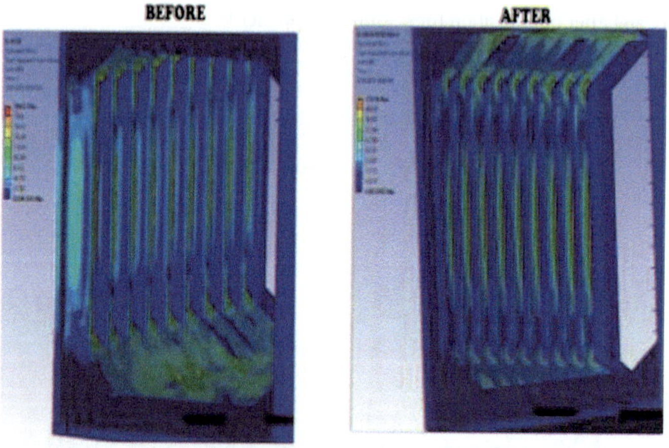

Fig. 10 Before and after optimization for position 4 of model 1

6 Conclusion

In this study, the simulation has been conducted using OWC model with environmental data related to reference actual condition. The objective of the study is successfully achieved. Based on analysis and results obtained in this project, the following conclusions can be made:

1. The Computational Aid Design (CAD) which is AutoCAD and Solidworks is use to model the OWC model and detail structure based on reference article [7]
2. The maximum wave pressure acting on OWC model is successfully identify using Ansys AQWA software

3. Structural strength of OWC is analyze by maximum equivalent stress (von Mises) and total deformation
4. Maximum equivalent stress (von Mises) and total deformation of OWC is successfully conduct and determine using Ansys Static Structural
5. Based on result and discussion, detail structure of Model 1 has less stress impact on structure
6. Redesign detail structure of Model 1 to optimize the maximum equivalent stress (von Mises) and total deformation
7. Optimization process decreased on structure up to 20–70% at different position for maximum equivalent stress (von Mises). Total deformation also decreased on structure up to 25–80% at different position.

References

1. ANSYS (2012) AQWA user manual. Release 14 Oct, pp 724–746
2. ANSYS (2013) ANSYS Mechanical user guide. Release 15 Nov, pp 724–746
3. Ashlin SJ, Sannasiraj SA, Sundar V (2015) Wave forces on an oscillating water column device. Procedia Eng 116:1019–1026
4. Bellamy NW, Bucchi A, Hearn GE (2016) Analysis of the SEA-OWC-Clam wave energy device—part A: historical development, hydro-dynamic and motion response formulations & solutions. Renew Energy 44:1–40
5. Bhattacharya R (1978) Dynamics of marine vehicles. Wiley, New York
6. Brusca S, Cucinotta F, Galvagno A, Lanzafame R, Mauro S, Messi-na M (2015) Oscillating water column wave energy converter by means of straight-bladed darrieus turbine. Energy Procedia 82:766–773
7. Bucchi A, Hearn GE (2016) Analysis of the SEA-OWC-Clam wave energy device part B: structural integrity analysis. Renew Energy 99:253–269
8. Bull D, Smith C, Jenne DS, Jacob P, Copping A, Willits S, Jepsen R (2014) Reference model 6 (RM6): oscillating wave energy converter. Sandia National Laboratories, 6 Sept
9. Falcão AFO, Henriques JCC (2016) Oscillating-water-column wave energy converters and air turbines: a review. Renew Energy 85:1391–1424
10. Kurowski PM (2012) What is calculated in FEA? 1–3. Retrieved from http://debis.deu.edu.tr
11. Park S, Choi E, Wang B, Kim H, Kwak S (2016) Design of spar-type floating substructures for a 2.5 MW-class wind turbine using fluid-structure coupled analysis, pp 3536–3548
12. Parkinson DB (2008) Function variability optimization in parameter design. Qual Technol Quant Manag 5(3):263–270

Main Propulsion Marine Diesel Engine Condition Based Maintenance Monitoring Using Ultrasound Signal

Mohd Naim Awang, Zakiman Zali, Nor Ashimy Mohd Noor and Ridwan Saputra Nursal

Abstract Operating a ship is very critical especially when the bad weather conditions. Therefore, a good and durable engine is especially important when the weather is bad. It is important to make periodic care or maintenance based condition of the engine to ensure that engine capabilities can help facilitate the operation of the vessel. This paper presents study of 2-stroke main propulsion marine diesel engine characteristic using ultrasound signal monitoring condition based maintenance (CBM). The increment of operation and maintenance cost was significantly increase the life cycle cost and it might interrupt the efficiently and return of investment in any particular asset, as in this project is the main propulsion 2-stroke marine diesel engine. The condition based maintenance (CBM) could be one of the option to reduce the maintenance cost and increase the efficiently of the main propulsion engine. The condition based maintenance (CBM) is the option for ship's owner to gain the profit margin with shrinking the performance of the engine. CBM is the proposed method for performing operation or maintenance and it was applied in various industries such as aviation industries, marine industries and power plant sector. Wave spectrum analysis using ultrasonic signal and wave sensor is one of the CBM method that can be applied for early identifying the excessive friction. In this study, both method condition based maintenance and wave spectrum analysis be utilized together for identifying the main engine performance without involving major and complex procedure. Most important of criterion for boost the profit margin are the reliability, availability, repairing, installation costs, operating costs, flexibility and

M. N. Awang
Politeknik Ungku Omar, 31400 Ipoh, Perak, Malaysia
e-mail: mnaim@puo.edu.my

Z. Zali (✉) · N. A. Mohd Noor
Center of Technology in Marine Engineering, 31400 Ipoh, Perak, Malaysia
e-mail: zakizali@puo.edu.my

N. A. Mohd Noor
e-mail: ashimy@puo.edu.my

R. S. Nursal
Politeknik Bagan Datuk, 36100 Bagan Datuk, Perak, Malaysia
e-mail: r.s.nursal@gmail.com

C. L. Saw et al. (eds.), *Advancement in Emerging Technologies and Engineering Applications*, Lecture Notes in Mechanical Engineering,
https://doi.org/10.1007/978-981-15-0002-2_19

engine size. Failure to optimize the performance of the main engine will tremendously reduce the company profit margin. This study suggested that once the engine is operating within the operating limit and that wave spectrum and signal show the decrement of wave amplitude, the engine is in good condition and no major maintenance required. If the working parameter is out of the range, and significant friction detected using wave spectrum analysis, then the maintenance is required. Therefore, both condition based maintenance and wave spectrum analysis or signal monitoring is a good combined method to identify the engine performance and able to reduce the maintenance cost that using the conventional maintenance schedule.

Keywords Marine main propulsion · 2-Stroke diesel engine · Marine diesel engine · Condition based maintenance · Ultrasound signal monitoring

1 Introduction

1.1 Main Propulsion Marine Diesel Engine

The ability of a vessel to operate depends on the ability of the ship's main propulsion produces optimum power and translated the power through the transmission system to the propeller shaft. Most ships nowadays worldwide using marine diesel engine as the prime mover or main propulsion. The selection of a main engine is influenced by several factors such as; reliability, availability, maintenance, installation costs, operating costs, flexible and engine size. Failure to make the appropriate selection and optimum engine will cause the shipping company will pay a very high cost. However, the factors that most concern to all shipping companies now is to control the increase in operating costs, including the cost of maintenance and spare parts costs. Therefore, all shipping companies are now focusing on the development and improvement in analytical methods, operations, diagnosing fault conditions, monitoring performance, optimize performance and engine control [1]. In order to minimize the risk of failure of equipment in line with rising maintenance costs. Therefore, these factors have prompted efforts latest technology helps provide a method of monitoring or monitoring and evaluation of mechanical systems for the purpose of replacing conventional methods such as waiting until the failure of either local or comprehensive and replace parts according to the scheduled routine [2]. In marine diesel engine operates the fluctuations piston movement causing contact and friction between the cylinder block and piston rings mounted on the piston. This shift would produce sound waves in the range of ultrasonic waves [3]. Ultrasonic waves generated during the period of operation of the engine caused a wave propagating in the structure of the engine cylinder and the outer frame of the engine. Resonant pressure propagation effects of this structural framework distortion engine will experience a very artistic and cannot be identified with the naked eye. Through a combination of structure and propagation effects distortion framework will form the engine knocking noise in the engine [4]. In addition, the propagation of vibration on the cylinder structure also occurs as a result of noise when the combustion process occurs. This is because the resulting noise is caused by the explosion of spontaneous combustion. However, the frequency and extremely loud noise will cause damage to the engine, especially on the surface of the piston. Extremely loud noise also produces acoustic waves and is usually measured using a pressure signal [4].

1.2 Condition Based Maintenance (CBM)

Through this diagnostic method, the sensor is mounted on the cylinder jacket to measure and record vibration signals generated in the engine. However, this method still has the disadvantage caused by vibration and noise signals generated when a fault occurs when it is too late and the resulting sound is able to be detected by hearing human. Method using ultrasonic signal propagation structure is a method that is suitable for monitoring the operation of the engine. This is because the ultrasonic signal is parallel direction and then the result of the directive, the signal-noise ratio (SNR) obtained better next use appropriate signal filtering [2]. The increment of maintenance cost was significantly increase the life cycle cost and it might interrupt the efficiently and return of investment in any particular asset. The condition based maintenance could be one of the option to reduce the maintenance cost and increase the efficiently of the marine engine. Condition Based Maintenance (CBM) is a suggested maintenance strategy that uses the actual condition of the machinery or equipment to decide what kind of maintenance need to be done. CBM will dictate that maintenance should only be performed when certain indicator or certain value shows sign of decreasing performance or upcoming failure.

The deterioration and failures of any systems might incur high cost and safety hazard. For that reason, preventive maintenance is necessary in order to replace the deteriorated system before failure. If the deterioration of the system or a control parameter strongly correlated with the state of the system can be measured (such as vibration analysis, cumulative wear monitoring or etc.), it is more appropriate to base the maintenance decision on the actual deterioration state of the system rather than on its age. This lead to the choice of a condition based maintenance [5]. The objective of CBM is to minimize the total cost of inspection and repairs by collecting and interpreting intermittent or a continuous data related to the operating condition of critical components of an asset [6]. There are noted that certain signs, conditions, or indications precede 99% of all machine failures [7]. Therefore, by monitoring an asset, action can be taken prior to it having a serious effect on the performance of the organization. Thus, condition-based monitoring offers an alternative to the PM assumption of age-related failure mode. Moreover, with condition-based monitoring, managers can focus on just-in-time (JIT) replacement.

This study is based on the hypothesis that the ultrasonic signal propagation structure capable of being used as a media monitoring in the operation of the engine. Through this study, the signals in the frequency range of ultrasonic sound generated by the engine during operation can be discerned. The ultrasonic sound signals can be generated through contiguity, interactive and friction between the piston ring and the inner walls of the combustion chamber. The monitoring process is studied through the operation of marine diesel engine 2-stroke type scavenge aligned. Engines will be charged and improved load through hydro dynamometer periodically. Hence, the observation of ultrasonic signals will be done at different speeds in each load applied. Observations wave or ultrasonic signals was performed using instrumentation such as Ultraprobe 10,000 to record the signal in the form of time domain. The objectives of this project are to evaluate the engine performance based working parameter, to identify the ultrasonic wave signature using Ultraprobe 10,000 and to propose the solution to minimize the maintenance cost. The study on ultrasonic wave signature is based on the output result of Ultraprobe 10,000.

2 Methodology

2.1 *Experimental Equipment and Test Rig*

The equipment used is a two-stroke marine diesel engine type Akasaka Mitsubishi UE Diesel Engine, ultrasonic sensors Ultraprobe 10,000 and wave signal is recorded. During the testing of engine, the engine will start under load than without load 0% and increase 25% (928 kg) and 35% (1160 kg). Data collected and observed of the ultrasonic wave signal for the development of an engine database. In these experiments, observations performed at each load imposed on the engine up to 35% and at a different rotational speed of 185, 235 and 255 rpm. The methodology is shown in the data collection, processing flowchart and test rig of Fig. 1. This research start with defining problem statement, follow by literature review. Then the collected data were verified and correlated. Once the data verified, the discussion process stated and follow by recommendation and at last, with conclusion.

Two-stroke marine diesel engine used in these experiments were in the marine engineering workshop, Marine Engineering Department, Politeknik Ungku Omar, Ipoh. The engine used in the process of teaching and learning for Marine Engineering Workshop Practice courses are initially installed in 1976 and now has over 42 years old. Marine diesel engine is used every 3 times a week over a period of 5–8 h each use. The maintenance process was conducted in 2009; this is the first major overhaul since the engine was first installed. Two-stroke marine diesel engine is also equipped with hydro dynamometer equipment capable of functioning as imposing load on the engine during a performance test conducted. Hydro dynamometer is using water as a medium for load able to impose the burden of more than 1 ton. Table 1 shows the specifications for marine diesel engines, Two-stroke engine used in the experiment.

The equipment must be kept in good condition and regularly before use. Before use Ultraprobe 10,000, the setting is done by giving the name of the file to be easier to record data. This task was done use Ultratrend DMS software to help organize the data. Start the engine cooling water pump, engine cooling jacket, hydro dynamometer

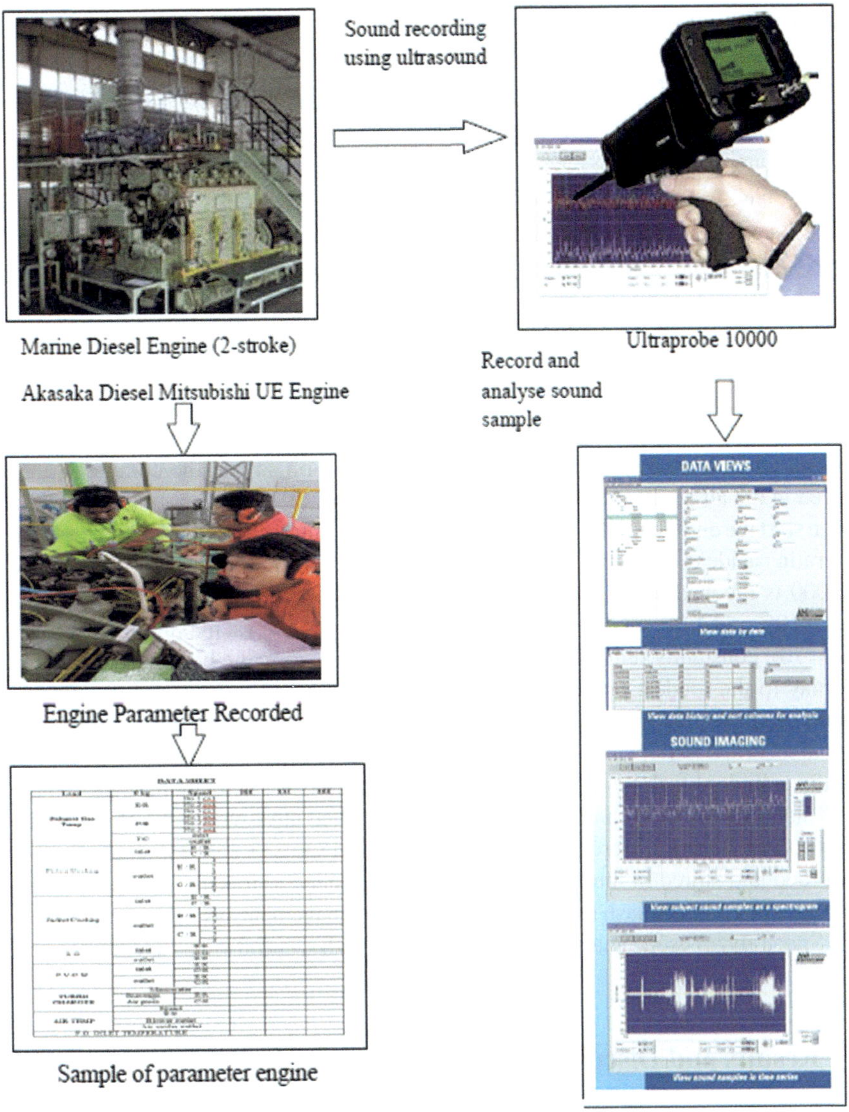

Fig. 1 Data collection flowchart and test rig

Table 1 Two-stroke marine diesel engine specification

No.	Item	Specification
1	Model/type	Akasaka diesel mitsubishi UE engine, uniflow scavenging, single acting, exhaust gas turbocharged, trunk piston
2	Number of cylinder	3 nos.
3	Bore	330 mm
4	Stroke length	550 mm
5	Max. output	750 PS
6	Max. speed	330 rpm
7	Starting system	Compressed air
8	Fuel system	BOSCH pump type

and oil lubrication engine. Air compressor is turned on so that the compressed air stored in the air storage tank located at a minimum of 20 bar for starting the engine. Then start the engine for the purpose of observing the ultrasonic waves. Let the engine operating at ideal and normal speed of around 180 rpm for 15 min. Nozzles Ultraprobe 10,000 is touched on the three marine diesel engine cylinder block on the engine loaded 0% and different speeds of 185, 235 and 255 rpm. Start at maximum sensitivity (S = 70), constantly reducing the sensitivity and following the bar graph amplitude display to the loudest point. Ultrasonic frequency sound can be heard or detected using headphones while waveforms can be seen on the screen ultraprobe itself. The signal recorded is a signal in the time domain. Accuracy step was repeated three times on the cylinder block to get a more accurate wave signal. There are periodic engine load starting with 0, 25, and 35%. At each load applied ultrasonic wave observations also recorded in 3 different rotational speeds of 185, 235 and 255 rpm. Data is recorded and stored in the compact flash card and need to transfer from ultraprobe through the compact flash card to ultratrend data management system. After transfer, UE Spectralyzer use to analyse sound samples in both spectral views and time series.

3 Results and Discussion

The experiment was carried out is to run observation ultrasonic wave propagation structure when the engine is operating in idle and loaded. Observation is done by touching the sensor nozzle Ultraprobe 10,000 on cylinder engine. There is a 3-cylinder engine, the reading frequency and magnitude of each cylinder wave recorded and recordable signal. Based on the Data Operating Parameter, the graphs for each loading condition for dedicated rpm are plotted on below graphs. For the ultrasonic wave analysis or wave spectrum, the discussion will be based on each load and each rpm but the graph were plotted together for 3 cylinders in one graph to ease the discussion process. The term early phase and end phase are referring to the time in x-axis. The term of low magnitude is referring to amplitude between −50 and 50 in y-axis. This is illustrated in Figs. 2, 3, 4, 5, 6, 7, 8, 9 and 10 respectively. For 0 kg load, 185 rpm, the graph had shown the early phase has high magnitude for cylinder 1 and cylinder 2. At the end phase shows low magnitude for cylinder 1 and 2. For 0 kg load, 235 rpm, the graph had shown the early phase has high magnitude for

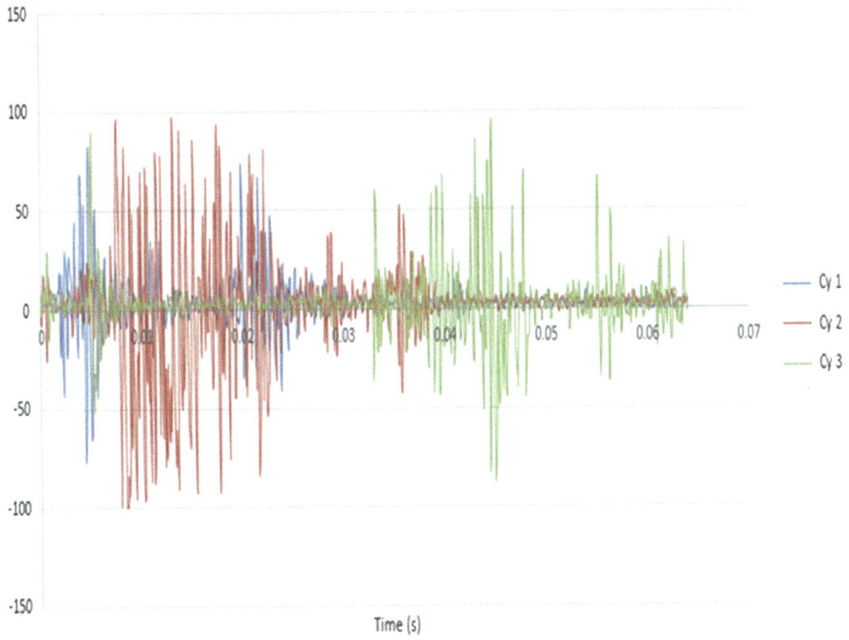

Fig. 2 Time domain analysis for load 0 kg at rotational speed of 185 rpm

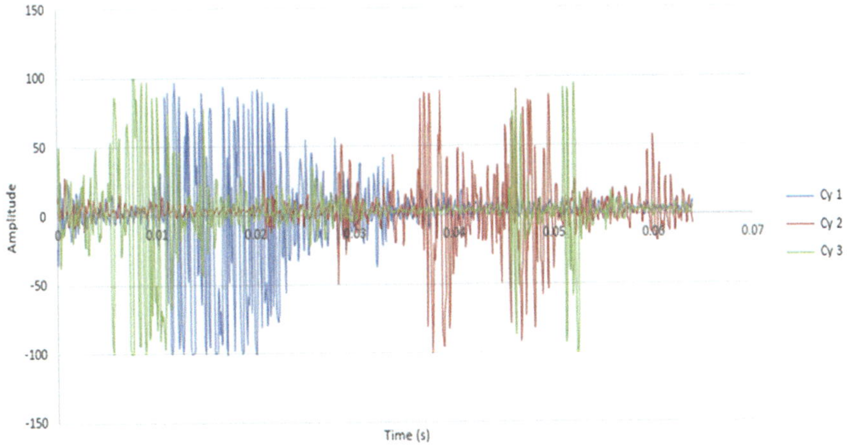

Fig. 3 Time domain analysis for load 0 kg at rotational speed of 235 rpm

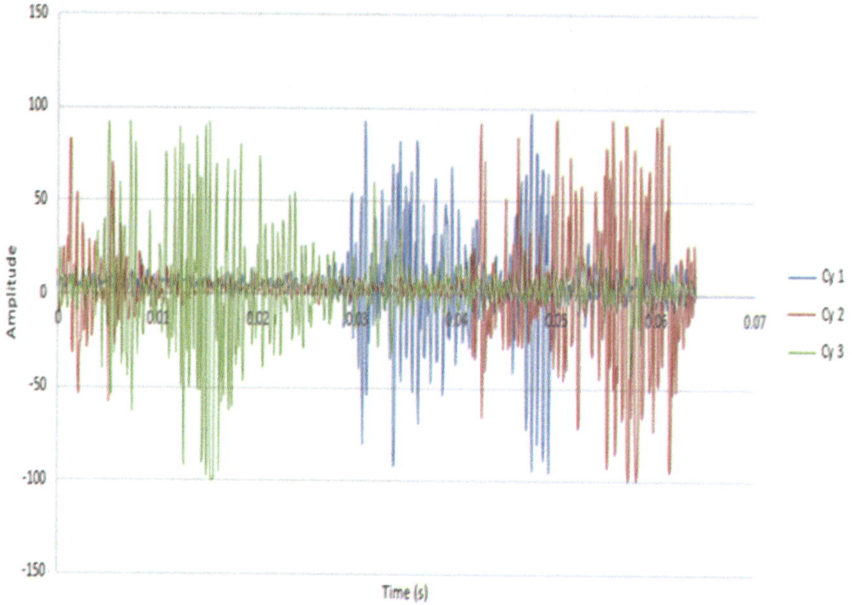

Fig. 4 Time domain analysis for load 0 kg at rotational speed of 255 rpm

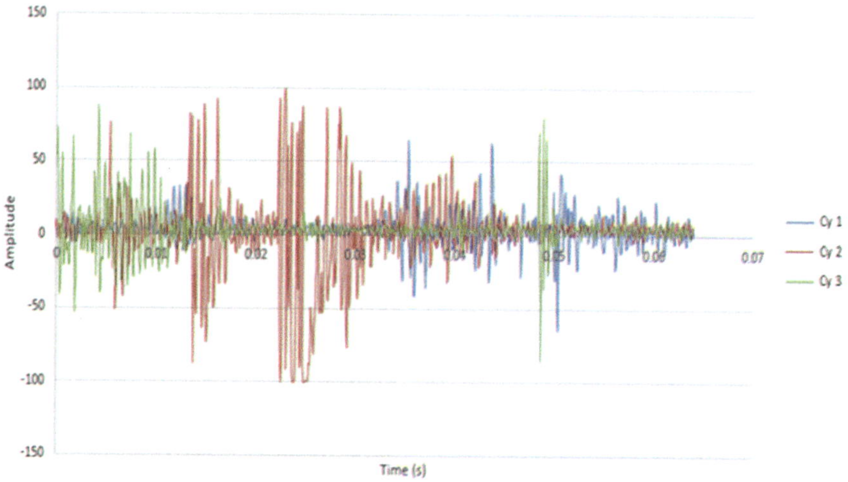

Fig. 5 Time domain analysis for load 928 kg at rotational speed of 185 rpm

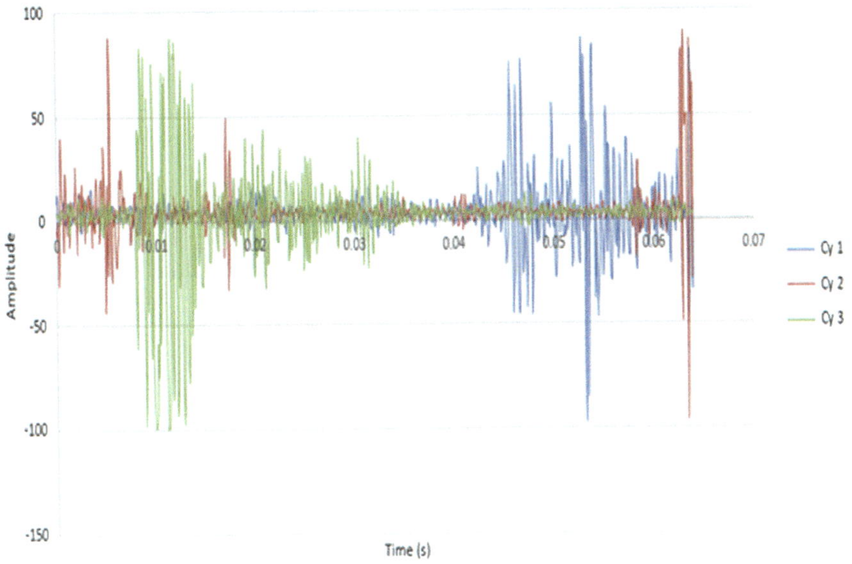

Fig. 6 Time domain analysis for load 928 kg at rotational speed of 235 rpm

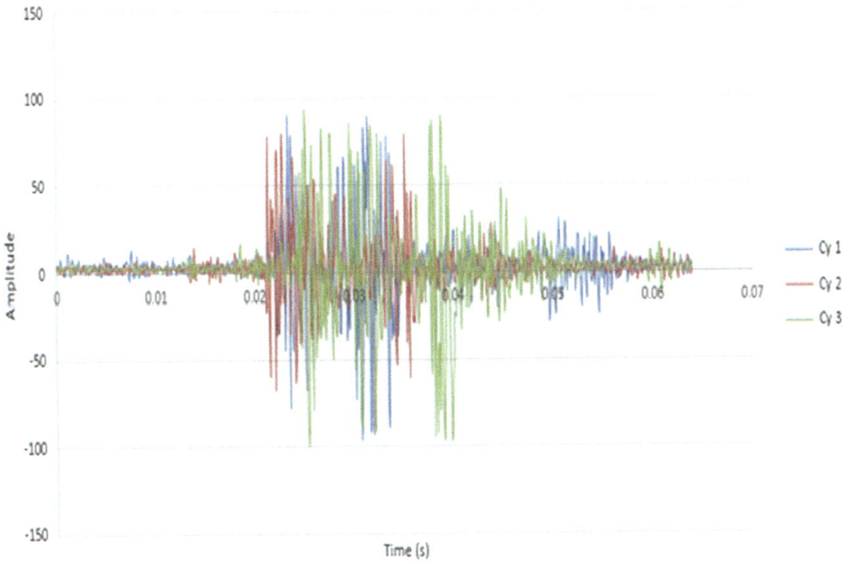

Fig. 7 Time domain analysis for load 928 kg at rotational speed of 255 rpm

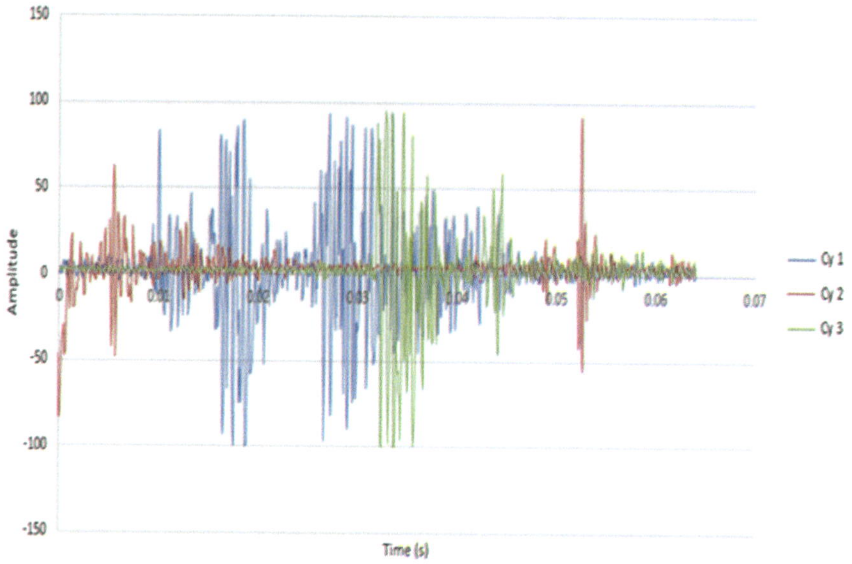

Fig. 8 Time domain analysis for load 1160 kg at rotational speed of 185 rpm

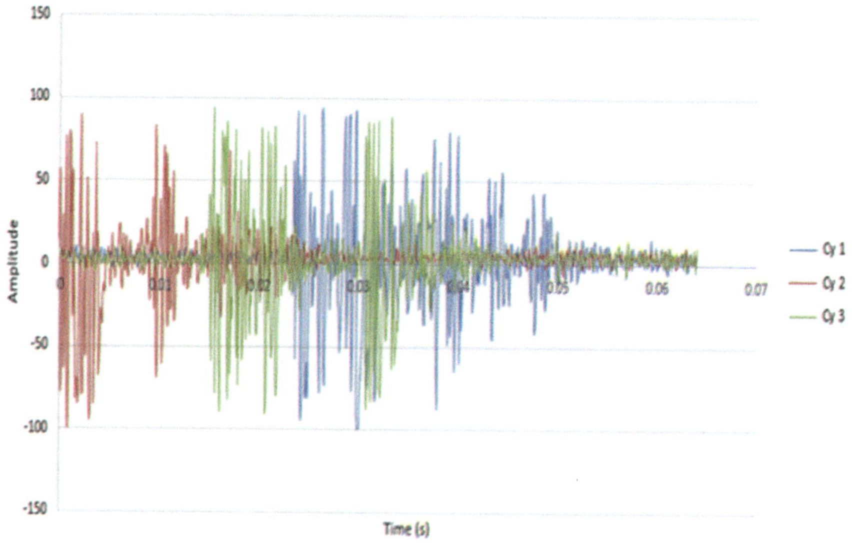

Fig. 9 Time domain analysis for load 1160 kg at rotational speed of 235 rpm

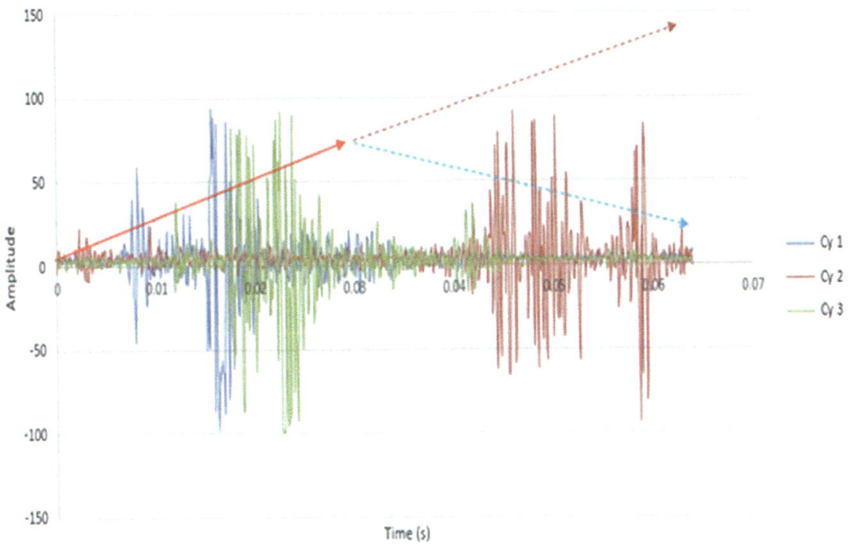

Fig. 10 Time domain analysis for load 1160 kg at rotational speed of 255 rpm

cylinder 1 and cylinder 3. At the end phase shows low magnitude for cylinder 2 and 3. For 0 kg load, 255 rpm, the graph had shown the early and end phase has high magnitude for all cylinders. For 0 kg load, it concludes that this load condition is not suitable because the amplitude recorded by Ultraprobe 10,000 shows the high magnitude. This is the main reason where running without load should be avoided. Light loading creates a condition termed wet stacking, indicating the presence of unburned fuel or carbon, or both, in the exhaust system. Its presence is readily indicated by the presence of continuous black smoke during engine-run operation. Carbon particles are deposited on top of the piston and in the injectors when fuel is not burned completely. Depending on the amount of time that the engine is run on low load, the engine may not be able to perform to its maximum rated load until these deposits have been burned off by operating the engine at higher loads. Piston rings are designed for optimum sealing under elevated combustion pressures. When these pressures are not reached due to the application of low loads, the fuel injected into the combustion chamber tends to get between the cylinder wall and the piston rings, causing dilution of the lubricating oil with fuel, with subsequent formation of acids and loss of lubricates [8].

For 928 kg load, 185 rpm, the graph had shown the early phase has high magnitude for cylinder 2 and cylinder 3. At the end phase shows low magnitude for all cylinders. For 928 kg load, 235 rpm, the graph had shown the early phase has high magnitude for cylinder 2 and cylinder 3. At the end phase shows low magnitude for cylinder 1. For 928 kg load, 255 rpm, the graph had shown both early and end phase has low magnitude for all cylinders but it shows high amplitude at the middle phase. For 928 kg load it concludes that the suitable load rpm for 928 kg load is at engine speed 255 rpm. It is because of the engine was stabilised at this speed. The middle phase shows the high magnitude because the engine is in the phase of achieving the optimum condition.

For 1160 kg load, 185 rpm, the graph had shown the early phase has high magnitude for cylinder 1. At the end phase shows low magnitude for all cylinders. For 1160 kg load, 235 rpm, the graph had shown the early phase has high magnitude for cylinder 2 and cylinder 3. At the end phase shows low magnitude for all cylinders. For 1160 kg load, 255 rpm, the graph had shown the early phase has high magnitude for cylinder 1 and cylinder 3. At the end phase shows low magnitude for cylinder 1 and 3. For 1160 kg load, it concludes the cylinder is running in good condition and no sign of excessive friction. If the graphs show the increment of amplitude consistently with the load and rpm, it will be the main reason where the maintenance work should be done. As an example, for load 1160 kg and 255 rpm, it shows the increment of amplitude (refer to red arrow). But at the end phase, the amplitude decrease tremendously (refer to dash blue arrow). If the amplitude increases (refer to red dash arrow), it indicates the necessity of performing maintenance on the engine. This result also can be verified with the result from operating condition parameters.

4 Conclusion

This study shows the capability of condition based maintenance to evaluate the marine engine performance and predict the maintenance required. The performance of Two-stroke marine diesel engine can be predicted by utilizing the normal operating condition and analysis on ultrasonic wave spectrum able to identify qualitatively the friction effect. The conclusion this research are Evaluating the operating parameter based on normal operating condition able to predict the engine performance. Recorded data should be in place for future reference. Ultrasonic probe 10,000 can be used to analyse the ultrasonic wave spectrum. The wave spectrum able to show the friction qualitatively. In order to minimize the maintenance cost, wave spectrum analysis can be used to identify the friction effect. The friction effect during 0 kg load can be the base line that show the maximum level of friction before maintenance work required. It concurred that the most friction on marine diesel engine occurred during engine started at 0 kg load or idling. The results of operating parameter and wave spectrum need to be analysis together because the operating parameters shows the standard parameter need to be obliged by the user and the value was given from the OEM and the wave spectrum able to identify the friction effect in the internal combustion cylinder.

References

1. Abdul-Kader HM (2009) Fuzzy knowledge base system for fault tracing of marine diesel engine. Commun IBIMA 11. ISSN 1943-7765
2. Nuawi MZ (2008) Pembangunan Sistem Pemantauan Keadaan Perkakas Mesin Menggunakan Kaedah Isyarat Ultrasonik Bawaan Udara. Universiti Kebangsaan Malaysia, pp 1–2
3. Nuawi MZ, Abdullah S, Ismail AR, Zulkifli R, Zakaria MK, Hussin MFH (2008) A study on ultrasonic signals processing generated from automobile engine block using statistical analysis. World Sci Eng Acad Soc Trans Signal Proc 4(5):279–288. ISSN 1790–5052
4. Behrens SC, Bohme JF (2001) Applying time-frequency methods to pressure and structure–borne sound for combustion diagnosis. In: Sixth international, symposium on signal processing and its application, vol 1, pp 256–259
5. Grall A, Berenguer C, Dieulle L (2001) A condition-based maintenance policy for stochastically deterioration systems. Reliab Eng Syst Saf 76:167–180
6. Knapp GM, Wang HP (1992) Machine fault classification: a neural network approach. Int J Prod Res 30(4):811–823
7. Bloch HP, Geitner FK (1983) Practical machinery management for process plants. Volume 2: Machinery failure analysis and troubleshooting. Exxon Chemical Co., Baytown, TX
8. Chishlim SD (2014). http://ecmweb.com/blog/running-without-load

Investigation on Performance and Emissions of a Single Cylinder Marine Diesel Engine Fuelled with Diesel and Crude Palm Biodiesel Oil

Shahril Jalil, Zamri Yusoff, Hairi Haizri Che Amat and Ridwan Saputra Nursal

Abstract Numbers of studies has been executed globally in order to seek the alternative way to replace the current fossil-fuel with regards to the rising of fuel-energy demand, running-out of petroleum reserves and to attain the cleaner atmosphere. For that reason, the exploration and studies on Biodiesel Fuel (BDF) has been one of the potential alternative. Thus, a research has been carried out to investigate the effects on performance and exhaust gas emissions of a single cylinder, 4-stroke marine diesel engine with capacity of 638 cc using biodiesel blends derived from Crude Palm Oil (CPO) along with neat diesel (DSL) that driven as standard reference. The CPO fuel was blended with DSL in various concentration to produce blended fuels that comprise CPO5, CPO10, CPO15 and CPO20. The CPO biodiesel was chosen in the study due to the availability and readiness of its feedstock sources in Malaysia all over the year. Moreover this research was focusing on marine engine compared to the most today's research which commonly engaged with automotive engines. The outcomes of this research found that the used of CPO blends fuels promotes to some extent of better performance of marine diesel engine and comparable exhaust gas emissions. The finding from this research also verified that the different levels of blending concentration have its pros and cons on the effects of engine

S. Jalil (✉) · H. H. Che Amat
Department of Marine Engineering, Politeknik Ungku Omar, 31400 Ipoh, Perak, Malaysia
e-mail: shahjalil@puo.edu.my

H. H. Che Amat
e-mail: haizri@puo.edu.my

Z. Yusoff · R. S. Nursal
Department of Marine Engineering, Politeknik Bagan Datuk, 36400 Hutan Melintang, Perak, Malaysia
e-mail: zamyusoff@pbd.edu.my

R. S. Nursal
e-mail: r.s.nursal@gmail.com

R. S. Nursal
Faculty of Mechanical and Manufacturing Engineering, Universiti Tun Hussein Onn Malaysia, 84600 Parit Raja, Johor, Malaysia

C. L. Saw et al. (eds.), *Advancement in Emerging Technologies and Engineering Applications*, Lecture Notes in Mechanical Engineering,
https://doi.org/10.1007/978-981-15-0002-2_20

performance and gas emissions. The used of crude palm biodiesel fuels however do good to the marine diesel engines which it can be ran without adjustments in equatorial regions.

Keywords Engine performance · Exhaust emission · Biodiesel · Crude palm oil

1 Introduction

From the word of 'bio', universally biodiesel is well known as a versatile biofuel that is renewable, biodegradable and environmentally beneficial in the sense that combustion discharges only biogenic carbon to the atmosphere [1–5]. The production of biodiesel undergo the transesterification process which converts triglycerides, like vegetable oil into fatty acid methyl esters or by the esterification of fatty acids [6–9]. The majority of biodiesel today is produced by alkali-catalysed transesterification with methanol, which results in a relatively short reaction time which means that lesser time is required to produce biodiesel at enormous volume [10–13]. The main advantages of biodiesel discussed by researchers includes its domestic origin, its potential for reducing a given economy's dependency on imported petroleum, biodegradability, high flash point, and inherent lubricity in the neat form [4, 7, 14–16].

Apart from those advantages, Malaysia as the leading of biodiesel producer in the world had bring many good returns to the country such as economic strengthened by exportations, job opportunities, environment quality as well as preparations towards the status of developed country. As the availability of the feedstocks is concerned for production purposes, the feedstocks should fulfill two main requirements i.e. low production costs and large production scale. Moreover, the latest policy of Malaysia's government is to mandate on the sale of commercial biodiesel blends from B5 by year 2009 to B10 and B7 by 2017 [9]. This blending ratio will keep increase by the authority as long as the consuming of the biodiesel blends fuels is verified to be matching with the local vehicles and machineries.

However, the crucial problem in using crude palm oil (CPO) based fuel is stoichiometric point, biofuel composition, oxidation stability, antioxidants on the degradation and much oxygen with comparing to fossil diesel oil. The application of the biodiesel fuel (BDF) in the compression ignition (CI) engines offer not only attractive and more economical fuel but also creates problems of higher emissions compared with petroleum based diesel specifically on the problem of emitting NO_x and Particulate Matter (PM) into the atmosphere. Thus, the improvement of emissions exhausted from BDF engines is urgently required to meet the future stringent emission regulations.

2 Experimental Setup

The fuels tested in the marine diesel engine were neat diesel with premium Euro 2 grade diesel, denoted as DSL along with Crude Palm Biodiesel Oil (CPO) blends fuel. Whilst, the purified crude oil methyl ester of oil palm was blended using specific blending machine in lab (see Fig. 1) with standard diesel (DSL) in numerous proportions of 5, 10, 15 and 20% by fuel. These blended fuels are denoted as CPO5, CPO10, CPO15 and CPO20 in further discussion. During the blending process, the scale blending machine will be operated at 70 °C and the mixture will be stirred at 60 °C for 1 h. The mixture will be stirred continuously with rotating blender blade speed was set to maintain at 250 rpm.

The blended CPO fuels properties were measured in order to meet the MS Standard requirements [17]. The density of fuel was measured by Metter Toledo Diamond Scale model JB703-C/AF. The kinematic viscosity of biodiesel was measured by Hydromation Viscolite 700 model VL700-T15 viscometer. The flash point of biodiesel was measured by using Pensky-Martens model PMA 4 flash point analyzer. The fuel emits enough vapours to ignite and is set to minimum flash point at 43 °C. The water content and acid value in biodiesel sample were measured using Volumetric KF Titrator model v20 and undergo titration process. During acid value

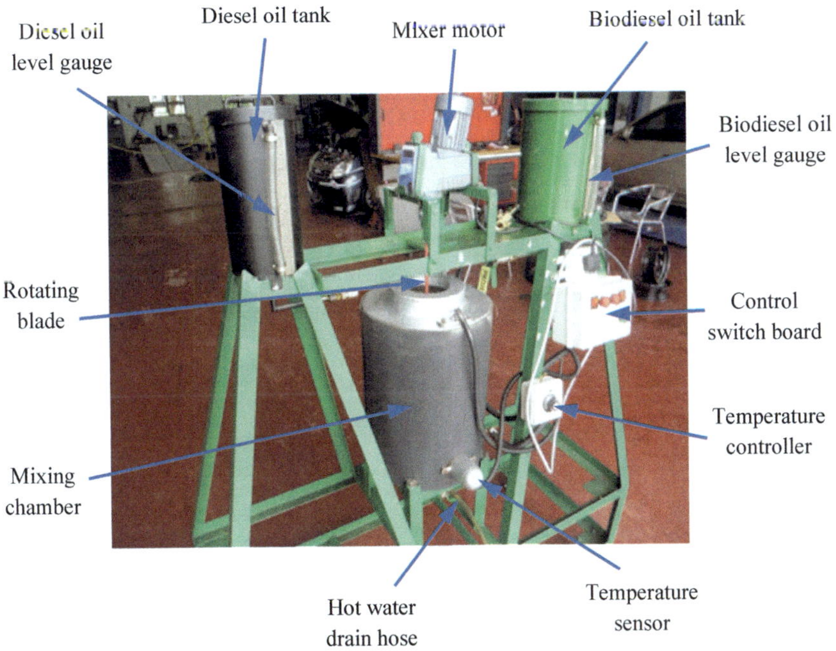

Fig. 1 Laboratory scale biodiesel fuel blending equipment

measurement, the biodiesel sample is titrated with alcoholic KOH using phenolphthalein as indicator [18, 19].

Table 1 gives the summary of the tested fuel properties.

The testing engine specifications are summarized in Table 2 whilst the experimental apparatus was arranged as in Fig. 2. All fuels were tested using a single cylinder marine diesel engine, model TF120-ML with 0.638 l capacity, horizontal type, 4-cycle/4-stroke with combination of water and air-cooled system made by Yanmar Motor. The engine was design with compact dimensions and in all applications with a power requirement up to 8.8 kW at 1 h due to its low weight. The dyno test load and speeds were controlled manually during experiment, whilst the overall engine operation was automatically monitored by set of a Supervisory Control and

Table 1 The properties of tested fuels

Fuel properties/testing method	Fuel type				
	DSL	CPO5	CPO10	CPO15	CPO20
Density (kg/m^3) ASTM D1298 (88 max)	833.74	837.05	837.66	840.43	841.17
Kinematic viscosity (mm^2/s) ASTM D445 (1.9–6.0)	3.01	3.01	2.9	3.03	3.1
Flash point (°C) ASTM D93 (43 min)	83	91.5	92	93.5	94.5
Water content (mg/kg) ASTM E203 (500 max.)	79.6	120.1	158.6	219	294.7
Acid value (mgKOH/g) ASTM D664 (0.8 max.)	0.423	0.259	0.662	0.322	0.364

Table 2 The specification of test engine

Item	Specification
Engine model	TF120-ML
Type	1 cylinder, Horizontal, 4 stroke
Combustion system	Direct injection
Aspiration	Natural aspiration
Cylinder bore × stroke	92 mm × 96 mm
Displacement	0.638 l
Maximum output power	12 Ps (8.8 kW) at 2400 rpm
Starting system	Manual and auto starting
Lubrication system	Complete enclosed forced lubricating system
Cooling system	Water-cooled, air-cooled
Electrical system	Alternator, 12 V–45 W
Fuel tank capacity	11 l
Engine oil capacity	2.8 l

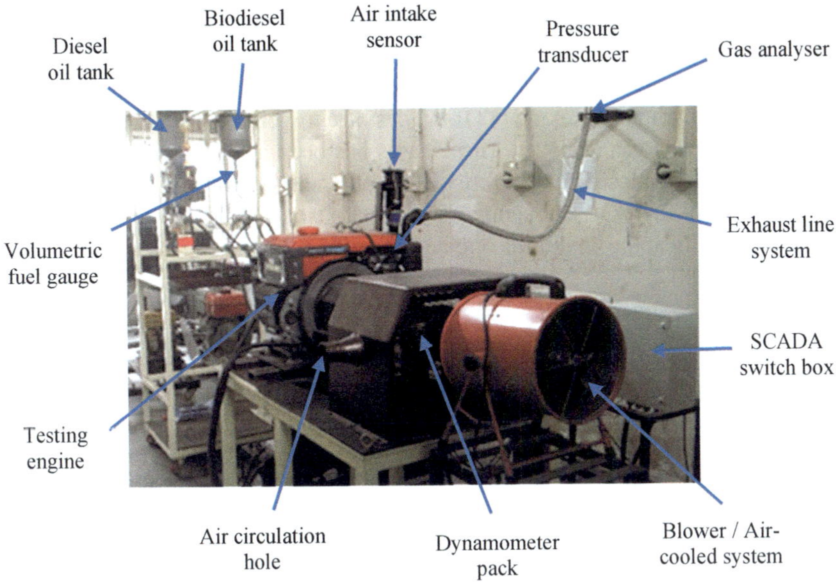

Fig. 2 Experimental equipment assembly

Table 3 The specification of gas analyzer

Gas measured	Range	Resolution
CO_2	0–9.999%	0.001%
CO	0–20.00%	0.01%
HC	0–10,000 ppm	1 ppm
O_2	0–25.00%	0.01%
NO_x	0–5000 ppm	1 ppm
Lambda	0.50–3.00	0.01
Other specs		
Power Supply	AC 90–270 V, 50/60 Hz	
	DC 12 V inverter (use in mobile testing PUC centre)	
Dimension	270 (D) × 370 (D) × 170 (H) mm	
Accessories	Probe, hose, fuse, leak test, dust and carbon filter	

Data Acquisition unit (SCADA) unit and a computer. The DEWESOFTv7.11 application is used to monitor the input parameters and measure the performance of the engine while the DYNO-MAX 2010 application is used to analyse the combustion behaviour.

For this research, the engine was operated under dynamometer loads at 50% load condition throughout the various engine speeds were simulated at 800, 1200, 1600 and 2000 rpm. The measurements data comprised brake power (P_B), Torque (τ), brake specific fuel consumption (BSFC), brake mean effective pressure (BMEP), brake thermal efficiency (η_{BT}) as well as exhaust gas temperature (T_{EG}).

In conjunction, the measurements of exhaust gas emissions comprised carbon dioxide (CO_2), carbon monoxide (CO), oxides of nitrogen (NO_x) and hydrocarbon (HC) using exhaust gas analyzer. These gases has been analysed and captured by using exhaust gas analyser model Airrex HG-540. The specifications of gas analyser are describes in Table 3. The sampling probes of smoke and gas analyser were mounted centrally at the end of the engine exhaust pipe.

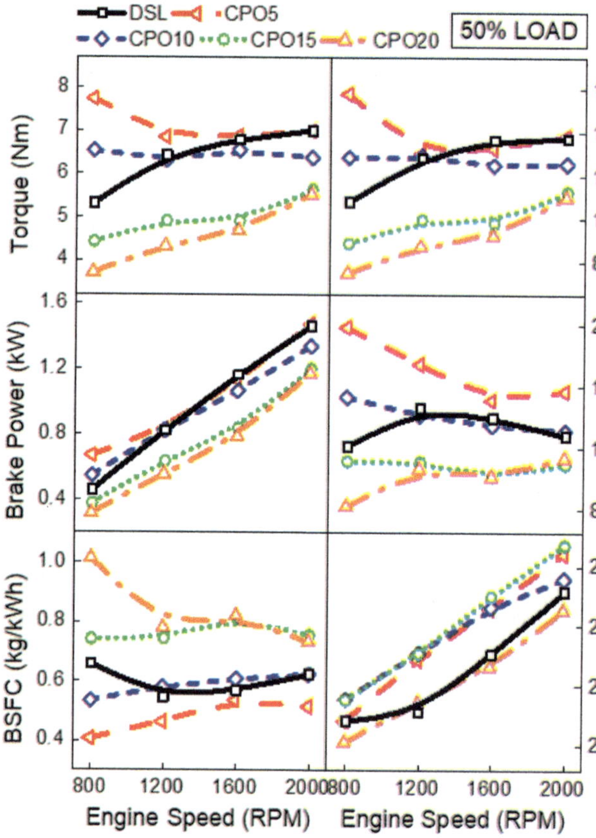

Fig. 3 The performance of marine diesel engine throughout engine speed from 800 to 2000 rpm under 50% rated load

3 Results and Discussions

Figure 3 shows the performance of marine diesel engine throughout the range of engine speed comprises of brake power (P_B), Torque (τ), brake specific fuel consumption (BSFC), brake mean effective pressure (BMEP), brake thermal efficiency (η_{BT}) and exhaust gas temperature (T_{EG}) by biodiesel blends CPO5, CPO10, CPO15, CPO20 and neat diesel (DSL) under 50% load condition. According to the figure, it is clearly shows the P_B and T_{EG} increased as the engine speed increased. However it is remarked that biodiesels given a slight lower of P_B and slightly higher T_{EG} compared to that of DSL. While the changes in τ by DSL increased as engine speed was increased while CPO5 and CPO10 showed contrary pattern over the changes of engine speed compared to DSL under the rated load. From the figure also, it is observed that there was small reductions of η_{BT} at higher engine speed for all biodiesel blends as well as DSL except for CPO15 and CPO20. Furthermore, the variant of η_{BT} between DSL and biodiesels blends was greater at low speed particularly for CPO5 and CPO10. It also observed that BSFC of all tested fuels were gradually increased as engine speed increased except for CPO20. The offset margin by CPO20 from DSL also a bit greater. This clearing up that the power produced during low speeds is low and the main part of fuel is consumed to overcome the engine friction [20–24]. Aside, the results prescribes that changes of BMEP throughout the engine speeds had similar behavior to the τ result of by all tested fuels.

Figures 4 and 5 indicates the characteristic of fuel combustion of all tested fuels inside the combustion chamber under conditions 50% load at running speed of 1200 and 1600 rpm respectively. These figures comprised the plots for the changes of cylinder pressure, heat release and cumulative heat release (CHR) with respect to the different crank angle degree (° CA). From Fig. 4, it is remarked that through engine speed of 1200 rpm under 50% load exerted, the maximum cylinder pressure for CPO15 and CPO20 blends fuel are 57.73 bar at 2° Crank Angle (CA) after top dead centre (aTDC) and 60.43 bar at 2° CA aTDC respectively. Whilst, the cylinder peak pressure for DSL was measure as 56.6 bar at 3° CA aTDC as depicts in Fig. 4. These results clarified that the highest cylinder peak pressure is remarked by CPO20 while slightly lower for CPO15 due to the reason of oxygen content, higher BSFC, cetane number and boiling points of CPO20 [25–28]. On the other hands, the maximum cylinder pressure recorded for DSL was slightly lower compared to that of all biodiesel blends fuel. Peak pressure depends upon the burning rate after the start of ignition during uncontrolled combustion stage.

The major aspect that determines the uncontrolled or diffusion combustion phase is the mixture preparation rate and atomization which is influenced by viscosity and volatility of the fuel [29–33]. In conjunction, all tested fuels shows similarity in initial heat release with slightly higher peak value of the premixed combustion by CPO20 while slightly by CPO15 relative to that of DSL as illustrates in heat released plots in Fig. 4.

It is considered that the differences in the heat release rate pattern influenced the local flame temperatures and resulted in the NO_x formation [34–38]. Ignition delay is

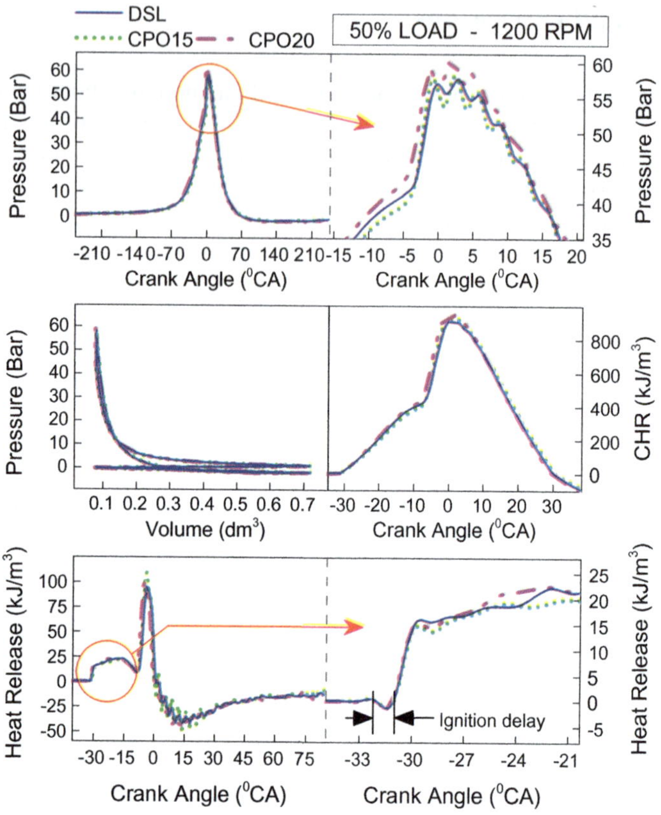

Fig. 4 Combustion characteristic of all tested fuels under 1200 rpm engine speed

one of the important parameters in combustion process. The ignition delay is defined as the time interval between the start of injection and start of combustion [39–42]. It is observed that the ignition delay by all fuels was approximately similar at 31° CA before top dead centre (bTDC). Besides, the start of combustion (SOC) timing is observed to occur simultaneously to the end of ignition delay means the combustion begin just after 31° CA bTDC as well. Aside, Fig. 5 indicates the characteristic of fuel combustion inside the combustion chamber by CPO under conditions of 50% load through 1600 rpm engine speed, along with that of DSL for comparison analysis. It is observed that CPO15, CPO20 and DSL exhibits 61.67 bar of cylinder peak pressure at 1° CA aTDC, 64.49 bar at TDC and 63.12 bar at 2° CA aTDC respectively. The SOC timing and ignition delay is observed to occur at 31° CA bTDC demonstrates by biodiesel blends fuels and DSL as well. Although all tested fuels have the same ignition delay, the slope for heat release at the initial of SOC by CPO20 is remarked was greater than that of CPO15 and DSL which was an indication of faster burning rate. Therefore, this was attributed to the higher rate of heat release for CPO20 than

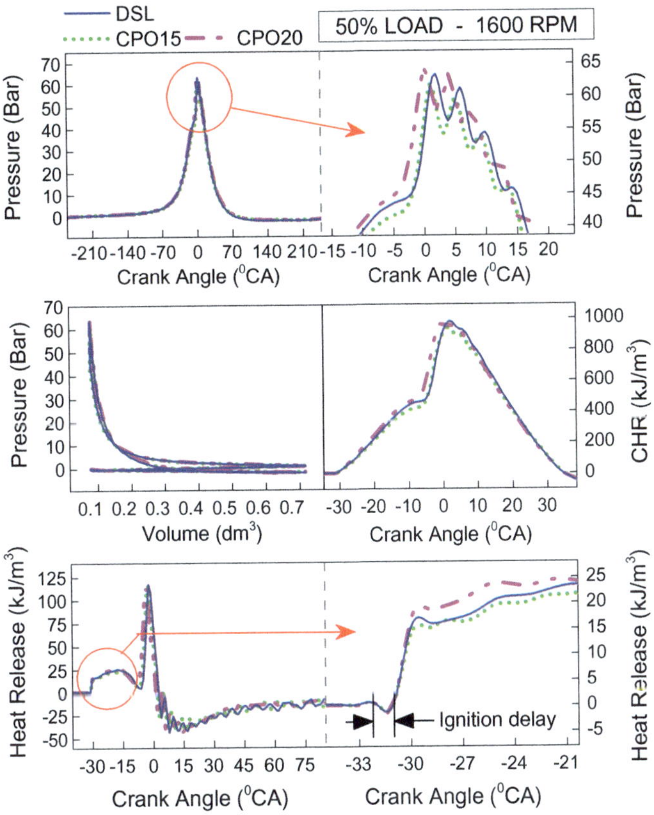

Fig. 5 Combustion characteristic of all tested fuels under 1600 rpm engine speed

that of DSL and significance to the quicker exceeds the cumulative heat released as depicts in the CHR plots in Fig. 5.

Figure 6 shows the emissions effects by CO_2, CO, NO_x and HC from diesel engine fuelled by biodiesel blends CPO5, CPO10, CPO15 and CPO20 as well as standard diesel (DSL) corresponding to the changes of engine speed under 50% load condition. This emissions results is clearly shows that CPO5 and CPO20 produced lower CO_2 emission as compared to that of DSL. However, the CO_2 emission by CPO10 decreased gradually proportional to the increasing of speeds. Aside, from all type of tested biofuels, only CPO20 among the biofuels that released lower CO gas emission than that of DSL. Figure 6 also dictates that 20% biodiesel blends fuel shows about similar graph pattern and contributes lower NO_x emission relative to that of standard diesel as well as CPO5 through running speeds of 1200–1600 rpm. Whereas CPO15 was remarked to have highest NO_x among the biodiesel blend fuel test. The higher of NO_x production from diesel engine is an indicator of higher burning rate which leads to higher peak combustion temperature [43–46]. On the other hand, the combustion of CPO5 also remarked having tolerable HC emission

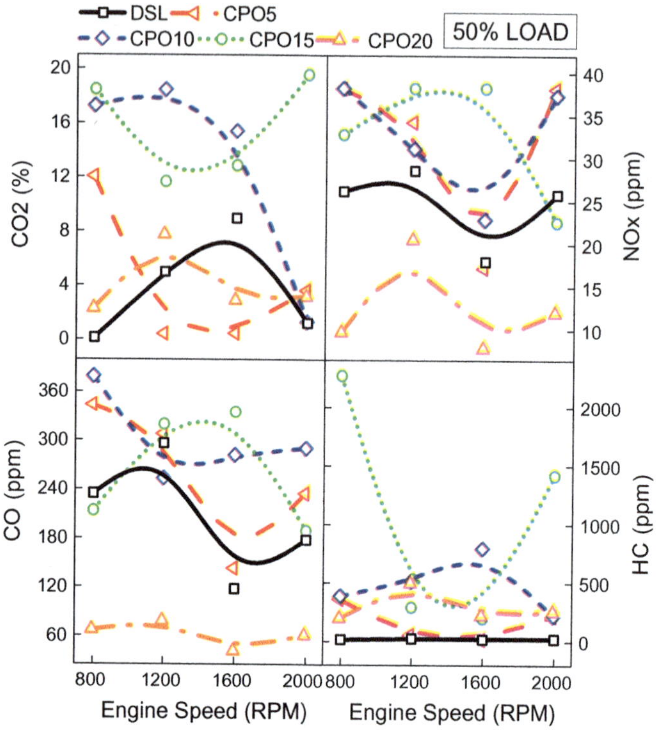

Fig. 6 Effects of exhaust gas emissions throughout engine speed from 800 to 2000 rpm under 50% rated load

while CPO10 and CPO20 were observed to have higher HC emission at a slight relative to that of DSL. Meanwhile, CPO15 exhibits the greatest of HC emissions which indicates that incomplete combustion occurs during either over-lean and over-rich fuel-air mixtures [47–51]. In overall, it can be concludes that the combustion of CPO20 promotes lower exhaust gas emissions of all gases whilst CPO5, CPO10 and CPO15 emit higher exhaust gas emissions as compared to that of petroleum diesel.

4 Conclusion

In this research, the CPO biodiesel fuels with different blending ratio (5, 10, 15 and 20%) were successfully tested in single cylinder marine diesel engine through engine speed designed at 800, 1200, 1600 and 2000 rpm. It can be concluded that the use of CPO5 promotes significant improvement of engine performance among the biofuels. While a clear drop of engine performance was experienced by CPO15 and CPO20 by comparing to diesel. Conversely, a significant reduction of gas emission was donated by CPO20, while the combustions of CPO10 and CPO15 contributed to somewhat higher gas emissions relative to standard diesel. This study also verified that there is better ignition capability by CPO biodiesel blends fuel due to the higher level of molecular saturation of it contains. This means a lower number of double bonds in the molecules directed to higher ignition quality in compression ignition (CI) engine. Aside, this also may leads to a higher cloud point which makes them difficult to be used in cold weather unless certain cold flow additives are being added. It also virtuous recommendation to perform onboard vessel test as well as numerical test/CFD [52] thus further the analysis and significance of biodiesel utilization particularly in marine diesel engine can be discovered more. Nevertheless, the CPO biodiesel blends fuel was proven can be used in marine diesel engine without engine alterations or modification.

Acknowledgements The authors would like to oblige Department of Polytechnic and Community College Education, Ministry of Education Malaysia for supporting this research. The authors also want to oblige Department of Marine Engineering, Politeknik Ungku Omar and Centre for Energy and Industrial Environment Studies, Universiti Tun Hussein Onn Malaysia for providing facilities, expertise and resources to complete the research.

References

1. Gasparatos A, Doll CNH, Esteban M, Ahmed A, Olang TA (2017) Renewable energy and biodiversity: implications for transitioning to a green economy. Renew Sustain Energy Rev 70(4):13
2. Ong HC, Mahlia TMI, Masjuki HH, Norhasyima RS (2011) Comparison of palm oil, *jatropha curcas* and calophyllum inophyllum for biodiesel: a review. Renew Sustain Energy Rev 15(8):3501–3515
3. Marella ER, Holkenbrink C, Siewers V, Borodina I (2018) Engineering microbial fatty acid metabolism for biofuels and biochemicals. Curr Opin Biotechnol 50:39–46
4. Gaikwad PU, Balan KL, Sarawade SS (2017) Biodiesel fuel—a review. IOSR J Mech Civ Eng 3:66–70
5. Khalid A, Nursal RS, Tajuddin ASA, Hadi SA (2016) Performance and emissions characteristics of alternative biodiesel fuel on small diesel engine. ARPN J Eng Appl Sci 11(12):7424–7430
6. Liu G (2017) An overview of biodiesel production and refining technologies. Curr Altern Energy 1(1):3–19
7. Bateni H, Saraeian A, Able C (2017) A comprehensive review on biodiesel purification and upgrading. Biofuel Res J 15:668–690
8. Ugwu BI, Nnaji JC, Simonyan KJ, Onukwuli OD (2018) Advances in biodiesel synthesis: the role of various catalysts. Open J Eng Sci 1(1):31–52
9. Nursal RS, Zali Z, Amat HHC, Ariffin SAS, Khalid A (2017) Comparative study of the performance and exhaust gas emissions of biodiesels derived from three different feedstocks with diesel on marine auxiliary diesel engine. ARPN J Eng Appl Sci 12(6):2017–2028
10. Atabani AE, Silitonga AS, Anjum I, Mahlia TMI, Masjuki HH, Mekhilef S (2012) A comprehensive review on biodiesel as an alternative energy resource and its characteristics. Renew Sustain Energy Rev 16(4):2070–2093
11. Koh MY, Idaty T, Ghazi M (2011) A review of biodiesel production from *Jatropha curcas* L. oil. Renew Sustain Energy Rev 15(5):2240–2251

12. Vasudevan PT, Briggs M (2008) Biodiesel production–current state of the art and challenges. J Ind Microbiol Biotechnol 35(5):421–430
13. Ramachandran K, Suganya T, Gandhi NN, Renganathan S (2013) Recent developments for biodiesel production by ultrasonic assist transesterification using different heterogeneous catalyst: a review. Renew Sustain Energy Rev 22:410–418
14. Nursal RS, Zali Z, Jalil S, Khalid A, Hadi SA (2017) Experimental study of the bio-additives effects in biodiesel fuel on performance, emissions and combustions characteristics of diesel engine. ARPN J Eng Appl Sci 12(6):1997–2005
15. Kathirvel S, Layek A, Muthuraman S (2016) Exploration of waste cooking oil methyl esters (WCOME) as fuel in compression ignition engines: a critical review. Eng Sci Technol Int J 19(2):1018–1026
16. Velmurugan K, Sathiyagnanam AP (2016) Impact of antioxidants on NO_x emissions from a mango seed biodiesel powered DI diesel engine. Alex Eng J 55(1):715–722
17. Department of Standards Malaysia (2008) MS 2008:2008 automotive fuels—Palm Methyl Esters (PME) for diesel engines—requirements and test methods
18. Khalid A, Tamaldin N, Jaat MNM, Ali MFM, Manshoor B (2013) Impacts of biodiesel storage duration on fuel properties and emissions. Procedia Eng 68:225–230
19. Novalia R, Sianipar R, Ariyani D, Fatyasari I (2017) Conversion of palm oil sludge to biodiesel using alum and KOH as catalysts. Sustain Environ Res 27(6):291–295
20. Hariram V, Bharathwaaj R (2016) Application of zero-dimensional thermodynamic model for predicting combustion parameters of CI engine fuelled with biodiesel-diesel blends. Alex Eng J 55(4):3345–3354
21. Manu PV, Sunil A, Jayaraj S (2016) Experimental investigation using an on-board dry cell electrolyzer in a CI engine working on dual fuel mode. Energy Procedia 90:209–216
22. Khalid A, Mudin A, Jaat MNM, Mustaffa N, Manshoor B, Fawzi M, Razali MA, Ngali Z (2014) Effects of biodiesel derived by waste cooking oil on fuel consumption and performance of diesel engine. Mech Mater Eng 554:520–525
23. Demirbas A (2009) Progress and recent trends in biodiesel fuels. Energy Convers Manag 50(1):14–34
24. Nursal RS, Hashim AH, Nordin NI, Abdul MAH, Danuri MR (2017) CFD analysis on the effects of exhaust backpressure generated by four-stroke marine diesel generator after modification of silencer and exhaust flow design. ARPN J Eng Appl Sci 12(4):1271–1280
25. Magno A, Mancaruso E, Vaglieco BM (2016) In-cylinder combustion process analysis of a small compression ignition engine fuelled with both blended and pure biofuel. Energy Procedia 66:29–32
26. Lapuerta M, Rodríguez-fernández J, Fernández-rodríguez D (2018) Cold flow and filterability properties of n-butanol and ethanol blends with diesel and biodiesel fuels. Fuel 224:552–559
27. Abed KA, Morsi AKEl, Sayed MM, Shaib AAEl, Gad MS (2018) Effect of waste cooking-oil biodiesel on performance and exhaust emissions of a diesel engine. Egypt J Pet
28. Khalid A, Tajuddin ASA, Jaat MNM, Manshoor B, Zaman I, Hadi SAA, Nursal RS (2017) Performance and emissions of diesel engine fuelled with preheated biodiesel fuel derived from crude palm, jatropha, and waste cooking oils. Int J Automot Mech Eng 14(2):4273–4284
29. Hosamani BR, Katti VV (2018) Experimental analysis of combustion characteristics of CI DI VCR engine using mixture of two biodiesel blend with diesel. Eng Sci Technol Int J
30. Kumar MS, Jaikumar M (2014) Studies on the effect of hydrogen induction on performance, emission and combustion behaviour of a WCO emulsion based dual fuel engine. Int J Hydrogen Energy 93(32):18440–18450
31. Nursal RS, Zali Z, Zainol I, Sabri MNM (2018) Experimental study of direct injected marine auxiliary diesel engine performance, emission and cylinder pressure using biodiesel fuels derived from *jatropha curcas* oil. Adv Struct Mater 85:115–135
32. Manieniyan V, Sivaprakasam S (2008) Investigation of diesel engine using bio-diesel (methyl ester of jatropha oil) for various injection timing and injection pressure. SAE Technical Paper Series, 2008-01-1577

33. Anand R, Kannan GR, Nagarajan S, Velmathi S (2010) Performance emission and combustion characteristics of a diesel engine fueled with biodiesel produced from waste cooking oil. In: SAE international paper, 2010-01-0478
34. Deivajothi P, Manieniyan V, Sivaprakasam S (2017) An impact of ethyl esters of groundnut acid oil (vegetable oil refinery waste) used as emerging fuel in DI diesel engine. Alex Eng J
35. Kannan D, Nabi N, Hustad JE (2009) Influence of ethanol blend addition on compression ignition engine performance and emissions operated with diesel and jatropha methyl ester, 4970
36. Yoshimoto Y (2009) Performance and emissions of a diesel engine fueled by biodiesel derived from different vegetable oils and the characteristics of combustion of single droplets. SAE Int J Fuels Lubr 2(1):827–838
37. Mat Yasin HM, Mamat R, Yusop AF, Daing Idris DMN, Yusaf T, Rasul M, Najafi G (2017) Study of a diesel engine performance with exhaust gas recirculation (EGR) system fuelled with palm biodiesel. Energy Procedia 110(12):26–31
38. Kumar MV, Babu AV, Kumar PR (2018) The impacts on combustion, performance and emissions of biodiesel by using additives in direct injection diesel engine. Alex Eng J 57(1):509–516
39. Madiwale S, Bhojwani V (2016) An overview on production, properties, performance and emission analysis of blends of biodiesel. Procedia Technol 25:963–973
40. Atabani AE, Silitonga AS, Ong HC, Mahlia TMI, Masjuki HH (2013) Non-edible vegetable oils : a critical evaluation of oil extraction, fatty acid compositions, biodiesel production, characteristics, engine performance and emissions production. Renew Sustain Energy Rev 18:211–245
41. Moser BR (2009) Biodiesel production, properties, and feedstocks. In Vitro Cell Dev Biol Plant 45:229–266
42. Knothe G (2010) Biodiesel: current trends and properties. Top Catal 53:714–720
43. Kumar MS, Ramesh A, Nagalingam B (2001) Experimental investigations on a jatropha oil methanol dual fuel engine. SAE Technical Paper Series, 2001 01 0153
44. Devarajan Y, Munuswamy DB, Mahalingam A, Nagappan B (2017) Performance, combustion and emission analysis of neat palm oil bio-diesel and higher alcohol blends in diesel engine. Energy Fuels 31(12):13796–13801
45. Gad MS, El-Araby R, Abed KA, El-Ibiari NN, Mors AKEl, El-diwani GI (2017) Performance and emissions characteristics of C.I. engine fueled with palm oil/palm oil methyl ester blended with diesel fuel. Egypt J Pet
46. Naik B, Radha K (2016) Performance and emission of VCR-CI engine with palm kernel and eucalyptus blends. Perspect Sci 8:195–197
47. Yusop AF, Mamat R, Yusaf T, Najafi G (2018) Analysis of particulate matter (pm) emissions in diesel engines using palm oil biodiesel blended with diesel fuel. Energies 11(1039):1–25
48. Sastry GRK, Deb M, Panda JK (2015) Effect of fuel injection pressure, isobutanol and ethanol addition on performance of diesel-biodiesel fuelled D.I. diesel engine. Energy Procedia 66:81–84
49. Abedin MJ, Masjuki HH, Kalam MA, Sanjid A, Rahman SMA, Fattah IMR (2014) Performance, emissions, and heat losses of palm and jatropha biodiesel blends in a diesel engine. Ind Crops Prod 59:96–104
50. Shahir VK, Jawahar CP, Suresh PR, Vinod V (2017) Experimental investigation on performance and emission characteristics of a common rail direct injection engine using animal fat biodiesel blends. Energy Procedia 117:283–290
51. Sivaramakrishnan K (2017) Investigation on performance and emission characteristics of a variable compression multi fuel engine fuelled with Karanja biodiesel–diesel blend. Egypt J Pet 1–10
52. Nursal RS, Nordin NI, Mohammad M, Afandi M, Hamid A, Awang MN, Ariffin AAM (2017) Investigation on resistance effects through towing test and CFD analysis on catamaran boat hull form. ARPN J Eng Appl Sci 12(4):1334–1343

A CFD Analysis of Ventilation System Improvement in the Two Stroke Marine Diesel Engine Workshop

M. R. Danuri and S. A. Rashid

Abstract The study was carried out to improve by identifying and determining the level of thermal comfort and ventilation system at the 2-stroke Marine Engine Diesel Engine, Department of Marine Engineering, Politeknik Ungku Omar, Ipoh, Perak, Malaysia. Computational Fluid Dynamic (CFD) software has been used to study and suggest appropriate working conditions in the workshop. The three-dimensional model for the workshop has been analyzed using the CFD software. Results obtained from the simulation using the natural ventilation system indicate that this system is not capable of providing thermal comfort in the working area. Thermal comfort levels and ventilation systems are defined in reference to the American Society of Heating, Refrigerating and Air Conditioning Engineers (AHSRAE) standards. The improvement involves the use of ventilation device to further enhance the efficiency of the naturally ventilated system that has been used in the workshop.

Keywords Ventilation system · Thermal comfort · Computational fluid dynamic

1 Introduction

Ventilation plays an important role in providing a comfortable and productive work environment. Makhbul [1] shows that health factors, workspace designs, shifting work, ventilation systems and working hours have significant relationship with stressful consequences [1]. The ergonomic workstation component also demonstrates that human and environmental factors have significant relationships with the effects of stress. Makhbul et al. [1] in his previous study found that 62.9% of stress variation was dependent on its own factors such as chair, work area, sound, lighting, working time and humidity [1]. Seats and office seats, working hours and air humidity are found to have significant relationships with working pressure. While the remaining

M. R. Danuri (✉) · S. A. Rashid
Politeknik Ungku Omar, Ipoh, Perak, Malaysia
e-mail: Zuwan.poli@1govuc.gov.my

S. A. Rashid
e-mail: manshid@puo.edu.my

C. L. Saw et al. (eds.), *Advancement in Emerging Technologies and Engineering Applications*, Lecture Notes in Mechanical Engineering,
https://doi.org/10.1007/978-981-15-0002-2_21

Table 1 ASHRAE standard

No.	Parameter	ASHRAE standard
1	Dry bulb temperature	20–27 °C
2	Air velocity	>0.2 m/s

factors, such as workplace design, sound and lighting, are not significantly affected. However, these factors need to be addressed by management in the interest of the institution in terms of minimizing working pressure. The design of the workstation from an ergonomic perspective can effectively stimulate productivity and minimize stress through the relationship between various components in the system.

The American Society for Heating, Refrigeration and Air Conditioning Engineering (ASHRAE) has identified the permissible indoor air quality as "*the air in which no contaminant is known as a hazardous concentration as determined by the authorities and a large majority (80% or more) of the people who are exposed do not feel dislike*" [2]. According to Parsons [3] thermal comfort has a great influence on productivity and satisfaction of occupants of closed buildings [3]. Thermal comfort is very difficult to value, this is because we need to take into account the various environmental and individual factors when deciding on the temperature and ventilation system that will make it feel comfortable. So the best method is to look forward to the condition of the thermal environment satisfying the majority of people at work, or in other words 'reasonable comfort'. The comfortable zone set by ASHRAE [2], for summer is in the following range Table 1.

According to Cândido et al. [4], in warm and humid climates, the addition of indoor air velocity increases the indoor temperature of the building that is the most comfortable building occupants [4]. However, the distribution of air velocities measured during field studies has tended to be relatively low values. According to Maallick [5], many previous studies have attempted to determine when and where air movement is either desirable or improper [5]. Thermal comfort research information indicates that indoor air speeds in hot climate areas should be set between 0.2–1.50 m/s, but 0.2 m/s was considered in ASHRAE Standard as the upper limit of the perception draft threshold allowed in the air-conditioned buildings in where the residents have no direct control over their environment [6]. No previous research was clearly addressed air reception of the reception, rather than focusing mostly on overall heat sensation and comfort [7]. According to Lau and Chen [8], based on their investigation on the performance of the level-of-supply ventilation shifts with swirl or perforated diffusers under high panel cooling loads [8]. This investigation uses a CFD program that solves the powerful equation for changes in mass, momentum, heat, pollutant concentration and quantity of turbulence. The results of this study show that workshops with floor shifting ventilation systems can improve indoor air quality because pollutant concentrations in the respiratory zone are lower than the combined system.

2 Methodology

2.1 *Model of Study*

The model shape of the 2-stroke marine engine diesel engine workshop is illustrated in Fig. 1a (natural ventilation system model) with is the current ventilation system and Fig. 1b (force ventilation system model). This model is a model diagram that has been used for CFD simulation. There are two doors and five exhaust fans in the model for ventilation purposes which use the concept of air pushing outward. The length, width and height of the model are 22.592 m × 12.274 m × 10.869 m. The exhaust fan size is 1.5 m × 1.5 m. Both doors are 2 m long and 2 m wide.

2.2 *CFD Simulations*

This study has been using the CFD software for modelling and simulating thermal comfort and airflow in the workshop that is being studied. The software allows modelling and simulation of fluid flow, mass and heat transfer in geometries complex. It has a high flexibility feature when doing canting or meshing process which is a very important process in the analysis of complex geometric shapes. The program structure shown in Fig. 2.

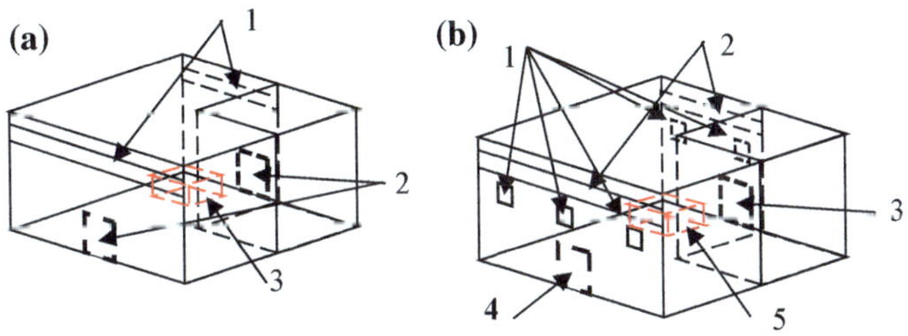

Fig. 1 **a** A computer simulation analysis model Sketches, the ventilation window (1), door (2) and the heat source (3). **b** A computer simulation analysis model Sketches, the exhaust fan (1), the ventilation window (2), door 1 (3), door 2 (4) and the heat source (5)

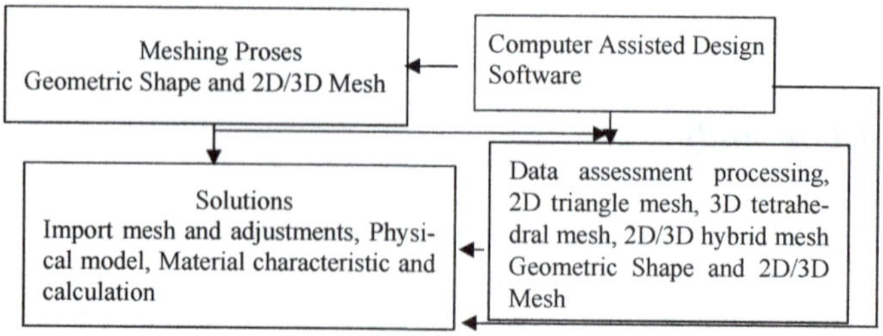

Fig. 2 The program structure

3 Result and Discussion

3.1 Comparison of Simulation Results Between Natural Ventilation Systems and Force Ventilation Systems Using 5 Sets of Exhaust Fan

Simulation of temperature distribution in a 2-stroke diesel engine workshop on the plane x-z, y = 1.6 m (Average human height).

Figure 3 shows the level of heat receiver and heat release from the heat source of the 2-stroke diesel engine. In Fig. 4, it shows the level of heat distribution on the

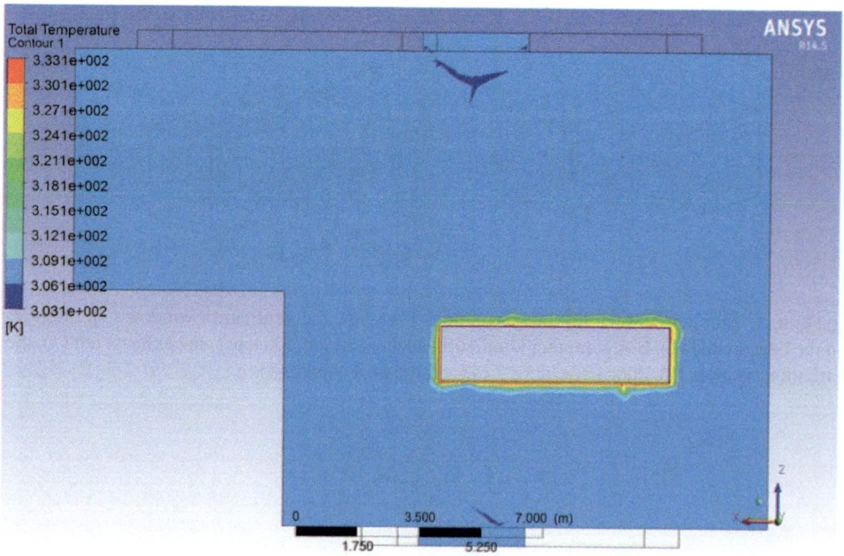

Fig. 3 Contour of heat flow distribution at plane x-z, y = 1.6 m (average human height) by using natural ventilation system

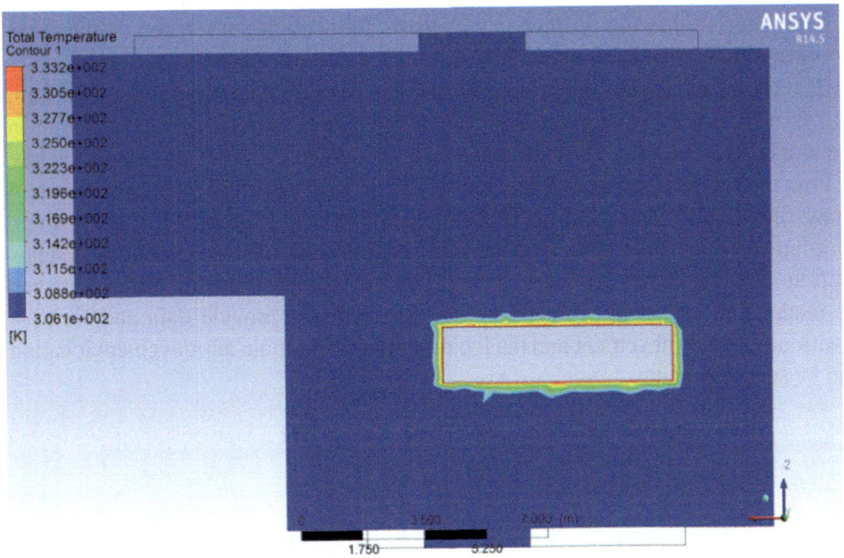

Fig. 4 Contour heat distribution on the plane x-z, y = 1.6 m from the floor (ventilation window height) by using force ventilation system

workshop when using the force ventilation system. This height limit is important to represent the heat received by employees and students in the workshop.

Table 2 shows the results of the analysis on temperature distribution using CFD simulation program at a height of 1.6 m from the floor level for existing ventilation system which is naturally and a force ventilation system. Comparison of CFD simulation results is able to provide conclusions and recommendations for the level of effort of the existing ventilation system. At a height limit of 1.6 m is ideal for visualizing heat distribution at average height of worker and student who is working in the workshop.

Table 2 Comparison on temperature distribution between natural ventilation systems and ASHRAE standards

Location	ASHRAE standard (a)	Temperature distribution (b)	Differences (b–a)
Respiratory height: 1.6 m from floor level (plane x-z, y = 1.6 m)	20–27 °C	Natural ventilation system Average value: 312 K (38.85 °C)	11.85 °C
		Force ventilation system Average value: 306 K (32.85 °C)	5.85 °C (15% reduce)

Simulated *ventilation air flow* in a 2-stroke diesel engine workshop on the plane x-z, y = 1.6 m (Average human height).

The use of the CFD simulation program has been implemented at a height limit of 1.6 m from the floor to visualize the air movement at average human height in the workshop. The use of colour contours is able to give a good and proper picture of how the air movement conditions in the 2-stroke diesel engine engines. Figure 5 shows the colour contour of air movement at 1.6 m from floor when using force ventilation system. Evaluation at the same level for available ventilation systems with an already-sustained system is important so comparisons and judgments can be made. The use of air velocity colour contours can provide data and location positioning of air movement and the location of the minimum air movement location can be seen and value.

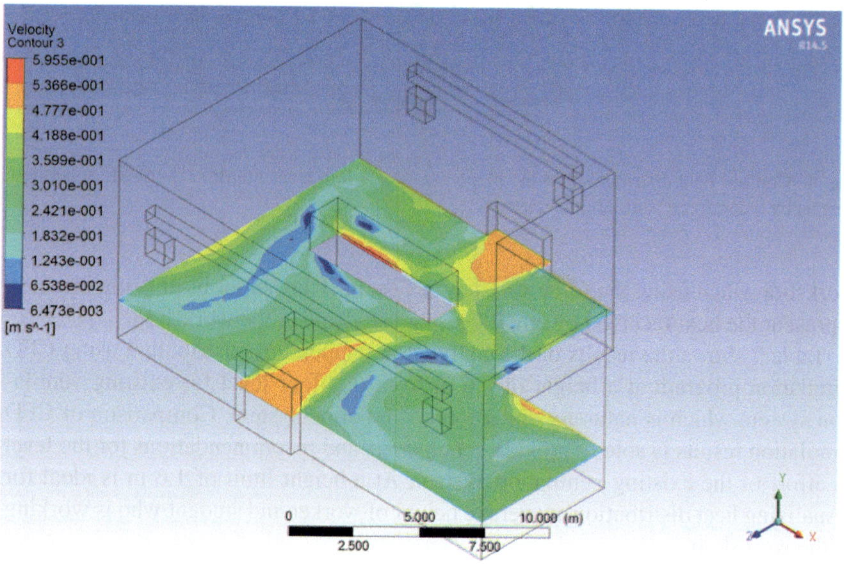

Fig. 5 Contour of air flow distribution on the plane x-z, y = 1.6 m (average human height) by using natural ventilation system

Table 3 Comparison of the value of the ventilation flow distribution across the ventilation system with the forced ventilation system in the workshop

Location	Natural ventilation system (c)	Force ventilation system (b)	Differences (c–b)
Respiratory height: 1.6 m from floor level (plane x-z, y = 1.6 m)	Maximum value: 0.480 ms^{-1} Average value: 0.192 ms^{-1}	Maximum value: 0.536 ms^{-1} Average value: 0.242 ms^{-1}	0.056 ms^{-1} (26% increase)

Table 4 Comparison between ventilation systems using exhaust fan with ASHRAE standard temperature distribution

Location	ASHRAE standard (a)	Ventilation system (b)	Differences (a–b)
Respiratory height: 1.6 m from floor level (plane x-z, y = 1.6 m)	>0.2 m/s	Natural ventilation system Average value: 0.192 ms^{-1}	0.042 ms^{-1}
		Force ventilation system Average value: 0.242 ms^{-1}	0.054 ms^{-1} (26% increased)

Comparison of aerial velocity simulation analysis (Table 3) at a height limit of 1.6 m from floor level is aimed at assessing and comparing data differences between natural ventilation systems are used in the 2-stroke diesel engine and the force ventilation system by using 5 sets of exhaust fan. This comparison is aimed at looking at the airflow at the average height limit of the worker and the student doing any activity in the 2-stroke diesel engine workshop. The colour contour that represents the air velocity flow is to see precisely the position of the air velocity that you most want to make any improvements. Comparison of the results of the analysis simulation on the movement of air velocities between the existing ventilation system and the ventilated system can be seen through Table 3. These differences are aimed at assessing and evaluating the effectiveness of ventilated systems that are significantly above the existing ventilation system.

Table 4 shows the comparison of the results of the CFD simulation analysis obtained with ASHRAE standards is aimed to see the impact that can be afforded from an aerated ventilation system to the thermal comfort level in a 2-stroke diesel engine engineering workshop. The comparison is that the ASHRAE airflow standard in the workshop is above 0.2 m/s against the maximum value obtained from the existing ventilation system and the ventilated system which has been at the level of effort.

4 Conclusion

CFD simulation is run on a model that represents a 2-stroke diesel engine workshop. Using this method, we can see the comparisons and changes that occur to the temperature distribution and airflow conditions in the workshop when it comes with ventilation systems. Observation shows that the temperature distribution of all ventilation workshops by the exhaust fan will still be outside the thermal comfort level range determined by ASHRAE guidelines. However, simulations on developed models have shown that the use of exhaust fans mounted on the walls can affect the flow conditions and reduce minimal internal temperature heat. By using displacement ventilation in all marine engineering workshop models, increased air velocity significantly compared to the exhaust fan system. The inclusion of displacement ventilation systems has been shown to increase the level of thermal comfort in the workplace. For this study, we can conclude that, the displacement ventilation system is useful for the marine engineering workshop ventilation.

References

1. Makhbul ZM, Idrus D, Abdul Rani MR (2008) Perhubungan Antara Stesen Kerja Ergonomik dan Stres: Kajian terhadap Operator Pengeluaran di Organisasi Multinasional di Malaysia. Technol J 48(E):29–69
2. American Society of Heating, Refrigerating and Air-Conditioning Engineers (2017)ASHRAE handbook: fundamentals, SI edn. ASHRAE (2017)
3. Parson KC (2000) Environment ergonomics: a review of principle, methods and models. Fundam Rev Appl Ergon 31(6):581–594
4. Cândido C, De Dear R, Lamberts R, Bittencourt L (2008) Natural ventilation and thermal comfort: air movement acceptability inside naturally ventilated buildings in Brazilian hot humid zone. In: Proceedings of conference: air conditioning and the low carbon cooling challenge. Cumberland Lodge, Windsor, UK
5. Mallick FH (1996) Thermal confort and building design in the tropical climates. In: Meir IA, Erell E, Etzion Y, Pearlmutter D, Cook J (eds) PLEA'94 international conference, vol 23(3). Elsevier, Netherlands, pp 161–311
6. De Dear R (2004) Thermal comfort in practice. Int J Indoor Environ Health 14(7):32–39
7. Toftum J (2004) Air movement—good or bad? Int J Indoor Environ Health 14(7):40–45
8. Lau J, Chen Q (2007) Floor-supply displacement ventilation for workshop. Build Environ 42(4):1555–1868

Thermal Comfort in Marine Workshop: Environmental Ergonomics Measurement Approach

M. H. Abd Razak and Nor Ashimy Mohd Noor

Abstract A field study conducted was to identify and determine the environmental ergonomics factor level in Marine 2-stroke Engines Workshop and Multifuel engines workshop. Thermal comfort is considered the main factor which is often emphasized by the occupant of the accommodation unit. The ambient temperature and relative humidity will lead to thermal comfort. Thermal comfort were calculated using the PMV (*Predicted Mean Vote*) and PPD (*Predicted Percentage of Dissatisfied*) according to the ISO 7730. Measurement data for all factors were collected by mean of "*Thermal Comfort Multistation*" before the data is then been transferred to *Microsoft Excel* for further analyze. Through this study, there are six environmental factors quantified by utilizing Thermal Comfort Multi-station (TCM) contrivance. The results were compared with standard the National Institute of Occupational Safety and Health (NIOSH) Malaysia, ASHRAE (American Society of Heating, Refrigerating and Air Conditioning Engineers) and ISO (International Standard Organization). The results show that the PMV at the Marine 2-stroke Engine's Workshop are in the range of 1.50–3. Its mean that the workshop can be conclude as not comfort and the thermal sensation is warm.

Keywords Thermal comfort multi station · Ergonomic · Predicted mean vote · Predicted percentage of dissatisfied

1 Introduction

Research development of thermal comfort in a workshop is part of ergonomics evolved in recent times. With rising standards of living and economy in the country, ergonomic factors are also given attention so that they meet the needs and comfort of workers and residents in the country. According to [1], stated that the "best that you can realistically hope to achieve is a thermal environment that satisfies the majority of people in the workplace, or put more simply, 'reasonable comfort'.

M. H. Abd Razak (✉) · N. A. Mohd Noor
PoliteknikUngku Omar, Ipoh, Perak, Malaysia
e-mail: shimzak@puo.edu.my

C. L. Saw et al. (eds.), *Advancement in Emerging Technologies and Engineering Applications*, Lecture Notes in Mechanical Engineering,
https://doi.org/10.1007/978-981-15-0002-2_22

This research will revealed the actual reading of the parameters that affect the studying environment at the Ungku Omar Polytechnic workshop. The environmental factors such as levels of temperature, humidity, noise and lighting are important because they influence the effectiveness of an activity. According to [2], open ventilation does not help in energy savings due to the need for room cooling system and it is a relationship between indoor and outdoor environments. While the importance of air quality are of concern, the release of excess carbon dioxide will cause degradation of work performance and productivity. Therefore, controlling the release of carbon dioxide levels is very important, according to [3], good ventilation should be adapt even though with the presence of only 1% carbon dioxide in the closed space.

Adequate lighting for work area will also contribute to employee performance. Lighting in office spaces need to be of the daylight while in open spaces the requirement for lighting are less. In the industry sector controllable task-lighting is not standard fixtures. Normally if task-lighting system is present, the user can easily switch it on or off. As compare to fixed-level task-lighting, the investment costs of the controllable task-lighting are higher.

Plant which involves machinery and equipment operation will produce noises. Noise is the sound that should be avoided, especially through mass development in developing country, [4]. The sources of noise can be in many ways and types such as the operation of machinery and equipment. The effects of occupational and environmental noise on a variety of cardiovascular risks such as hypertension, usage of anti-hypertension drugs, consultation with a general practitioner or medical specialist, usage of cardiovascular medicines, angina pectoris, myocardial infarction and prevalence of ischemic heart disease was review by recent meta-analysis, [5].

2 Methodology

This study started by doing literature review on the environmental ergonomic aspects with regard to temperature, humidity, noise and lighting. Initial studies focused on the studies done earlier in the same field. This study was conducted in eight level document preparation, TCM set up, data record/collection, data verification, data analyse, conclusion and recommendation. The brief description is as in Fig. 1.

This study is done in 2-stroke diesel engines workshop and multifuel engines workshop at Ungku Omar Polytechnic, Malaysia. The focusing activities is teaching and learning process. The studies were conducted on day where there are lecture or practical activities involved. The TCM measurement device is installed from 8.45 am to 4.00 pm. Although the practical activities ended around 1.00 pm, the TCM measurement device are left to measure, observe and validate the changes in the parameters measured. Thermal Comfort Multistation (TCM) is placed at the sitting point 600 mm. The data will be recorded in the computer using ADAM VIEW software.

Through the practical training the instructors and students will be exposed to the heat generated from several sources such as, the operation of several machineries,

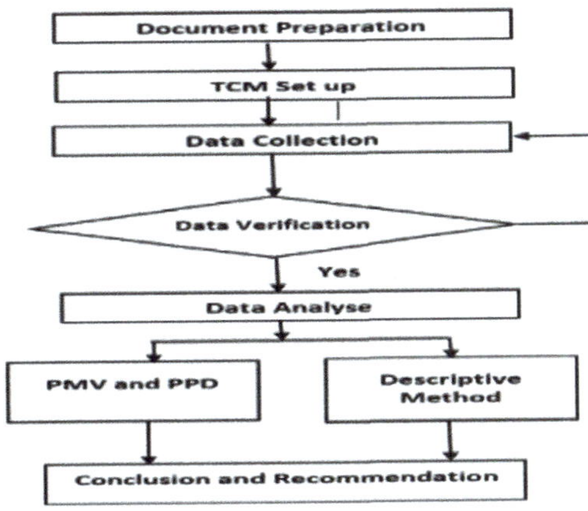

Fig. 1 Flow chart of methodology in the study

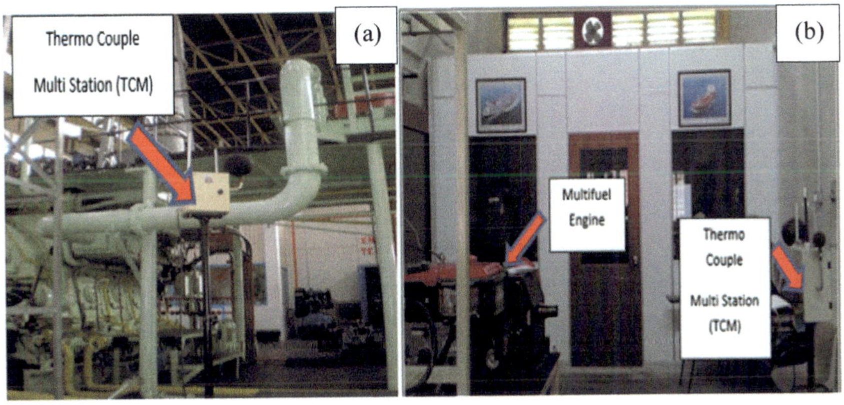

Fig. 2 Location of TCM in the workshop. **a** 2-stroke engine, **b** multifuel engine

auxiliary machineries and surrounding ambient air humidity. Figure 2a shows the location of the 2-stroke diesel engines workshop and Fig. 2b shows the location of the multifuel engines workshop.

3 Results and Discussion

After measurements were conducted for three sessions on different days, the results of the reading level of air velocity, ambient temperature, relative humidity, predicted percentage of dissatisfaction (PPD), Predicted mean vote (PMV), illuminance and noise is as shown in Table 1.

Table 1 List of results

Item	2-stroke diesel engines			Standard	Multifuel engines		
	Minimum	Maximum	Mean		Mini mum	Maximum	Mean
Air velocity (m/s)	0	1.08	0.11	WHO 0.25 m/s	0	0.86	0.3
Ambient Temperature, t_a (°C)	25.39	30.48	28.14	DOSH (23–26 °C)	26.72	34.87	30.93
Relative humidity, RH (%)	49.47	83.71	62.40	ASHRAE 62-2001 (30–60%)	42.99	85.6	60.12
PMV	1.5	2.60	2.07	Fanger −2 to +2	1.7	3	2.4
PPD (%)	50.4	95.3	78.66	Fanger (5–80)%	61.8	99.1	88.94
Noise (dBA)	43.34	90.27	72.16	DOSH 92 dBA for 6 h	43.6	82	62.06
Illuminance (lux)	229.90	912.30	477.91	DIN (250 lx) and IES (540 lx)	97.2	414.4	239.9

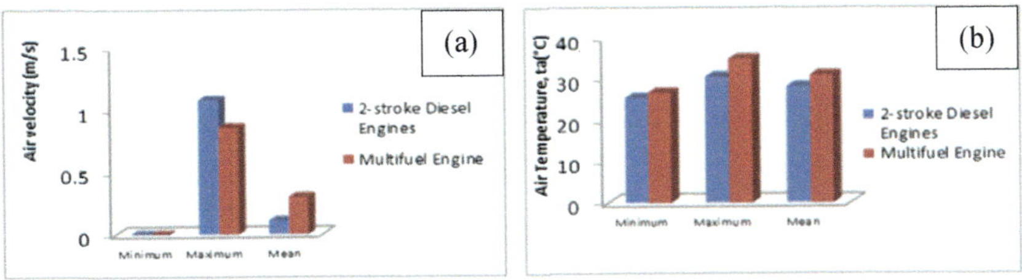

Fig. 3 Measurement taken in marine workshop. **a** Air velocity, **b** air temperature

Referring to Fig. 3a, the overall results of air velocity, air temperature and relative humidity for the two workshops, there is a significant difference in the readings. In general, the movement of air workshop multifuel engine with an average value of 0.3 m/s have better air movement compared 2-stroke diesel engine's workshop with an average value of just 0.11 m/s. Air movement in the multifuel engine's workshop helped by a fan located quite close to the wall near the site of multifuel engines. Air movement in the 2-stroke diesel engine's workshop are heavily dependent movement of air from outside the building.

Referring to Fig. 3b, the multifuel workshop has the highest readings and almost reached a temperature of 35 °C. For multifuel workshop, maximum reading only about 30.48 °C. Among the factors to be taken into account, the size of the 2-stroke engine workshop area is more spacious, air movement in and out of the building more open.

Refers to the rate of humidity (Fig. 4), the average reading of 2-stroke diesel engines have better reading than reading the workshop multifuel. The average reading Workshop 2-stroke engine is 62.40%, while reading in the workshop multi fuel is 60.12%. Among the factors that should be considered is the size of the area of the workshop and air movement in and out of the workshop. The relative humidity of compliance standards referred to ASHRAE Standard 62-2001, which has determined that the appropriate humidity in the building is 30–60%. It indicates the level of humidity both workshops located at acceptable levels.

Figure 5a, b shows the PMV and PPD index in workshop between two workshops. Referring to Fig. 5a, from the graph is plotted, PMV index for this workshop for three days in between 1.50 and 2.07 for two-stroke diesel engine and 1.7–2.4 for multifuel

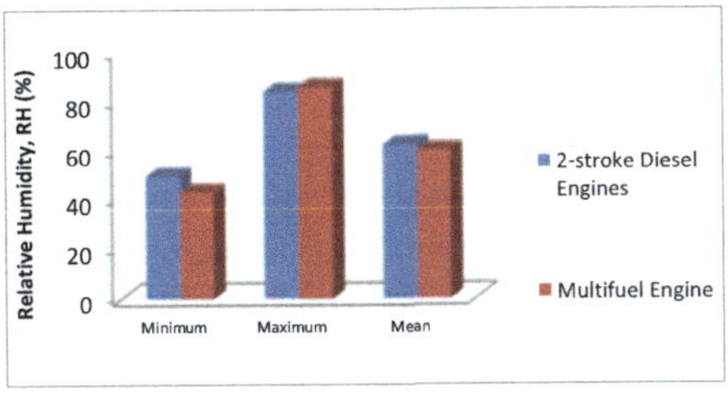

Fig. 4 Measurement of the relative humidity in marine workshop

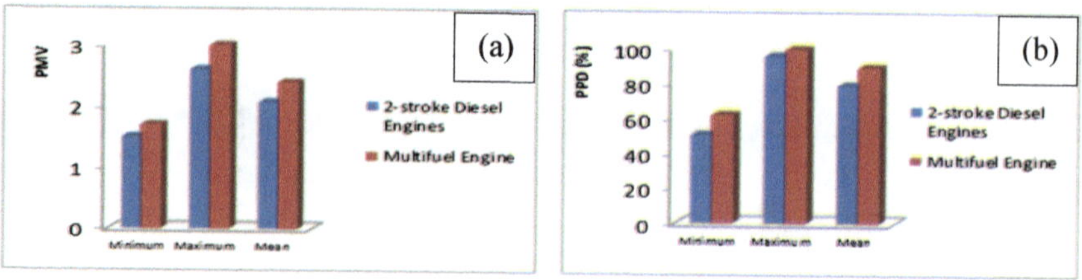

Fig. 5 Measurement taken in marine workshop. **a** PMV, **b** PPD

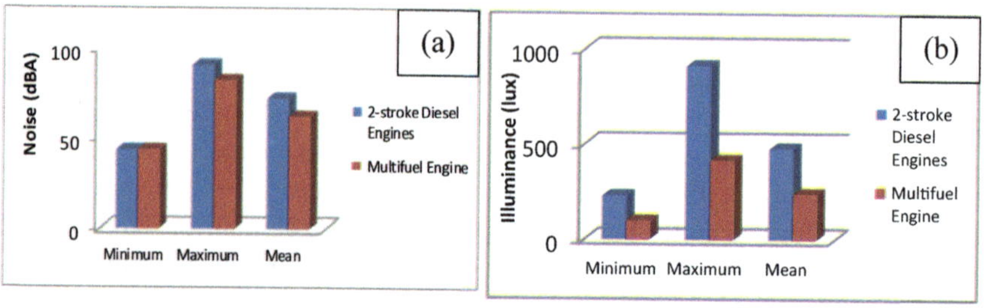

Fig. 6 Measurement taken in marine workshop. **a** Noise, **b** Illuminance

engine's workshop. It shows both the workshop environment is categorized as slightly warm (+1) and warm (+2) which refers to the model uncomfortable by Fanger thermal comfort. PPD also gives an analysis of the results were not much different where it falls in the range of 50.4–95.3%, which significantly only 4.7% are satisfied with the environmental terms of 2-stroke diesel engine. Referring to data analysis workshop multifuel engines, the PMV is between 1.7 and 3. The level 3 is the most hot and uncomfortable conditions exist for students in this workshop. The recorded value of PPD between 67 and 99.1%. This data shows that only 0.9% are satisfied with the conditions in the workshop environment.

Referring to the Fig. 6a, measurement of the sound level is the highest recorded 90.27 dBA for 2 stroke diesel engine's workshop. While for multifuel workshop, the noise level recorded the highest is 82 dBA. For information, exposure to noise for both workshops at a level not exceeding a period of 3 h.

According to OSHA limited to an average exposure over a period of 8 h is 90 dB. Institute for Occupational Safety and Health (National Institute of Occupational Safety and Health, NIOSH, 1972) also has suggested the level of exposure for 8-h work day is 85 dBA. Therefore, the sound level of the second workshop is still under the control of form.

Suitability changes in the workshop and office lighting set between 500 and 2000 lx; in areas where lighting is required. Appropriateness of changes in the workshop and office lighting set of between 500 and 2000 lx; according to the area where lighting is required. As a result in Fig. 6b, the level of lighting in workshop multi fuel is uncomfortable in terms of vision and need additional lighting to ensure improved visibility and comfort during practice runs.

4 Conclusion

Based on the results of the analysis of PMV and PPD as it has been reviewed, the results tend to states that the workshop environment is warm and uncomfortable. From the analysis, it was found that the level of comfort for both the workshops was a warm environment and not comfortable with the students and lecturers. The environmental conditions in the workshops monitored did not satisfy all recommended criteria, the impact of this is on the students' learning.

In general, the level of brightness in the 2-stroke diesel engine workshop, during engine operation is carried out is 500 lx or more. This value is sufficient, safe and good for practical activities conducted as mentioned in the German DIN Standard Lamp (DIN 5035 1972), American Illuminating Engineering Society (IES 1972), IES Lighting Handbook, 1981 or Lighting Standards Department Occupational Safety and Health (Safety, Health and Welfare) 1970. But the level of brightness of the lights in the multifuel engine workshop is not healthy i.e. it was below 500 lx during the practical training session.

Generally, the 2-stroke engine workshop have a noise level that is not fixed and it depends on the full engine operating speeds. The noise readings recorded a range between 43.34 and 90.27 dBA. Under DOSH guidelines, the allowable noise exposure is 92 dBA for a continuous period of 6 h. But real-time practical activities is approximately 5 h. On average, the level of noise generated in the past four days was 71.16 dBA. This value is safe and has no negative effect on student's who are doing practical under normal schedule in about 5 h. Furthermore, students are provided with ear muff's during practical activities in the workshop.

The multi fuel engine workshop also had noise levels that are not fixed and it depends on the full engine operating speeds. The noise readings recorded were ranged between 43.6 and 80 dBA. But real-time practical activities is approximately 5 h. On average, the level of noise generated in the past three days was 62.06 dBA with a period of continuous activity not exceeding 5 h.

References

1. Health & safety executive. Thermal Comfort Home Page. http://www.hse.gov.uk/temperature/thermal/index.htm. Last accessed on 04 Oct 2018
2. Lipin W, Hien WN (2007) Applying natural ventilation for thermal comfort in residential buildings in Singapore. Architect Sci Rev 50(3):224–233
3. Jonathan LS, David GK, Randal JK (2009) Occupational hazards of carbon dioxide exposure. J Chem Health Safety 16:18–22
4. Banerjee D, Chakraborty KS, Bhattacharyya S, Gangopadhyay A (2009) Apprisal and mapping the spatial-temporal distribution of urban road traffic noise. Int J Environ Sci Tech 6(2):325–335
5. Van Kempen EE, Kruize H, Boshuizen HC, Ameling CB, Staatsen BA, de Hollander AE (2002) The association between noise exposure and blood pressure and ischemic heart disease: a meta-analysis. Environ Health Perspect 110(3):307–317

A Study of Carbon Dioxide and Temperature

S. A. Rashid, M. R. Danuri and Nor Ashimy Mohd Noor

Abstract This scientific project is conducted to identify and determine the comfort level in the lecturer room, Center of Technology in Marine Engineering (CTME), Politeknik Ungku Omar. Air quality control and comfortable room are important to ensure that occupants feel at ease during everyday activities especially in the lecturer's room. Comfortable conditions will contribute towards the effectiveness of daily work processes. This level of comfort refers to the ISO 7730 Standard. The parameters measured and used in the calculation on this study are air temperature, mean radial temperature, relative humidity and Carbon Dioxide (CO_2) level in the Lecture room. Measurements were carried out using the Thermal Comfort Multistation (TCM). The data obtained were analyzed descriptively through the plotting of the relevant graphs. This study also presents the importance of temperature and content of Carbon dioxide (CO_2) around the human comfort especially in indoor space. This factor encourages the worker to be in a safe, comfortable environment and thereby enhances job satisfaction and performance.

Keywords Heat · Temperature · CO_2

1 Introduction

Ergonomics is a field of human biology science related to engineering sciences among workers with their environment. Therefore, it is very relevant between human self-adaptation and working requirements. It is important to ensure the safety, health, comfort and quality of work throughout the duration of eight (8) hours of work. In achieving high n working performance, an ergonomic space should be considered.

S. A. Rashid (✉) · M. R. Danuri · N. A. Mohd Noor
Politeknik Ungku Omar, Ipoh, Perak, Malaysia
e-mail: manshid@puo.edu.my

M. R. Danuri
e-mail: zuwan@puo.edu.my

N. A. Mohd Noor
e-mail: ashimy@puo.edu.my

C. L. Saw et al. (eds.), *Advancement in Emerging Technologies and Engineering Applications*, Lecture Notes in Mechanical Engineering,
https://doi.org/10.1007/978-981-15-0002-2_23

Ergonomic design and work space not only provide comfort but it also reduces risk for long term or short term. In general, according to [1] the station or workspace should be designed and adjusted to the body and the working condition and mind. This study was conducted to examine the work environment ergonomic in the lecturer room of the CTME, Politeknik Ungku Omar. This is because ergonomic comfort is also influenced by environmental conditions such as temperature quality, aeration quality and lighting quality. Thermal comfort is a measurement that assesses the level of heat or coldness of an object or place. The heat generated from the metabolism of the human body during work can provide a balance of body temperature to the surrounding space environment. In addition, the quality of the workplace's outdoor temperature can also affect the ambient temperature of the room. Carbon dioxide is one of the gas components found in the atmospheric surface. The carbon dioxide content (CO_2) found in neutral air environment is 400 ppm. Excessive CO_2 presence in the air will cause negative symptoms to humans such as tiredness, irritability and less comfort.

2 Literature Review

In a study conducted by Nakamurab [2] found that sensation thermal and thermal comfort have a close relationship under which conditions continuous and uniform as well as thermal comfort for thermal sensation is selected at the level warm. Under constant but non-uniform conditions, thermal sensation changes according to the space identified as an important factor in determining thermal comfort. Under dynamic conditions, the change of thermal sensation with time has a clear effect on thermal comfort. According to the standard ASHRAE Standard 55, human thermal comfort is a condition where the mind shows satisfaction with the environment. Thermal comfort is detected by convective heat, irradiation and evaporation. Thermal comfort is permanent when heat produced by human metabolism can be removed, and it will in turn maintain a thermal balance with the environment. Any loss or heat increase will result in discomfort. Hot or cold feeling is not dependent on temperature only, but it also involves environmental factors and other individuals. Early contact between thermal sensation and thermal comfort was first investigated [3] who have collected and analyzed the response man to various environmental conditions with temperature changes between 12 and 48 °C which in turn finds that the "Comfort" domain can be projected from the sensory temperature of the "neutral" stage to cold or warm stages. Overall it can be concluded that there are six main variables to consider in terms of thermal comfort for a person or group of individuals: Metabolic rate, Clothing, Air temperature, Min of radial temperature, Relative humidity and Air velocity.

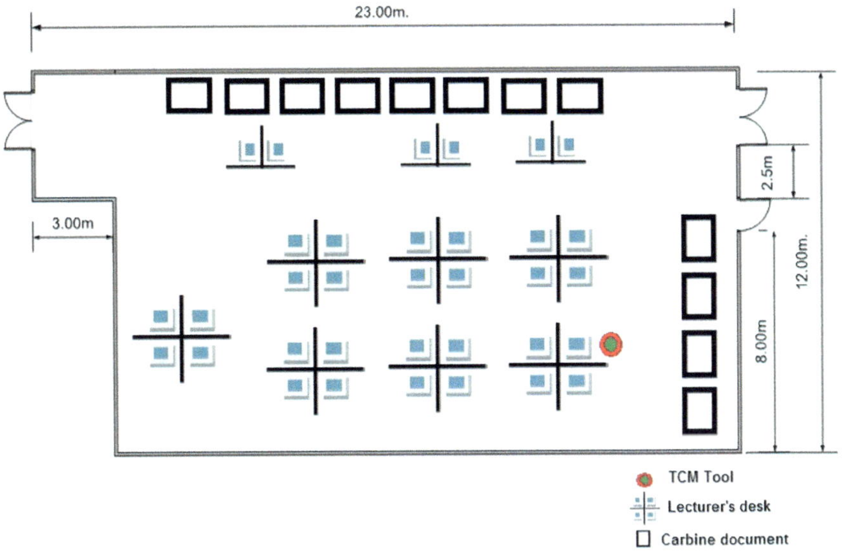

Fig. 1 Layout lecture room in CTME

3 Methodology

This study was conducted in the CTME lecturer room. Activities carried out is mainly office work. During the lecture session the lecturer will be directly exposed to the heat source as a result of the use of 34 fluorescent lamps, 24 computer screens and individual metabolism as well as the surrounding air humidity in the room. In addition, lecturers are exposed to the release of Carbon Dioxide (CO_2) in the room. Inside there are 4 air conditioning system with 2HP and 14 window frames and 2 door frames. Figure 1 shows the position of TCM tools to collect all data such as room temperature, Carbon dioxide, humidity, wind speed and lighting.

This study was conducted in seven level of preparation, TCM set up, data record/collection, data verification, data analyse, conclusion and recommendation. The brief description is as in Fig. 2.

Figure 3 show the exact position of the tool used to record all the data required in this study.

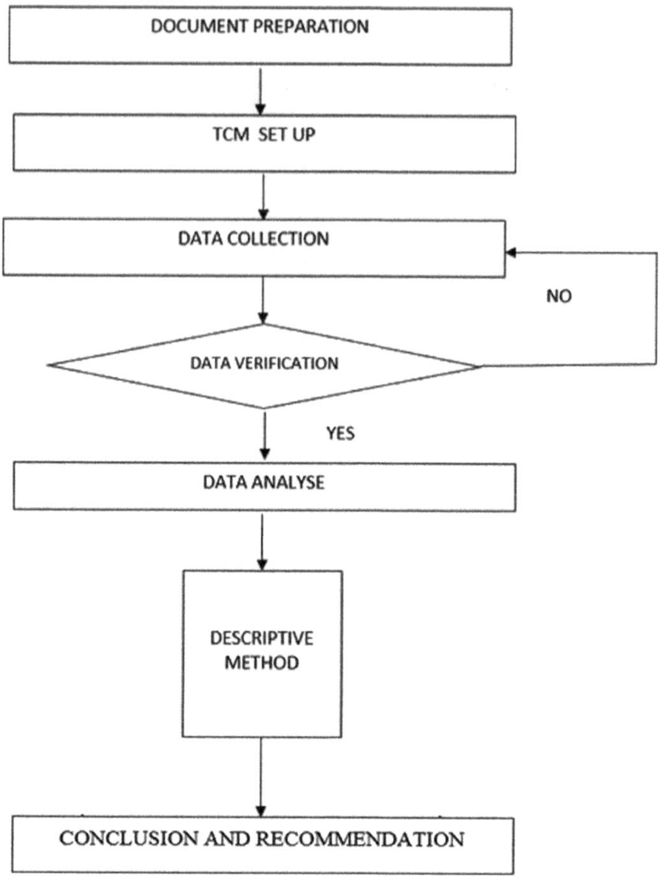

Fig. 2 Flow chart of studies in CTME lecture room

4 Results

Tables 1 and 2 show data that has been collected for 4 days in the CTME lecturer room. The data collection time is from 8.40 am to 5.20 pm covering working hours.

5 Discussions

Carbon dioxide as well (CO_2) and other greenhouse gases in the atmosphere play an important role in the global climate change environment [4]. At low concentrations, greenhouse gases serve as a medium of temperature control of the earth by trapping the wave of hawk. However, increasing greenhouse gases from time to time causes the earth's temperature to rise. This phenomenon is known as a greenhouse effect.

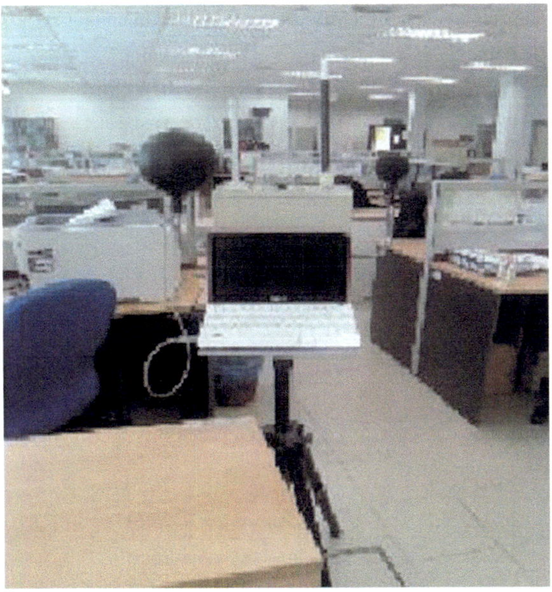

Fig. 3 Thermal comfort multistation (TCM)

The United Nations mission through the Kyoto Protocol to control human activities affecting the quality of the environment should be achieved to ensure a balanced CO_2 cycle to the atmosphere [4]. The development of industrial activity and land use change are among the factors that affect the balance and concentration of greenhouse gasses, especially Carbon dioxide (CO_2), Methane (CH_4), Nitric oxide (N_2O) and Chlorofluorocarbon [4]

However, every development has their own bad impact on the environment. Among them is the reduction of land use and the release of dirty gases into the air. Carbon dioxide (CO_2) is one of the causes of dirty gas emissions resulting from today's development and technology. The technology that contributes significantly to the level of carbon dioxide generation is derived from the nuclear industry, the burning of fossil materials and the use of turbines to produce electricity. Carbon dioxide is also one of the components of the gas found on the surface of the earth's atmosphere. The carbon dioxide content (CO_2) found in neutral air environment is 0.04% or 400 ppm. Excessive carbon dioxide (CO_2) in the air may cause negative symptoms to humans such as rapid tiredness, irritability and discomfort. Through this project the measurement and monitoring of carbon dioxide content in the air in closed areas such as the CTME office's room.

The use of electricity is one of the sources of the increasing of (CO_2) as well as transport, waste and other development. The release of carbon dioxide is from the process of producing electricity and this means that the less energy the less energy is consumed (CO_2) is released. There are many sources that can produce electricity. One source is the burning of fossil material. The heat energy generated from the

Table 1 Room temperature data of CTME lecturer measured in units °C

Time	Point 1		Point 2	
	1/14/2016	1/15/2016	1/16/2016	1/17/2016
	Day 1	Day 2	Day 3	Day 4
08:44:13	26.575	26.069	27.05	26.4
08:54:13	26.369	25.869	27.269	26.213
09:04:13	26.256	26.006	27.3	26.194
09:14:13	26.269	25.65	27.419	26.231
09:24:13	26.237	25.631	27.481	26.725
09:34:13	26.081	25.606	26.806	26.775
09:44:13	26.444	25.344	26.662	26.288
09:54:13	26.388	25.094	26.419	26.487
10:04:13	26.231	24.994	26.531	25.806
10:14:13	26.412	24.994	26.244	25.331
10:24:13	26.319	25.225	26.613	25.213
10:34:13	26.281	25.344	26.519	25.756
10:44:13	26.3	25.594	26.262	25.994
10:54:13	26.319	25.556	26.55	25.925
11:04:13	26.156	25.612	26.375	25.85
11:14:13	26.294	25.606	26.231	26.138
11:24:13	26.262	25.594	26.35	26.237
11:34:13	26.225	25.519	26.6	26.231
11:44:13	26.15	25.444	26.219	25.925
11:54:13	26.394	25.2	26.325	26.094
12:04:13	26.344	25.594	26.819	26.225
12:14:13	26.281	25.656	26.35	26.563
12:24:13	26.587	25.619	26.894	26.563
12:34:13	26.656	25.475	27.094	26.594
12:44:13	26.7	25.675	26.794	26.337
12:44:13	26.7	25.756	27.3	26.337
12:54:13	26.638	25.487	27.256	26.669
13:04:13	26.431	25.313	26.806	26.306
13:14:13	26.731	25.2	26.819	26.575
13:24:13	26.806	25.037	27.25	26.844
13:34:13	26.825	25.006	26.831	26.8
13:44:13	26.638	25.625	26.837	26.831
13:54:13	26.369	25.356	26.563	26.8
14:04:13	26.35	25.019	26.837	26.794

(continued)

Table 1 (continued)

Time	Point 1		Point 2	
	1/14/2016	1/15/2016	1/16/2016	1/17/2016
	Day 1	Day 2	Day 3	Day 4
14:14:13	26.363	25.494	26.656	26.531
14:24:13	26.55	26.681	26.775	26.469
14:34:13	26.625	27.113	26.844	26.206
14:44:13	26.575	27.475	27.063	26.563
14:54:13	26.6	27.556	26.831	26.837
15:04:13	26.581	26.844	27.369	26.219
15:14:13	26.481	26.688	26.825	26
15:24:13	26.681	26.694	27.475	25.794
15:34:13	26.863	26.775	27.356	26.469
15:44:13	26.781	26.856	26.9	26.213
15:54:13	26.831	26.925	26.85	26.231
16:04:13	26.819	26.894	27.356	26.231
16:14:13	26.863	26.906	26.8	26.244
16:24:13	26.875	26.888	27.487	26.219
16:34:13	26.837	26.837	27.412	26.813
16:44:13	27.237	26.837	27.412	26.85
16:44:13	27.237	26.837	27.412	26.85
16:54:13	26.869	26.744	27.263	26.694
17:04:13	26.831	26.775	27.469	26.85
16:14:13	26.969	26.162	27.469	
17:24:13	26.706	25.888	27.475	

fuel is converted into electricity. However, the effects of burning fossil material will produce carbon dioxide. Referring to the TCM readings obtained for all four days of measurement, the highest reading of carbon dioxide content in the air in the CTME office's room was on the first day of about 970 ppm.

According to 2010 Air Quality Industry Code of Practice (Department of Safety and Health Malaysia), the level of indication of carbon dioxide gas induction is 1000 ppm and should not exceed at any time. Accordingly the recorded reading is approaching the maximum limit of CO_2 gas content in the room. The reading record was recorded at 12.40 pm with an average temperature of about 26.5 °C and was the highest average temperature for the four days. This factor is fueled by the use of high air conditioning at relatively hot temperatures. Overall the level of carbon dioxide content in the CTME room is still within the permissible limit as well as the control requirements for the use of the machine and electrical equipment in the room (Fig 4).

The actual results of the study are based on measurement results and analysis for all four days of measurement. This is the purpose of obtaining results which is more

Table 2 CTME lecturer room CO_2 data measured in ppm units

Time	Point 1		Point 2	
	1/14/2016	1/15/2016	1/16/2016	1/17/2016
	Day 1	Day 2	Day 3	Day 4
08:44:13	26.575	26.069	27.05	26.4
08:54:13	26.369	25.869	27.269	26.213
09:04:13	26.256	26.006	27.3	26.194
09:14:13	26.269	25.65	27.419	26.231
09:24:13	26.237	25.631	27.481	26.725
09:34:13	26.081	25.606	26.806	26.775
09:44:13	26.444	25.344	26.662	26.288
09:54:13	26.388	25.094	26.419	26.487
10:04:13	26.231	24.994	26.531	25.806
10:14:13	26.412	24.994	26.244	25.331
10:24:13	26.319	25.225	26.613	25.213
10:34:13	26.281	25.344	26.519	25.756
10:44:13	26.3	25.594	26.262	25.994
10:54:13	26.319	25.556	26.55	25.925
11:04:13	26.156	25.612	26.375	25.85
11:14:13	26.294	25.606	26.231	26.138
11:24:13	26.262	25.594	26.35	26.237
11:34:13	26.225	25.519	26.6	26.231
11:44:13	26.15	25.444	26.219	25.925
11:54:13	26.394	25.2	26.325	26.094
12:04:13	26.344	25.594	26.819	26.225
12:14:13	26.281	25.656	26.35	26.563
12:24:13	26.587	25.619	26.894	26.563
12:34:13	26.656	25.475	27.094	26.594
12:44:13	26.7	25.675	26.794	26.337
12:44:13	26.7	25.675	26.794	26.337
12:54:13	26.638	25.756	27.3	26.669
13:04:13	26.431	25.487	27.256	26.306
13:14:13	26.731	25.313	26.806	26.575
13:24:13	26.806	25.2	26.819	26.844
13:34:13	26.825	25.037	27.25	26.8
13:44:13	26.638	25.006	26.831	26.831
13:54:13	26.369	25.625	26.837	26.8
14:04:13	26.35	25.356	26.563	26.794

(continued)

Table 2 (continued)

Time	Point 1		Point 2	
	1/14/2016	1/15/2016	1/16/2016	1/17/2016
	Day 1	Day 2	Day 3	Day 4
14:14:13	26.363	25.019	26.837	26.531
14:24:13	26.55	25.494	26.656	26.469
14:34:13	26.625	26.681	26.775	26.206
14:44:13	26.575	27.113	26.844	26.563
14:54:13	26.6	27.475	27.063	26.837
15:04:13	26.581	27.556	26.831	26.219
15:14:13	26.481	26.844	27.369	26
15:24:13	26.681	26.688	26.825	25.794
15:34:13	26.863	26.694	27.475	26.469
15:44:13	26.781	26.775	27.356	26.213
15:54:13	26.831	26.856	26.9	26.231
16:04:13	26.819	26.925	26.85	26.231
16:14:13	26.863	26.894	27.356	26.244
16:24:13	26.875	26.906	26.8	26.219
16:34:13	26.837	26.888	27.487	26.813
16:44:13	27.237	26.837	27.412	26.85
16:44:13	27.237	26.837	27.412	26.85
16:54:13	26.869	26.744	27.263	26.694
17:04:13	26.831	26.775	27.469	26.85
17:14:13	26.969	26.162	27.469	
17:24:13	26.706	25.888	27.475	

accurate and precise to describe the actual situation of the environment in the CTME office.

Figure 5 shows the temperature environment of the CTME lecturer room against the selected four days. From the plotted graphs, the rated air temperature is equal to the time of the four days. Overall, the graph illustrates the temperature of this room influenced by the environmental activity and the individual capacity found in it. Malaysia's comfortable temperature is between 25.5 and 28.0 °C [5]. Thermal comfort basically occurs when a person is protected from direct sunlight other than the presence of wind generated from its surroundings either through natural ventilation systems or mechanical ventilation systems [6].

The maximum temperature recorded is 27.5 °C and the minimum is as low as 25.0 °C for all four days. The main goal of the study of thermal comfort is to create a "comfortable zone" or temperature range where most of its residents feel comfortable also defines "comfortable zone" is a situation in which human beings will be able to reduce energy production levels in their bodies to adapt themselves to

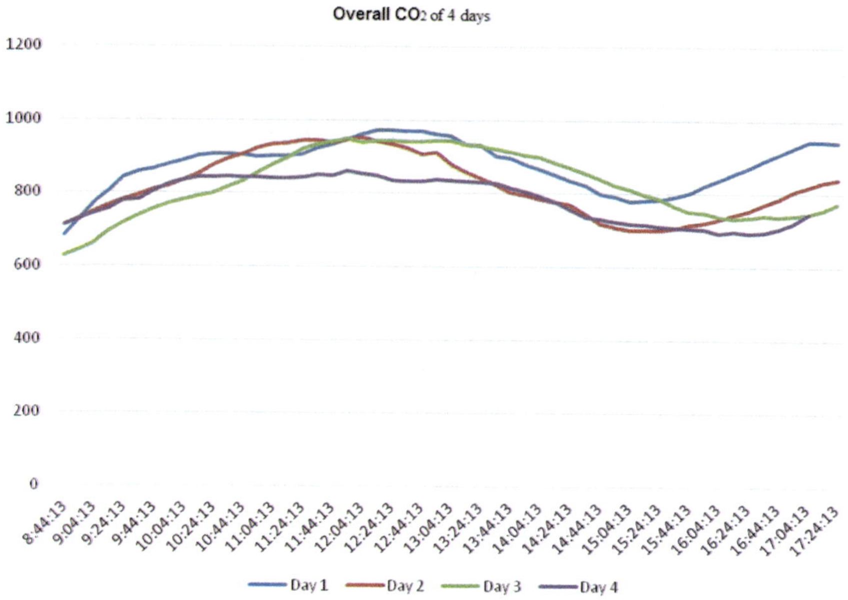

Fig. 4 CTME lecturer room CO_2 data measured in ppm units

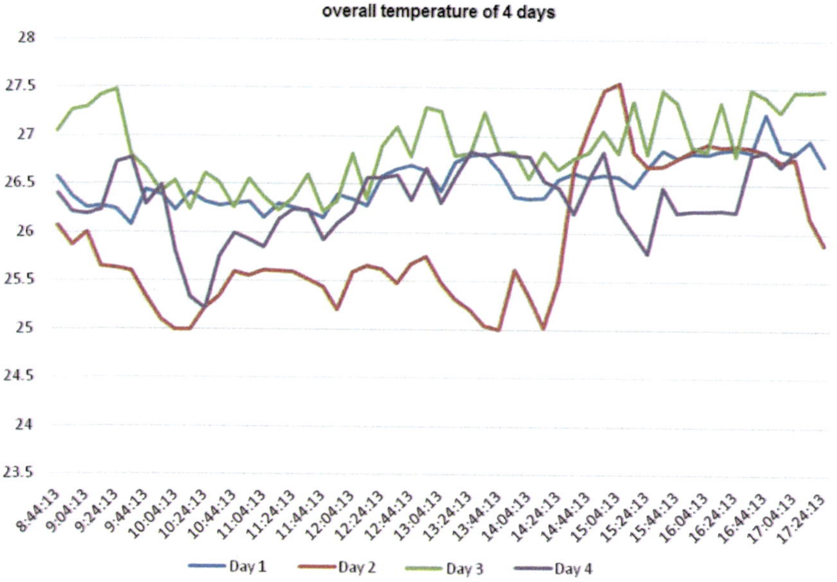

Fig. 5 CTME lecturer room temperature data measured in °C units

their environment. Overall, the measured measurements are somewhat compatible with individual comforts. Referring to the study [7] which states that temperature reduction can lead to increased productivity.

6 Conclusions

The findings show that maximum and minimum temperatures around the CTME lecturer room are suitable and comfortable. It shows similarities with the study of Shukor [8], the comfort temperature of Malaysians is delivered at 25.5–28.0 °C.

References

1. Auliciems A, Szokolay SV (1997) Thermal comfort department of architecture. University of Queensland, Australia
2. Nakamurab (2006) Effect of humidity on human comfort and productivity after step changes from warm and humid environment. Build Environ 42:4034–23, 404
3. Ismail RA, Jusoh N, Nuawi MZ, Deros BM, Makhtar NK, Rahman MNA (2009) Assessment of thermal comfort at manual car body assembly workstation. J World Acad Sci Eng Technol 54
4. Hoa JC, Xueb H, Taya KL (2003) A field study on determination of carbon monoxide level and thermal environment in an underground car park. Build Environ 39:67–75
5. Ismail AM (1993) Pengudaraan dan Alir Udara di dalam Bangunan serta Permasalahannya. Dicapai pada April 12, 2012 dari. http://www.hbp.usm.my/ventilation/pengudaraan_dan_permasalahannya.htm
6. Malek AR (2004) Low energy cooling technology for Malaysian homes. Universiti Sains Malaysia, Pulau Pinang
7. Sulong AB, Deros BM (2008) The relation between the discomfort level of automotive industries operators towards their workstation design and work environment. J Achievements Mater Manuf Eng 31(2):756–761
8. Abdul Shukor AM, Young A (1993) Thermal comfort study as an aid to determine energy savings in buildings in Malaysia

Fire Retardant and Physical Properties of Kapok-Chopped Strand Mats Fiberglass Composites for Marine Application

M. A. Mun'aim Idrus, Asmalina Mohamed Saat, Maisarah Mohamed Bazin and Redzuan Zoolfakar

Abstract In this study, kapok/Chop Strand Mat (CSM) reinforced unsaturated polyester hybrid composites were fabricated using hand lay-up method with a boat composite sandwich configuration laminate method. The glass CSM was used as a shell with kapok as a core, with an approximate total fiber content of 60%. Three different designated layers of kapok and weight ratios of (60/40)% and (40/60)% were used to fabricated for comparison purposes. Fire retardant properties were investigated using UL-94 which are vertical burning and horizontal burning test. The result showed that the hybrid composite with 3 layers of kapok took the longest time to burn. It has proved that the hybrid kapok/CSM reinforced displayed better in fire retardation than the raw CSM composites by using sandwich arrangement. It was found physical properties of kapok reinforced composite showed the increasing trend of water absorption and thickness swelling with the increasing storage time.

Keywords Kapok fiber · Composites · Fire retardant

1 Introduction

Over the last few decades, due to increasing global awareness of environmental issues, there has been great interest and motivation in research to develop natural fibre composites to replace glass fibre composites in certain applications. Marine industry especially in boat industry are looking forward for the new promising materials that are more environmentally friendly, strong as fiberglass and low cost. There are two types of fibers commonly used as reinforcement's polymer composites: synthetic fibers such as glass or carbon fibers and natural fibers including kenaf, jute, flax, hemp, sisal, etc. [1]. A fiber reinforced polymer (FRP) is a composite material

M. A. Mun'aim Idrus (✉) · A. M. Saat · R. Zoolfakar
Malaysian Institute of Marine Engineering Technology, Universiti Kuala Lumpur, Kuala Lumpur, Malaysia
e-mail: mamunaim@unikl.edu.my

M. M. Bazin
Malaysian France Institute, Universiti Kuala Lumpur, Kuala Lumpur, Malaysia

C. L. Saw et al. (eds.), *Advancement in Emerging Technologies and Engineering Applications*, Lecture Notes in Mechanical Engineering,
https://doi.org/10.1007/978-981-15-0002-2_24

consisting of a polymer matrix imbedded with high strength fibers, such as fiberglass, aramid, kevlar and carbon. Fiberglass is one of the most widely used materials in boat fabrications. Although the FRP follow structural and durability requirement, they have shortcomings, relatively high fiber density, non-biodegradable and the potential health hazards effect by glass fiber particulates [2].

Meanwhile natural fibers are naturally friendly and sustainable materials, which can be used to partially replace synthetic fibers in composite industry. The advantages of natural fibers over fiberglass are low cost, low density, comparable specific tensile properties, non-abrasive to the equipment's, non-irritation to the skin, less health risk and biodegradability. These composites materials are suitable applicable for aerospace, construction, sport and marine industries. The most popular natural fibers used in the composites include the bast, kenaf, hemp, jute, cotton, sisal and kapok fibers. According to Khalil et al. [3], the flexible high-performance applications of natural fiber composites can replace glass and carbon fibers.

One of the most important safety worries make it difficult for composites to meet the strict fire safety requirements practical to offshore oil and gas platforms is that most FRP materials have poor fire-resistant properties, such as short high rates of heat release, ignition time and, smoke production and flame spread, while it is generally recognized that composites have much lower thermal conductivity than metallic material [4].

This research based on natural fiber which Kapok is (ceiba petandra) as reinforce fiber. In Asia, the biggest production of kapok is from Indonesia and Thailand. In 2014, the production of kapok is led by Indonesia with 61,000 tons and followed by Thailand with 35,000 tons. Kapok fiber and its composite materials have been gaining increasing attention in recent years due to their unique features, kapok fiber-based materials have opened possibilities for various new research and application fields [5–8]. Kapok fiber can be mixed and blended with other natural fibers to achieve new characteristics for paper making and others [9].

In this research, kapok fiber/fiberglass hybrid composites were prepared and studied. The objective of this research is to study the effect of hybrid kapok fiber in chopped strand mat fiberglass composites in terms of the characteristic of fire retardant and physical properties.

2 Materials and Preparation

Raw kapok fibre (RKF) was purchased from Kuala Kangsar, Perak, Malaysia. The RKF was clean, keep in the polyether bag and sun dried. The polyester resins and Methyl-ethyl ketone peroxide (MEKP) used in this study was based on a commercially available supplied by Advance Moulding Chemical Sdn Bhd. The fiberglass was obtained from Unikl Mimet Fiberglass workshop.

2.1 Composites Preparation

Fiberglass type CSM was used to manufacture composite laminates with kapok natural fiber. Kapok fiber was obtained from kuala kangsar Perak. A hand lay-up method was used to produce the composites as it is widely used in fiberglass boat construction. The weight of the resin and the ratio of the kapok weight percentages was based on the fiberglass which are woven roving and csm types. The materials used in the preparation of composite flat panel represents in Table 1.

A combination of hand carder and hand lay-up was using to prepare the hybrid kapok/fiberglass composites. Hybrid kapok fiberglass composite is the mixture of kapok fibre, CSM fiberglass, catalyst and resin. Kapok hybrid composite is being produced by using different amount of kapok and CSM layer. Kapok first has been processing to be in a form of non-woven layer by using hand carder machine. The advantage of drum carder it can process fiber in a short time. Kapok mat then keep before being use in composite preparation. The average kapok mat weight is 20 g/mm^2 (Fig. 1).

Resin content for specimens depends on the calculation of fiber resin ratio. It is measured in weight and there are 5 varying fiber—resin ratio. In this research, 20 grams of kapok fiber used to fabricate a layer of sample. The total resin used for a layer will be $21 \times 10 = 210$ grams while the total resin used for a sample consist of 5 layers is $210 \times 5 = 1050$ grams per sample as shown in Table 1.

Table 1 Layers of raw composites materials for sample A, B, C, D and E

Specimen	Layer	Materials
A	5	csm/csm/csm/csm/csm
B	5	csm/kapok/kapok/kapok/csm
C	5	kapok/csm/kapok/csm/kapok
D	5	csm/kapok/csm/kapok/csm
E	5	csm/csm/kapok/csm/csm

(a) (b)

Fig. 1 **a**, **b** Show kapok/CSM hybrid composite

2.2 *Preparation of Raw Fiberglass and kapok/csm Hybrid Composites*

Raw fiberglass Chopped Stranded Matt (CSM) were cut and weight. 5 layers of CSM 300 were weight and cut into 33 cm × 33 cm mould dimension. A proper amount of polyester and catalyst were then prepared according to the ratio required. Mould were inspected and cleaned to ensure free from any form of particles. Wax were applied to the mould surface for best result of sample surfaces and easy to de-mould. A mixture of polyester was applied to the mould. 1 Layers of CSM were added layer by layer according to the weight ratio stated. Air bubble were released using roller. The mould were left for drying process at the room temperature.

Finally, samples were cut into 280 mm × 30 mm for mechanical testing. Kapok hybrid manufactured by layers of kapok and CSM according to the sample design in Table 1. A proper amount of resin, catalyst and kapok were then prepared according to the ratio required. Air bubble were released using roller. The mould were left for drying process at the room temperature about 24 h.

3 Experimental

3.1 *Water Absorption Test and Thickness Swelling*

Samples with dimension 10 cm × 10 cm based on ASTM D5229 were prepared for the water absorption test. prior to immersion in sea water, the samples were dried at room temperature until a constant weight is reached. The specimens were periodically taken out of the water, wiped with tissue paper to remove water on the surface, reweighted, re-measured and immediately put back into the water. Water absorption was calculated according to the following equation:

$$\text{Water absorption}\,\% = (\mathrm{W}_2 - \mathrm{W}_1 / \mathrm{W}_1) \times 100\% \tag{1}$$

where W_2 is the specimen weight after soaking and W_1 is the specimen weight before soaking.

Thickness swelling tests were carried out according to ASTM D5229. Composite samples were immersed in sea water during a time until the saturation was reached. Each specimen with dimension 10 cm × 10 cm was cut from composite panels. All the samples dried at room temperature for 24 h. The thickness of specimen before soaking (T_1) was taken. After 24 h, the samples were removed from the water and were measure the thickness using a digital scale immediately after they were dried with dry cloth. This process was repeated, to weigh the specimens regularly over 10 days of water immersion. Thickness swelling (TS) were calculated using the following equation:

$$TS = [(T_2 - T_1)/(T_1)] \times 100\% \quad (2)$$

where T_1 and T_2 is the thickness of the specimen before and after the test respectively.

3.2 Fire Retardant Test

Fire retardant properties were investigated using UL-94 which are vertical burning and horizontal burning test. Horizontal Testing. The specimen placed in a horizontal position supported by tripod. A flame is applied to the end of the specimen at 45° angle for 30 s. The burn lengths were recorded by using measuring tape. Three set of 3 specimens are tested. Meanwhile for vertical Testing, a specimen is supported in a vertical position and a flame is applied to the bottom of the specimen. The flame is applied for ten seconds and the removed until flaming stops at which time the flame is reapplied for another ten seconds and then removed. The burn lengths were recorded by using measuring tape. Three sets of three specimens are tested.

4 Results and Discussion

4.1 Water Absorption and Thickness Swelling Behavior

Figure 2 showed the effect of the storage time checked after 10 days with 5 samples respectively. It was found that kapok reinforced composite showed the increasing trend of water absorption with the increasing storage time. The lowest and highest water absorption recorded at sample A with 0.16% and sample B with 0.22%. Increase of water absorption caused by the increasing number of fibers because of fiber properties is hydrophilic group. After 7 days of test, the specimen remains constant because it already achieved maximum water absorption. Specimen D and E showed highest water absorption due to existence of voids in the sample.

Nevertheless, water absorption of different composite clearly increases because of the water absorption of kapok reinforced fiber matrix because the raw kapok fiber not treated with Sodium Hydroxide (NaOH) that change the physical structure, therefore

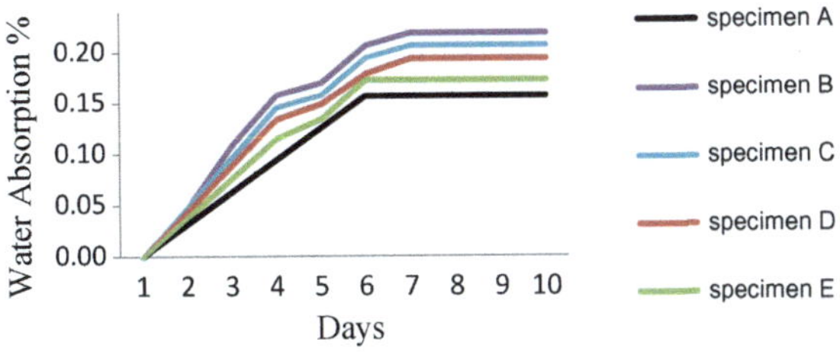

Fig. 2 Water absorption of the kapok/CSM hybrid composite

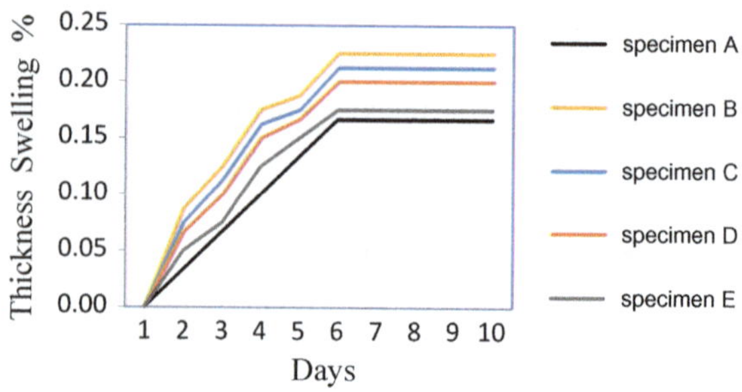

Fig. 3 Thickness swelling of the kapok/CSM hybrid composite

the hollow internal wall of the kapok fiber increase. Moreover, the weak adhesion between the matrix and cellulosic fibers also increase the free volume of the starch molecular chains and leader to the easy of water penetration [10]. It was expected that the water absorption of the kapok reinforce fiber should be higher than those of the fiberglass composite since kapok is strongly hygroscopic and readily absorbs moisture and due to the non-treatment effect.

From Fig. 3, it shows the thickness swelling of the composite coefficient for the composite which is varying depending upon the fiber loading. From observations, it shows that the thickness swelling is increased for all specimens. The highest thickness is recorded by sample B with 0.23% increment while the lowest was stated by specimen A with 0.17% increment. It shows that the thickness of raw kapok is higher than synthetic fiber. From the result obtained, the increase number of porosity lead to highest water absorption as the water filled void space in the composite structure. It also indicated that thickness swelling values increase with water absorption time until reach the maximum values. The highest number of kapok layer will increased the thickness swelling as observed in specimen B [11].

4.2 Vertical and Horizontal Burning Test Result

Figure 4 shows the vertical burning for fire retardant properties observe the highest length of burning by sample A which is 64 mm while the lowest by sample B which is 39 mm. This is because the specimen B consist of 3-layer kapok fibre while specimen A is a raw CSM fiberglass. It is clear shows that when increasing layer of kapok fibre it can reduce the length of burning effect to the composite sample. This is believed to occur due to decreased supply of oxygen to the inside of the composite layer. This in turn decreases the burning rate of the composite. Figure 5 shows the horizontal burning for fire retardant properties. It can be observed the highest length of burning by sample A which is 44 mm while the lowest by sample C which is 22 mm. This is because the specimen B consist of 3 layer kapok fibre while specimen A purely CSM fiberglass. This shows that increasing layer of kapok fibre can reduce the length of burning as discussed previously.

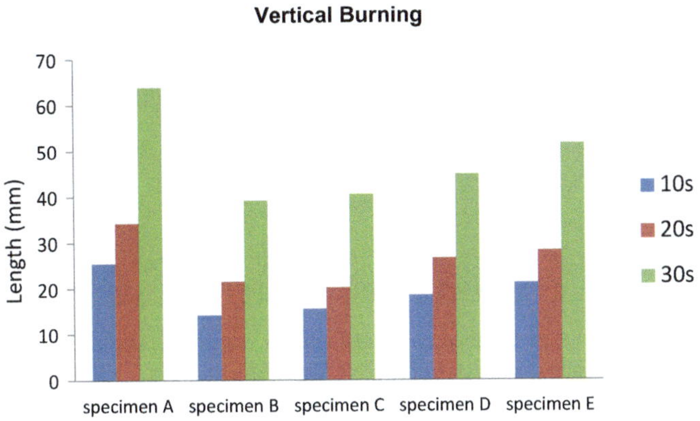

Fig. 4 Vertical Burning of the kapok/CSM hybrid composite

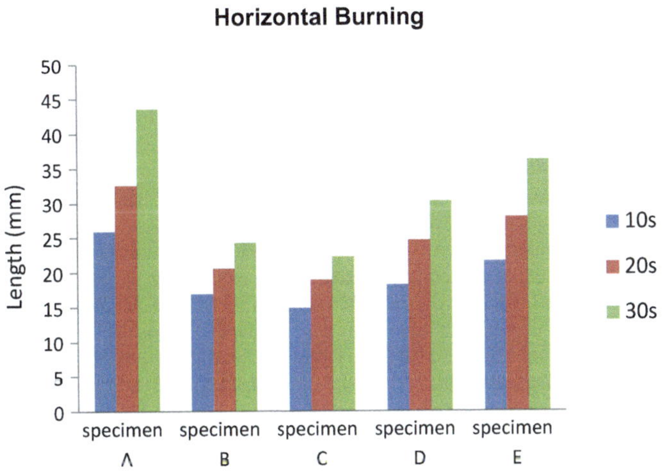

Fig. 5 Horizontal burning of the kapok/CSM hybrid composite

5 Conclusion

In this work hybrid composite structures non-woven kapok/CSM fiberglass composite was manufactured by using hand lay-up method and tested using physical testing and fire-retardant testing to assess the effect of kapok layer on the fire retardant, thickness swelling and water absorption properties. It was found that kapok reinforced composite showed the increasing trend of water absorption and thickness swelling with the increasing kapok layer and storage time. From the result, it shows that the increment layer of kapok fibre will lead to lower length of burning effect. The best sample is specimen B which showed the lowest burning effect. Two external layers of kapok involves the highest increases in lower the burning rate properties of hybrid laminates compared to those of CSM laminates. These results highlight both that the kapok fibres may be considered as a possible alternative of the glass ones in marine application and hybrid structures could be used in the manufacturing of a boat.

References

1. Saheb N, Jog NJ (2015) Natural fiber polymer composites: a review. Adv Polym Technol 2329:351–363
2. Shi J (2011) Kenaf Bast for fiber reinforced polymer composites by Jinshu Shi A dissertation submitted to the Faculty of Mississippi State University in Partial Fulfillment of the Requirements for the Degree of Doctor of Philosophy in Forest Products in the Department
3. Abdul Khalil HPS, Hanida SCW, Kang CW, Nik Fuaad NA (2007) Agro-hybrid composite: the effects on mechanical and physical properties of oil palm fiber (EFB)/glass hybrid reinforced polyester composites. J Reinf Plast Compos 26(2):203–218
4. Selvaraju S Ilaiyavel S (2011) Applications of composites in marine industry. J Eng Res Stud II(II):89–91
5. Venkata Reddy G, Venkata Naidu S, Shobha Rani T (2008) Kapok/glass polyester hybrid composites: Tensile and hardness properties. J Reinf Plast Compos 27(16–17):1775–1787
6. Venkata Reddy G, Venkata Naidu S, Shobha Rani T (2008) A study on hardness and flexural properties of kapok/sisal composites. J Reinf Plast Compos 28(16):2035–2044
7. Mwaikambo LY, Martuscelli E, Avella M (2000) Kapok/cotton fabric-polypropylene composites. Polym Test 19(8):905–918
8. Mwaikambo LY, Ansell MP (2002) Chemical modification of hemp, sisal, jute, and kapok fibers by alkalization. J Appl Polym Sci 84(12):2222–2234
9. Chaiarrekij S, Apirakchaiskul A, Suvarnakich K, Kiatkamjornwong S (2012) Kapok I: characteristics of kapok fiber as a potential pulp source for papermaking. BioResources 7(1):475–488
10. Prachayawarakorn J, Chaiwatyothin S, Mueangta S, Hanchana A (2013) Effect of jute and kapok fibers on properties of thermoplastic cassava starch composites. Mater Des 47:309–315
11. Idrus MAMM., Yang MFM, Ismail SB, Hamid, SEA, Abdullah, SFAS (2015) Development of treated kapok/fiberglass hybrid composite for marine application. Int J Mining Metall Mech Eng 3(3):117–121

Important Parameter Related to AFM Lithography for Fabrication of Silicon Nanowires

Ahmad Makarimi Abdullah, Khatijah Aisha Yaacob, Zainovia Lockman and Sabar Derita Hutagalung

Abstract Silicon nanowires (SiNWs) have been a great candidate to be used in development of electronic devices because it is easy to prepare in small dimension with high surface to volume ratio and its sensitivity of the carrier mobility to the variation in the electric field at their surface. Silicon nanowires (SiNWs) can be prepared using both "bottom-up" and "top-down" approached. AFM lithography is one of top-down approach used to fabricate the silicon nanowires by using the local anodic oxidation (LAO) process. The local anodic oxidation process involves the application of a positive voltage between the AFM tip and the surface of the silicon-on-insulator (SOI) wafer in the atmosphere with high relative humidity. The humidity in the ambient will generates a meniscus of water between the AFM tips and SOI wafer. The applied voltage on the tips will ionizes the water molecules and producing OH- ions, which will react with silicon on SOI wafer and form silicon oxide patterned on SOI wafer. Several AFM lithography important parameters in the patterning of the oxide mask, such as substrate surface roughness, applied voltage, writing speed and relative humidity were discussed. After the oxide pattern growth, the SOI wafers were undergo for chemical etching to remove silicon and SiNWs were formed.

Keywords Silicon nanowires · AFM lithography

A. M. Abdullah (✉) · K. A. Yaacob · Z. Lockman
School of Materials and Mineral Resources, Engineering Campus,
Universiti Sains Malaysia, 14300 Nibongtebal, Pulau Pinang, Malaysia
e-mail: mraisha@usm.my

A. M. Abdullah
Marine Electrical and Electronics Technology Section,
Malaysian Institute of Marine Engineering Technology (MEET),
Universiti Kuala Lumpur, 32000 Lumut, Perak Darul Ridzuan, Malaysia

S. D. Hutagalung
Physics Department, Jazan University, Jazan 45412, Saudi Arabia

C. L. Saw et al. (eds.), *Advancement in Emerging Technologies and Engineering Applications*, Lecture Notes in Mechanical Engineering,
https://doi.org/10.1007/978-981-15-0002-2_25

1 Introduction

The efforts are rapidly increase since last two decades to develop the nanodevice structures, which can utilize the fundamental physics (low-dimension structures, high surface areas, single charge effects, high density storage etc.) in nanotechnology applications. These works become more interesting when its can combine the technology between bottom-up technique and top-down technique for fabrication of nanodevices. However, there are spaces for improvements and enhancements to ensure the technology applied is well significant to apply in real industry application. Hence, this technology is quite stable to utilize for real industry scale, there are few methods still in lab scale for improvement studies.

Present era, is the generation of new device which focus on circuit patterning resolution and positioning accuracy which ranging beyond limit of fabrication processes. The consideration of the parameters in the lab scale plays the significant role as stepping stone to apply in industry scale. Conventionally the designing the circuits on semiconductor fabrication are using the various lithographical methods, which is highly developed and optimized. Most of these approaches are based on top-down techniques which require high operating cost, multiple-step processes and limited accessibility for nanofabrication applicability by using these techniques. In the other hand, an unconventional method provides simplest and cheaper technique in nanofabrication fabrication. Nano-imprint lithography (NIL), soft-lithography and atomic force microscopy (AFM) nanolithography are the novel methods which have been emerged the flexible alternatives in nanometer scale patterning and fabrication. All these techniques are potential to replicable for patterning process and fabrication since it can be reproducible based on the critical parameters set-up and environmental control. Then these methods can be a novel method for nanometer scale patterning and fabrication. Among those techniques mentioned, AFM nanolithography technique is one of the best method which can capable to fabricate and characterize the microarrays, nanometer scale structures, sensors and devices.

In general, this principle was applied in Scanning probe microscope (SPM) in different applying principles to induce modification of oxides, semiconductors and metals on the nanometer scale. There is low energy exposure of resists, nano manipulation [1, 2], thermo-mechanical writing [3], local oxidation [4, 5] and mechanical modification [6]. While for AFM application it can be generalized in three groups of operational principles. It can be categorized as force assisted, bias-assisted and thermal assisted AFM nanolithography. From all these operational principles listed, tip induced local anodic oxidation (LAO) as bias-assisted method is widely investigated for AFM nanolithography. This technique was observed for the first time in 1994 by Snow and Campbell [7] while they utilized the scanning tunneling microscope (STM) based method. Here, from their study shows that, the proper choice and special care of preparation of sample are the keys to ensure the success of LAO fabrication. Even though from their study mentioned the 'proper choice and preparation', it is too general to understand the limit and variant can be play around during the fabrication process.

Based on their study, the first experiments starts with Si as substrate and oxidize on the Si(111) surface. The electrical properties were remarkably reported by many researchers from the devices made by AFM nanolithography. Some recent works were performed to improve this method of AFM nanolithography. As discussed above, its shows only the half way of the whole process to gain the final structure devices. The pattern fabricated by that technique need to go through the etching process to get the desired final structures. Wet etching and dry etching are applicable techniques in fabrication of Si nanostructures. Dry etching which is top-down method need special attention due to mixtures of gasses needs to be at correct ratio to optimum the shapes of nanostructures needed. In addition, this method are quite costly and danger due to the gas used during the process [8, 9]. While wet etching is a bottom-up method is still applicable to optimize the technique and recipe for Si nanofabrication. Tetramethyl-amino-hydroxide (TMAH) and potassium hydroxide (KOH) saturated with Isopropyl alcohol (IPA) are two well-known etchants and widely use. This etching process is necessary to remove the unwanted oxide (which is thin layer mask) in nanostructures fabricated by LAO techniques. Wet etching method is simpler and effective cost compared to the dry etching method. In addition this method also safer compared to usage of gasses (as used in dry etching method) in the nanofabrication process. The nanometer scale structures are defined by well precise crystalline plains. During the process, the etching solution play an important role to terminate the anisotropic behavior on the substrate used. The controllable factors influence the cross section of the nanowire (trapezoidal cross section) during that process produce ultra-small and long nanowire [10].

In this paper, present the comprehensive review by elaborating the investigation of parameter variation impact on the fabrication of nanometer scale devices by AFM-LAO nanolithography followed by wet etching process using KOH-IPA solution. The investigation is aiming to give the complete information guide or comparison in details. It discuss the precise view of every single steps of (bias assisted) AFM-LAO by reviewing some of previous works to support the discussed data presented. The important factors contributing to the fabricating parameter were elaborated. It starts from preparation of SOI substrates wafer, AFM vector scan programming algorithms, voltage applied, ambient environment condition and etching process to the final extraction of the structure are presented as well.

2 Experimental Methodology

Refer to Fig. 1, the processes start with cutting the Si wafer substrate to size 1.5 cm × 1.5 cm. The wafer substrate was used is p-type B-doped (100) SOI wafer (Unibond International Ltd. Uxbridge, Middlesex, UK). The top silicon of 100 mm and ~200 nm buried oxide (BOX) thickness with a resistivity of 13.5–22.5 V cm (Soitech) as shown in the illustration in Fig. 2. This wafer will further clean with the standard Radio Cooperation of America (RCA) cleaning procedure. Then the native oxides were remove using hydrogen fluoride (HF). Parallel with this process, the pre-design structures

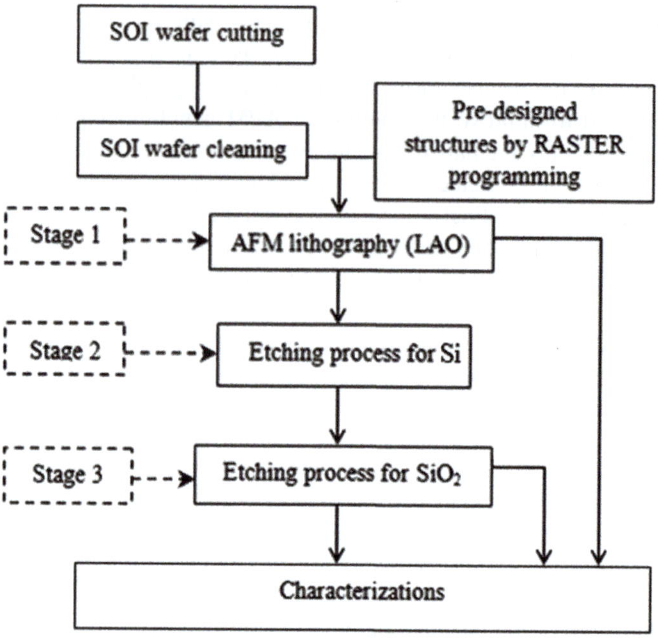

Fig. 1 Methodology of fabrication SiNWT by using AFM nanolithography

Fig. 2 Illustration of wafer used in the study supplied by Soitech

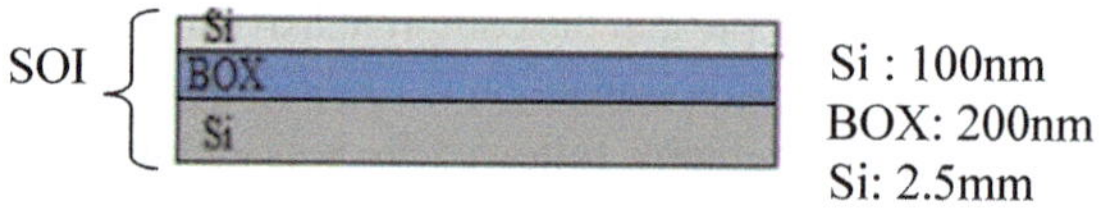

programming was set-up for vector scan/bias assisted LAO-AFM nanolithography. In the stage 1, nanolithography process was fabricated by scanning probe microscope machine model SPI3800N/4000, SII Nanotechnology Inc. Chiba-shi, Chiba, Japan. Contact mode was used in LAO procedure by employing the AFM conductive coated tips with the force constant at 0.2 Nm^2 with the resonance frequency of 13 kHz.

During the LAO process, the ambient environment in the room was controllable from 50 to 70% with the accuracy of the device is 1%. After performed the desired structure with LAO procedure, where the SiO_2 formed is considered as a mask on the top layer of the SOI surface as shown in Fig. 3a. Then the sample will proceed with the stage 2 of the process, here the sample will go through the etching process using KOH:IPA solution to remove the top layer of Si (~100 nm) as shown in Fig. 3b, c. Then the sample will proceed for the morphology characterization to ensure the undesired area was etched completely.

Stage 3 will follow with the second stage of etching by removing the top layer mask performed by LAO process (Fig. 3c). Here the top layer will remove by dilution of HF to reveal the patterned structure device as shown in Fig. 3d, e. The process

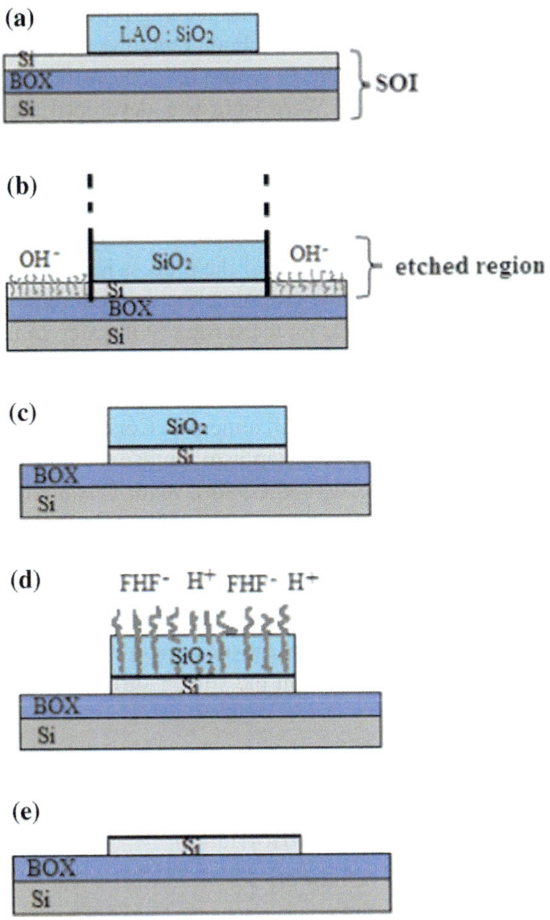

Fig. 3 Illustration of stage 1 process. **a** Mask performed by LAO process, **b** etched region by etching solution, **c** structure of the mask on top of Si layer, **d** mask remove by dilution of HF solution, **e** desired final nanometer scale structure

will follow with the morphology characterization before further process for electrical characteristic to test the performance of the devices.

3 Results and Discussion

3.1 Sample Preparation

The cleaning process is important in nanometer scale fabrication. In ambient environment the silicon wafer surface already covered by varieties of contamination, such as organic and inorganic contaminations and as well native oxide. To ensure the surface is purely clean as possible and nanometer flat oxide resistant, the wafers necessarily need to further the leaning processes. The contaminant elimination by native oxide removal and silicon dangling bonds saturation removed by hydrogen atoms as discussed by Ju et al. [11]. The workers of Radio Cooperation of America (RCA) in 1965, Wern Kern was developed the standard procedures for wafer cleaning; and these methods are known as RCA method [11]. However, this RCA method was modified to optimized to get the best output for SOI wafer which is have very thin layer of Si. Here, the parameter for the dilution, time and temperature must be exactly same ratio which accurately less than 5% different for each time of preparation. The summary of the process is shown in Table 1.

In organic cleaning (RCA 1), the solution removed the entire insoluble organic contaminants particle on the surface SOI wafer. In this process, the solvating action of ammonia/peroxide mixture oxidizes the silicon surface using hydrogen peroxide (H_2O_2), it will simultaneously removes the silicon oxides by means of ammonium hydroxide (NH_4OH) at alkaline pH. Concentration of NH_4OH was use is 25wt% in this formula. Second step is to remove the ionic contamination by using mixture of hydrogen peroxide/hydrochloric acid. This mixture will resemble bubbles to trap the contaminants by means of soluble chlorides. Here the metal ions and metal hydroxides were solubilized. In Ionic cleaning (RCA 2), the solution use to remove the ionic and heavy metal contamination. HCl generates a metal cat ion in solution to forms an ionic bond with the chloride ion to rinse entire metal tracks off on the silicon surface [12]. In both step, the temperature and time interval play the crucial role for the nanometer flatness on the silicon surface. Figure 4a shows the surface of SOI

Table 1 The modified steps of RCA method used for SOI wafer cleaning

Step 1—RCA 1: Organic cleaning
Solution of DIW: H_2O_2: NH_4OH (5:1:1)
Prepared in glass beaker. Heated at 75–78 °C
The temperature must maintain not more than 78 °C for 15 min
Use a magnetic stirrer bar to stir the solution at 400 rpm for uniform cleaning
Then the wafer rinsed with DIW for at least 1 min
Step 2—RCA 2: Ionic cleaning
Solution of DIW: H_2O_2: HCl (6:1:1)
Prepared in glass beaker. Heated at 75–78 °C
Use a magnetic stirrer bar to stir the solution at 400 rpm for uniform cleaning
The temperature must maintain not more than 78 °C for 15 min
Then the wafer rinsed with DIW for at least 1 min
Step 3—Oxide strip
Solution of HF: DIW ratio 1:80
Prepared in polypropelene beaker
Wafer was soaked in the solution until hydrophobic
Rinse with the DIW for at least 1 min

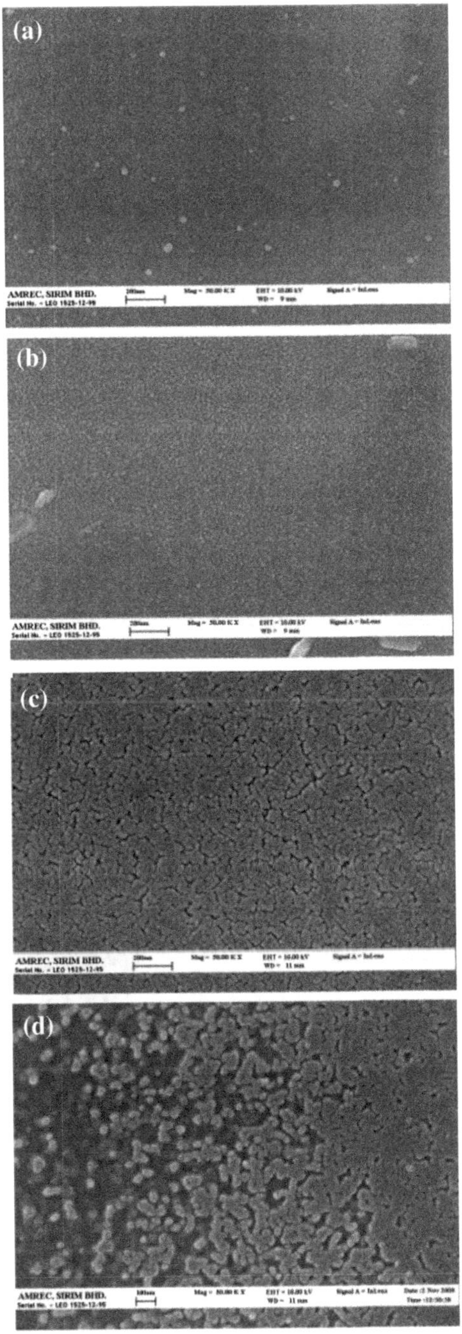

Fig. 4 SEM micrograph pictures showing the effects of cleaning parameters. **a** SOI wafer surface before cleaning process, **b** 12–15 min and 78–80 °C, **c** duration time interval ~20 min, and **d** high temperature ~90 °C

wafer before cleaning process. Figure 4b shows the perfect time interval 12–15 min at temperature 78–80 °C in RCA solution. Figure 4c shows the effect of the surface with improper time interval ~20 min in RCA solution, while Fig. 4d shows the high temperature solution effects at ~90 °C. Here clearly can see, when the time interval longer in the solution, the surface roughness becomes rougher. When the temperature arises more than 80 up to ~90 °C, the surface becomes not uniformity. Aqueous HF solution at 49% of purity was used to remove the native oxide. This aqueous solution diluted with formula of 1:80 (HF: DIW). This step is quite easy to see from naked eye. To ensure the native oxide is entirely removed, the surface is hydrophobic to de-ionized water (DIW).

3.2 *AFM Nanolithography*

First stage after cleaning process is to perform the oxide growth process using biased assisted local anodic oxidation using AFM. Here, the Si–H was formed on the surface of SOI. This happen, when the surface was de-passivated on the first layer of Si–Si becomes polarized due to the high electron negativity of the OH^- in the ambient environment in the AFM chamber. During the process, a biased AFM tip (10–215 V) oxidizes the surface based on the command of RASTER programming set-up. The gap between AFM tip and surface just a few nanometers would generate a field of 10^8–10^{10} V/m. The available water meniscuses on the surface are water bridge look-a-like will dissociated by the negative tip bias to react by performing the desired oxide growth. Figure 5 shows schematically the LAO-AFM process.

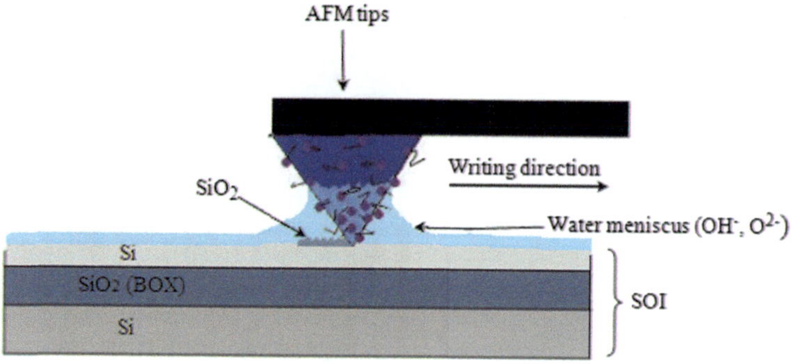

Fig. 5 Schematic process of LAO-AFM

3.3 AFM Tip and Exposure Time (Writing Speed)

In this study, the gold coated (Budget Sensors Au coated ContGB-G) was used. Parameter is set for the force constant (kc) of 0.2 Nm^{-1} and the resonance frequency of 13 kHz (tip radius, 25 nm). The LAO process should consider the speed writing and applied vaoltage to the tip during patterning process. Both parameters are crucial to control and ensure the oxide growth is performed uniformly. Based on the previous study on the LAO [1, 6] the oxide height decrease with the writing speed. The low writing speeds produced wider oxidation tracks on the substrate. If the faster writing speeds it might provide negligible track for constant applied voltage. As the writing speed increases, the numbers of diffusible ion into Si/SiO_2 interface decreases. In this study the voltage applied to the tip is between 7.0 and 8.4 V

3.4 Environment Relative Humidity and Temperature

Another parameter need to concern and play important role in LAO process is RH%. This factor play role to give necessity of water molecules in chamber environment during the oxidation process. Basically, the thickness of the oxidation is proportionally increased with the increase of humidity. Here to understand better, the presence of H^+ surrounding the AFM tip significantly activate the better growth of oxide on the surface [6] High concentration of oxyanions are coming from the effects of water bridge on the surface to give high electric filed strength to react with the SOI surface area. Figure 6a–e show the effects of RH% to the oxide growth on the SOI wafer. The results show that, it's proportionally increased by the increase of RH% as shows in Fig. 7.

The stage 3 process is mask removal or etching for the SiO_2 mask performed by LAO-AFM nanolithography. Many studies discussed details regarding these topics, either dry etching or wet etching for device fabrication. As result discussed, the hydrazine is one of the good etchants to negligible degradation during deep etching [9, 10], but extreme caution need to consider since it's highly toxic and have potential to explosive. However, this type of etching has great variety profiles ranging from isotropic etch to vertical sidewalls. Selectivity of etching for oxide mask and the silicon are poorer for the liquid etchants [12, 13]. Even though, the quality poorer, it can be utilized to optimal results for non-anneal oxide performed by LAO-AFM, since the density of oxide are lower and higher dielectric constant of 5.9 versus 3.9 [8]. More advantage this oxide also has a greater H_2O content by weight in comparison to the thermal case. Based on these factors, the selectivity agent to this oxide has high aspect ratio for the nanostructure, means that there is critical parameter need to be concern during the wet etching process. Mixture solution of KOH and IPA is a well-known etchant for the anisotropic etching technique. Many research teams elaborate and investigate deeply about the mechanisms. However, there are some matters need

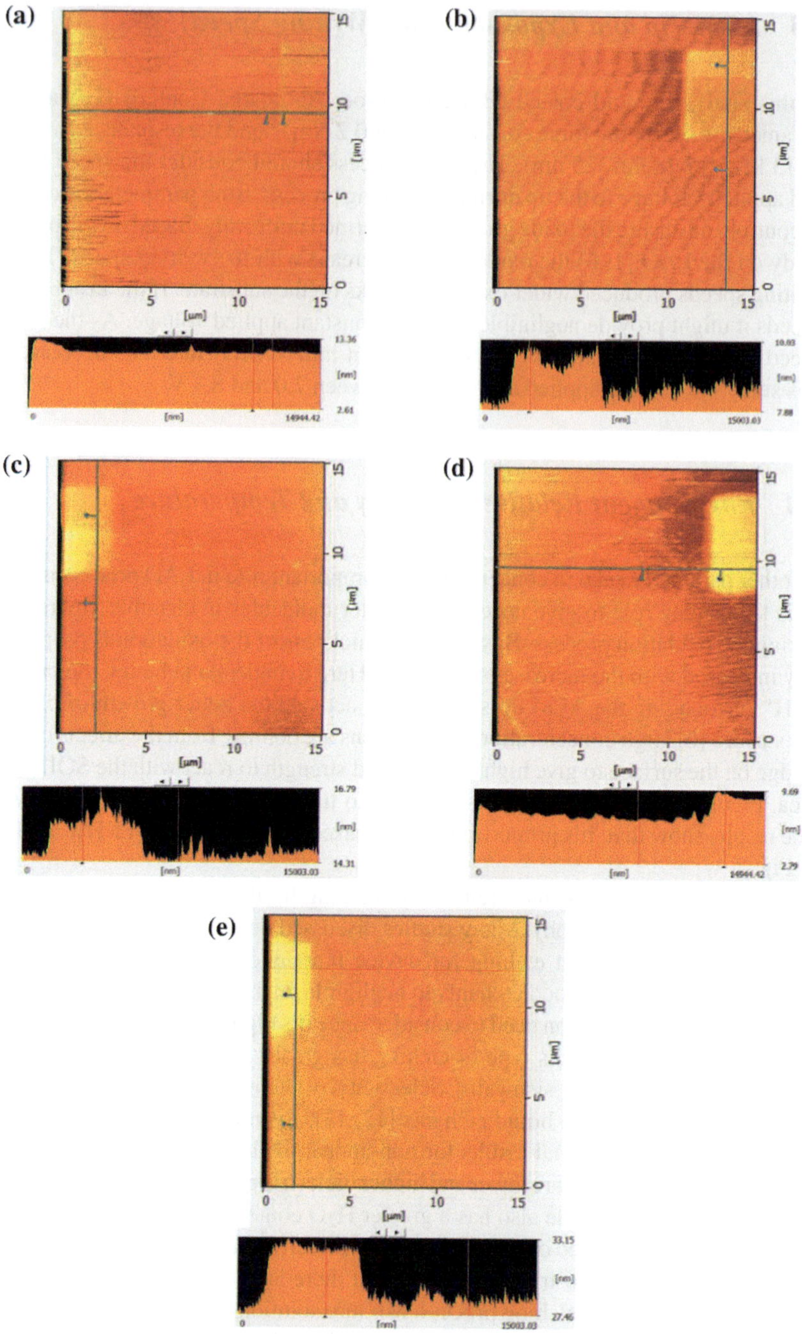

Fig. 6 Topographic image of fabricated area with 15 μm × 15 μm. **a** RH 55% with average thickness ~0.48 nm, **b** RH 60% with average thickness ~1.02 nm, **c** RH 65% with average thickness ~1.53 nm, **d** RH 70% with average thickness ~2.47 nm, **e** RH 72% with average thickness ~3.04 nm

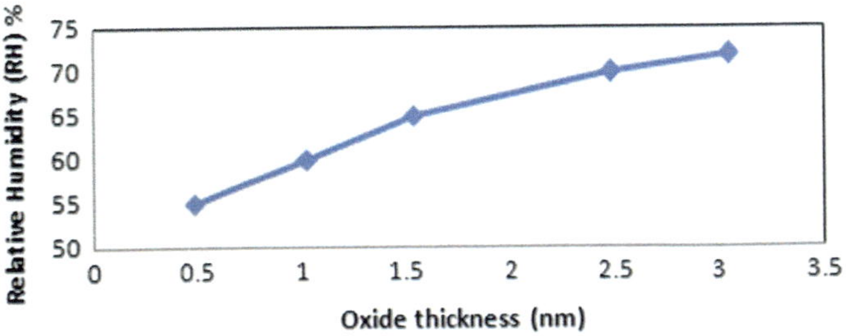

Fig. 7 Relation between Relative Humidity (RH) % and oxide thickness (nm)

to consider when it comes to nanometer scale structure with low density oxide and high H_2O content.

To extract the desired structure of the nanometer scale devices, the mask removal is a crucial step need to go through. Here in this study, use KOH-IPA with varieties of ratio to perform the optimum result of final device structure. Many possibilities could happen and hardly to avoid. The process could be either under etch, over etch or other contamination could affect to the final structure. Many precaution need to consider during the process of etching such as etch rate, temperature of solution and uniformity of the solution during etching process work. By considering the rate of Si removal by KOH etching as 0.28 μm/min [14], there are many attempts to optimize the final structure. Etch rate and concentration are independent for the temperature below 70 °C. Faster etch rate can be achieved by using high temperature reaction. The IPA will hinder the access of reactive OH- ion during the etching process from the bubbles occurred [9]. As the results it will smooth the surface etched area. Based on these finding, the works are done at temperature 65–70 °C for 20–22 s and the ratio for the mixture solution is 30wt% KOH + 10% IPA with stirred at 600 rpm. After etched, the sample should rinse with DIW for at least 1 min.

Figure 8 shows the results of surface pattern etched by KOH wt% from 10–40wt%. At KOH 10wt% which is low concentration of KOH, the surface is rougher and the formation of insoluble precipitates can be observed. And the roughness reduces when the concentration increases as reported by Shikida et al. [12], Noorhaniah et al. [15], Najihah Alias et al. [13].

As a finale result, Fig. 9 shows the results for variable temperature and time applied to the solution during etching process. It shows that, the nanowires structures and rectangle shape are over etched when the temperature arise up to 70 °C and immersed up to 22 s (Fig. 9a). The fabricated pattern not complete rectangle shape and the nanowires entirely etched at temperature 65 °C and immerse time up to 32 s in Fig. 9b. While the optimum results show in Fig. 9c, which the parameter at set up at 65 °C, immersed time 22 s and stirred at 600 rpm. Figure 10 shows the real time color image of the optimum finale structure which measured by 3D microscope. Form the

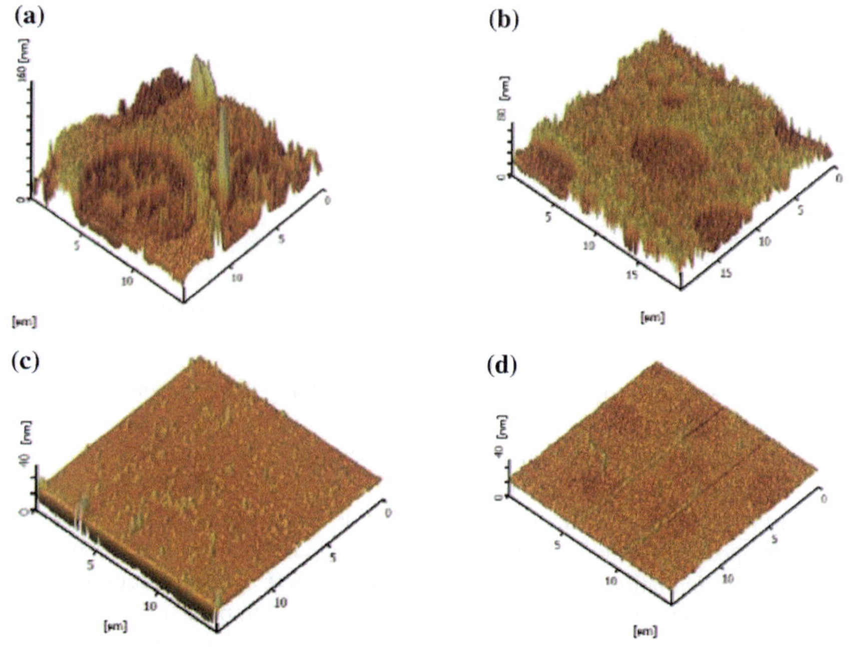

Fig. 8 AFM topography image of silicon surface etched with KOH by wt%. **a** KOH 10wt%, **b** KOH 20wt%, **c** KOH 30wt%, **d** KOH 40wt%

picture, there are no un-necessary marking and deformity. It shows the good quality final Si nanowire device with smooth, uniform and clean surface.

4 Conclusion

In this finding, different parameters were investigated in fabrication of nanostructure device on SOI wafer performed by AFM nanolithography and wet etching process. Sample cleaning preparation are follows the standard RCA procedure and improve by means of nanostructure flatness needed. The impact of LAO were mentioned the crucial parameters in type of tip, applied voltage, writing speed and relative humidity to perform the optimum structure. Etching procedure was investigated to remove the Si layer and SiO_2 (mask) are very critical by means of temperature, volume ratio and immersed time. The best parameter was finalized by using KOH 30wt% admixture with 10% IPA with etching time 22 s at 65 °C temperature and stirring at 600 rpm.

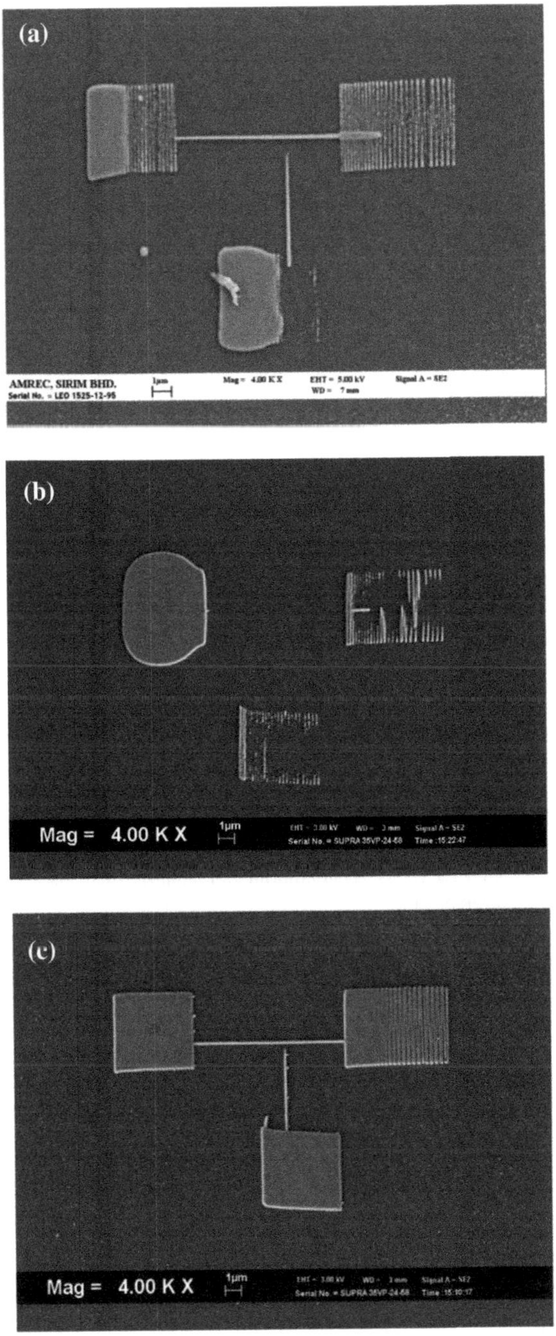

Fig. 9 SEM micrograph pictures for final structure of the device. **a** T = 70 °C, t = 22 s, and stirred at 600 rpm, **b** T = 65 °C, t = 32 s immerse time and stirred at 600 rpm **c** T = 65 °C, t = 22 s and stirred at 600 rpm

Fig. 10 Image of real time color of final structure measured by 3D laser microscope

Acknowledgements Authors would like to thank A.P Dr Sabar D. Hutagalung for his assistance in experimental work under support grant from MOSTI (2008). Author also would like to thank Dr Khatijah Aisha Yaacob for her support under Research university grant (RUI) Universiti Sains Malaysia under project number RU-8014205 in conference expenses and UniKL MIMET management for study sponsorship.

References

1. Garcia Ricardo, Knoll Armin W, Riedo Elisa (2014) Advanced scanning probe lithography. Nat Nanotechnol 9:577–587
2. Guo LJ (2007) Nanoimprint lithography: methods and material requirements. Adv Mater 19:495–513
3. Mikolajick T, Weber WM Silicon nanowires: fabrication and applications. anisotropic nanomaterials. NanoSci Technol 1–25 (2015)
4. Singh N, Buddharaju KD, Agarwal A, Rustagi SC, Lo CQ, Balasubramanian N, Kwong DL (2008) Fully gate-all-around silicon nanowire CMOS devices. Solid State Tech 51(5):34
5. Colinge JP, Lee CW, Afzalian A, Akhavan ND, Yan R, Ferain I, Razavi P, O'Neil B, Blake A, White M, Kelleher AM, McCarthy B, Murphy R (2010) Nanowire transistors without junctions. Nat Nanotechnol 5:225–229
6. Xie X, Chung H, Sow C, Wee A (2006) Nanoscale materials patterning and engineering by AFM nanolithography. Mater Sci Eng 54:1–48
7. Snow E, Campbell P (1994) Fabrication of Si nanostructures with an AFM. Appl Phys Lett 64:1932–1934
8. Zhang Y, Santhanam S, Liu J, Fedder G (2010) Active CMOS-MEMS AFM like conductive probes for field-emission assisted nano-scale fabrication. IEEE 336–339 (2010)
9. Camon H, Moktadir Z (1997) Simulation of silicon etching with KOH. Microelectron J 28:508–517
10. Chen LJ (2007) Silicon nanowires: the key building block for future electronic devices. J Mater Chem 17:4639–4643
11. Kern W, Puotinen DA (1970) Cleaning solutions based on hydrogen peroxide for use in silicon semiconductor technology. RCA Rev 31:187–206

12. Shikida M, Sato K, Tokoro K, Uchikawa D (2000) Differences in anisotropic etching properties of KOH and TMAH solutions. Sens Actuat A 80:179–188
13. Alias NN, Yaacob KA, Yusoh SN, Abdullah AM (2018) Comparison of KOH and TMAH etching on sinw arrays fabricated via AFM lithography. IOP Conf Ser: J Phys: Con Ser 1082, 012051
14. Ju C, Hesketh PJ (1992) High index plane selectivity of silicon anisotropic etching in aqueous potassium hydroxide and cesium hydroxide. Thin Solid Films 215:58–64
15. Yusoh SN, Yaacob KA (2016) Effect of tetramethylammonium hydroxide/isopropyl alcohol wet etching on geometry and surface roughness of siliconnanowires fabricated by AFM lithography. Beiltein J Nanotechnol 7:1461–1470

An Investigation on the Overall Efficiency of a Ship with Shaft Generator Using an Engine Room Simulator

Wilfredo Yutuc

Abstract The use of a marine shaft generator onboard is considered an alternative means to reduce fuel consumption and improve the energy efficiency of a ship. It is a type of an electric generator driven by the main propulsion diesel engine to supply the electric power requirement of the ship at sea. It is intended to take advantage of the low specific fuel consumption of the main propulsion diesel engine when engaged and in operation. Since it is a power take-off appliance, its operation causes an increase in load on the main propulsion diesel engine as compared to when a separate auxiliary diesel generator is running to generate and supply electricity onboard. To determine the effectiveness of this auxiliary equipment, this paper attempted to investigate and compare the overall plant efficiency of a ship when running on a shaft generator as against an auxiliary diesel generator for electric power generation, using an engine room simulator. Data were gathered from a series of simulations on a main propulsion diesel engine load of 50–100% with a steady power load on the running electric generator. The theoretical overall efficiency of the combined propulsion and electric power plant in both scenarios were calculated, compared and presented in this paper.

Keywords Diesel engine · Ship energy efficiency · Engine room simulator · Shaft generator

1 Introduction

A ship at sea is an autonomous system which generates its own propulsion and electric power. For the propulsion system, the prime mover may be a diesel engine, a gas turbine, a steam turbine or a combination of any of these. In merchant vessels, the diesel engine is still the most common due to its higher efficiency compared with the other types [1]. For electric power generation, separate generators (also termed 'alternators'), which may be driven by any type of prime mover mentioned earlier,

W. Yutuc (✉)
Universiti Kuala Lumpur, Malaysian Institute of Marine Engineering Technology (UniKL MIMET), 32200 Lumut, Perak, Malaysia
e-mail: wilfredo@unikl.edu.my

C. L. Saw et al. (eds.), *Advancement in Emerging Technologies and Engineering Applications*, Lecture Notes in Mechanical Engineering,
https://doi.org/10.1007/978-981-15-0002-2_26

running in a single unit or multiple units in parallel, supply the power demand. In some cases, the generator is coupled to the main propulsion engine either directly or through a gearbox, and if this is the case, it is called a shaft generator [1, 2]. While most generator prime movers rely on fossil fuels to generate electricity, the shaft generator is unique in such a way that it does not directly require the burning of fossil fuels since it is driven by the main propulsion engine. When engaged, full alternator power is available at a range between 45% and 105% of propeller speed, sufficient to supply the ship's electric power requirement [3–5]. For this reason, it has been considered as an alternative means to improve the ship's energy efficiency [6, 7], in addition to other green ship technologies which may include the use of renewable energy or alternative fuels.

A typical propulsion and electric power supply system employing a shaft generator on large conventional cargo vessels is shown in Fig. 1. A low-speed diesel engine provides propulsion power by directly driving a fixed pitch propeller, and the auxiliary diesel generators or the shaft generator provide the electric power requirement at sea. An emergency diesel generator is fitted to supply power to the emergency equipment in case of failure of the primary generators.

When in operation, the shaft generator is to take advantage of the main propulsion engine's low specific fuel oil consumption (SFC) and its use of less expensive heavy fuel oil (HFO). If the auxiliary diesel generator engines use marine diesel oil (MDO) instead of HFO, the improvement in energy efficiency is even greater [2, 6, 8]. Also, by running the ship with the shaft generator as the main source of electric power during long passages, the auxiliary diesel generator units operate for short periods only with a reduced maintenance requirement. However, shaft generator operation

Fig. 1 A typical propulsion and electric power plant system of a ship with a shaft generator

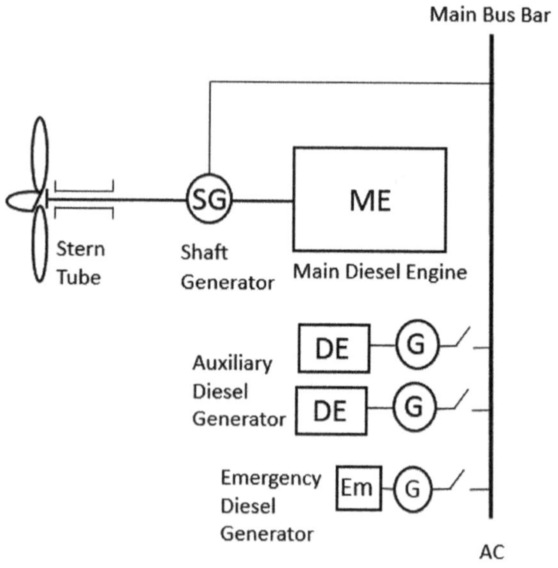

causes an increase in load on the main propulsion engine since the power to drive the alternator is supplied by the latter.

To investigate the effectiveness of the shaft generator, the overall efficiency of the combined propulsion and electric power plant of a ship with a similar configuration in Fig. 1 was investigated using an engine room simulator. With the advancement of computer technology, computer simulation and use of simulators nowadays has become more and more accurate in modeling physical processes and operations, and predicting the real-world performance [9]. In marine engineering, full-mission engine room simulators may utilize a combination of software and hardware to mimic the engine room environment and the system processes with much greater realism [10]. Its use may not only be limited to education and training purposes [11], but also in investigation and performance analysis, as in this paper.

2 Theory

In power plants, the overall efficiency of a system may be expressed as the total useful energy divided by the energy input [12, 13]. The useful energy is usually converted in the form of power from a source of heat energy which is commonly fuel. A simplified illustration showing the energy conversion flow with fuel as the main energy source to the rotating output shaft of an engine is shown in Fig. 2.

For most common ships, energy conversion mainly takes place in two systems. The first one is in the propulsion power plant wherein chemical energy contained in the fuel is converted into mechanical energy to drive the propeller. The second one is in the electric power plant wherein chemical energy is converted into electric energy to run the electric machinery onboard [1].

For a typical ship similar to what is shown in Fig. 1, with two separate systems, the overall efficiency, η_O may be taken as

$$\eta_O = \eta_P \cdot \eta_E \tag{1}$$

where η_P is the propulsion plant efficiency and η_E is the electric power plant efficiency.

To indicate the efficiency of an engine, the brake thermal efficiency is commonly used [12]. It is the ratio of the brake power produced to the heat input, and may be expressed as

$$\eta = P_B/Q_{IN} \tag{2}$$

Fig. 2 Energy conversion flow

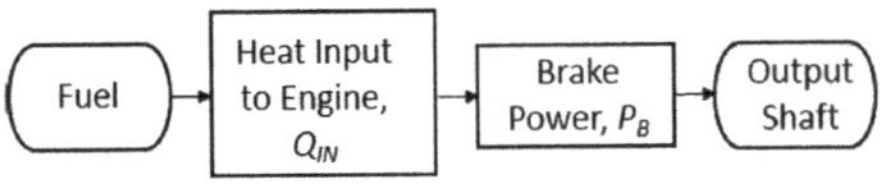

The fuel heat input to the engine Q_{IN}, may be calculated as

$$Q_{IN} = m_F \cdot \text{CV} \tag{3}$$

where m_F is the fuel mass flow rate and CV is the lower calorific value. As applied to ships, HFO and MDO are the common types of fuel with a CV of 40,200 MJ/kg and 42,700 MJ/kg, respectively [14].

If the brake power output and corresponding s0pecific fuel consumption is known, the energy input to the engine may be determined by

$$E_{IN} = SFC \cdot P_B \cdot CV \tag{4}$$

and the rate of fuel consumption may be obtained by

$$m_F = SFC \cdot P_B \tag{5}$$

For an electric generator driven by a diesel engine, the generator efficiency may be considered as

$$\eta_{GEN} = P_O / P_B \tag{6}$$

where P_O is the electric power output.

The overall efficiency of the combined propulsion and electric power plant efficiency in (1) may also be taken as the ratio of the total power developed by both the main and auxiliary diesel engines to the total heat input [13]. The expression may be written as

$$\eta_O = \Sigma P_B / \Sigma Q_{IN} \tag{7}$$

3 Methodology

To collect and analyze the ship's propulsion and electric power plant performance data, the 'Virtual Engine Room—Free Student Version 2.5 ©' engine room simulator, created by Dr. Stefan Kluj, was used. It is a computer-based full-mission engine room simulator developed to comply with the International Convention on Standards of Training, Certification and Watchkeeping for Seafarers (STCW), Section A-1/12 and Section B-1/12, and the International Safety Management (ISM) Code, Section 6 and Section 8 [5]. It simulates a typical ship's engine room which includes the main engine for propulsion, auxiliary systems, power plant and steam systems. The program runs on multiprocessing and code overlaying in order to ensure that the mathematical model will be able to operate on Pentium class computers, as what was used in this study.

3.1 Simulator Propulsion and Electric Power Plant

The engine room simulator's propulsion and electric power plant layout are similar to that of Fig. 1. A large slow speed, reversible two-stroke diesel engine drives a fixed pitch propeller with a remote control system. The electrical requirements are met either by a shaft generator mounted at the front part of the main engine directly driven by the crankshaft or by the separate auxiliary diesel generator engines.

The technical data of the simulated propulsion and power plant machinery is shown in Table 1.

The simulation program set-up 'Full Ahead at Sea' was used, wherein the main propulsion diesel engine runs at a nominal speed of 91 rpm. In this condition, the shaft generator is engaged and loaded with 0.41 MW with Auxiliary Diesel Generator No. 2 on standby mode. A clean ship's hull navigating in a calm sea was assumed in this program. A screenshot of the ship's electric power plant is shown in Fig. 3.

In this condition, the main engine fuel consumption in kg/h, SFC in g/kW-h and shaft power in kW, were recorded from the digital gauges on the 'Main Engine General View' window of the main engine system panel (Fig. 4).

To get data on various load conditions, the main propulsion engine speed was reduced at an interval equivalent to 5% reduction in load considering 91 rpm as 100% load. This was accomplished by moving the fuel control lever downwards at

Table 1 Simulator propulsion and power plant technical data [5]

Main diesel engine	
Type	Two-stroke
Number of cylinders	7
Number of turbochargers	2
Nominal output	19 670 kW
Nominal speed	91 rpm
Diameter of cylinder	700 mm
Length of stroke	2674 mm
Shaft generator	
Nominal output	600 kW
Auxiliary diesel generator	
Type	Four-stroke
Number of cylinders	6
Nominal output	600 kW × 2
Nominal speed	1000 rpm
Emergency diesel generator	
Type	Four-stroke
Nominal output	300 kW

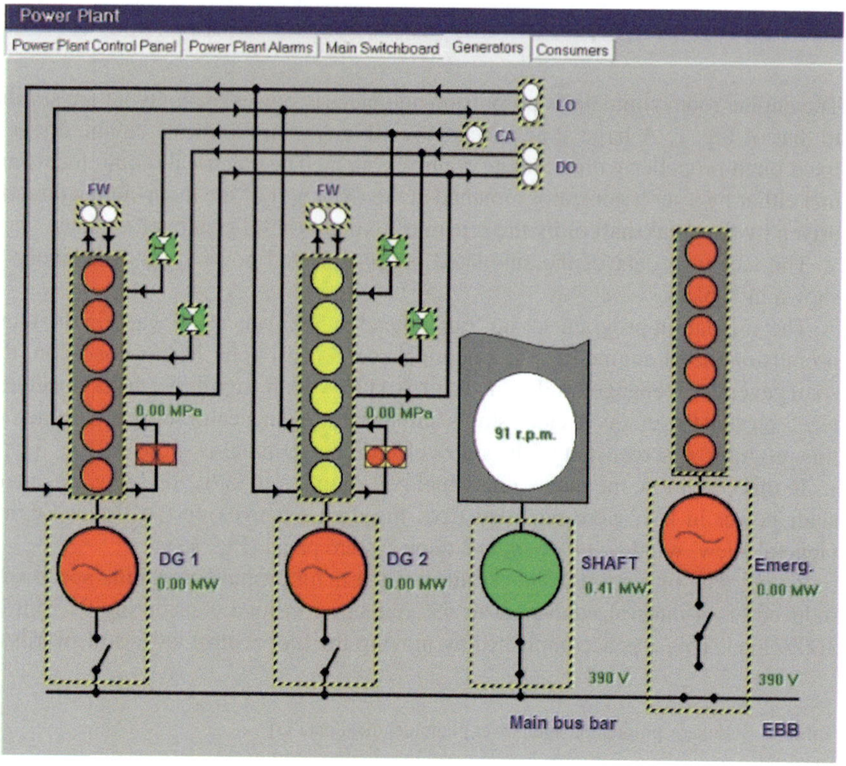

Fig. 3 The simulated electric power plant

the main engine panel of the 'Engine Control Panel' window, shown in Fig. 5. The shaft generator load was kept steady at 0.41 MW.

In simulating the condition wherein one auxiliary diesel generator is running with the shaft generator disengaged, Auxiliary Diesel Generator No. 2 was started and connected in line. A similar procedure was done in performance data recording as with when the shaft generator was engaged.

3.2 Fuel Consumption

The fuel consumption of the main propulsion diesel engine on every simulated speed and load was directly measured from the fuel oil flow meter at the 'Engine General View' window. It was running on HFO in the entirety of the simulation.

The auxiliary diesel generators use MDO with no fuel meter installed. To determine the fuel consumption of Auxiliary Diesel Generator No. 2, a generator efficiency

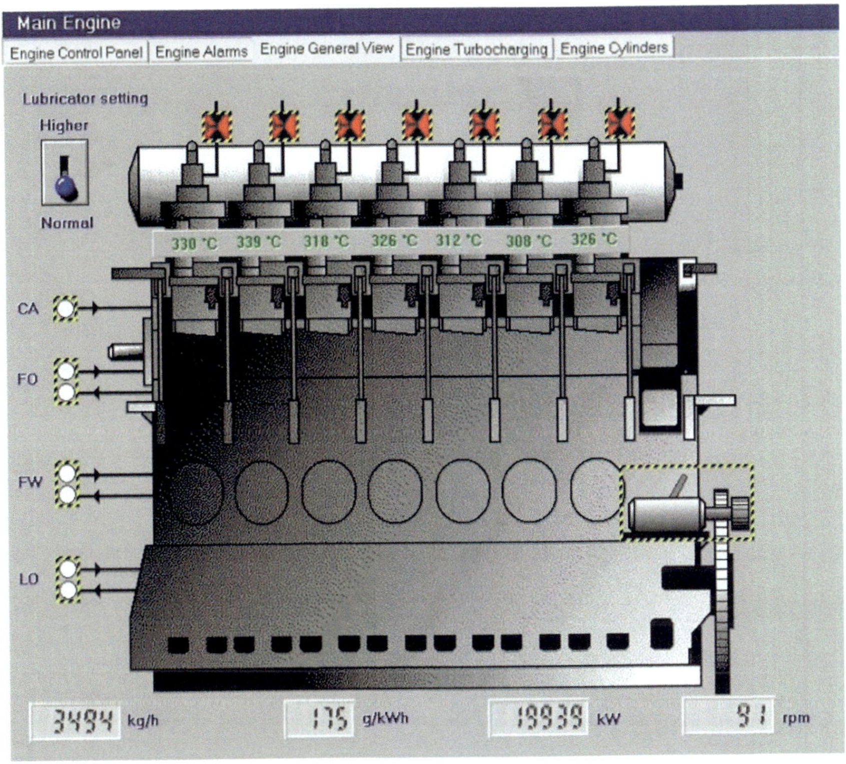

Fig. 4 Screenshot of 'Engine General View' window

of 95% was used [15, 16] and with a steady output of 0.41 MW, or 410 kW, the generator brake power was calculated using (6). (5) was thereafter used to calculate the generator fuel consumption with an SFC of 195 g/kW-h at 75% load.

Three simulations were done following the same procedures mentioned earlier for data recording. The average of these data were used in plotting and analysis of simulation results.

4 Results

A graphical representation of the simulation results of the main propulsion diesel engine power output on increasing load when running on either shaft generator or auxiliary diesel generator is shown in Fig. 6. When the shaft generator was engaged and running, the main propulsion diesel engine brake power output was higher due to the additional power required to run the shaft generator while maintaining propeller speed. However, no separate auxiliary diesel generator was running for electric power generation as the ship operated in this condition.

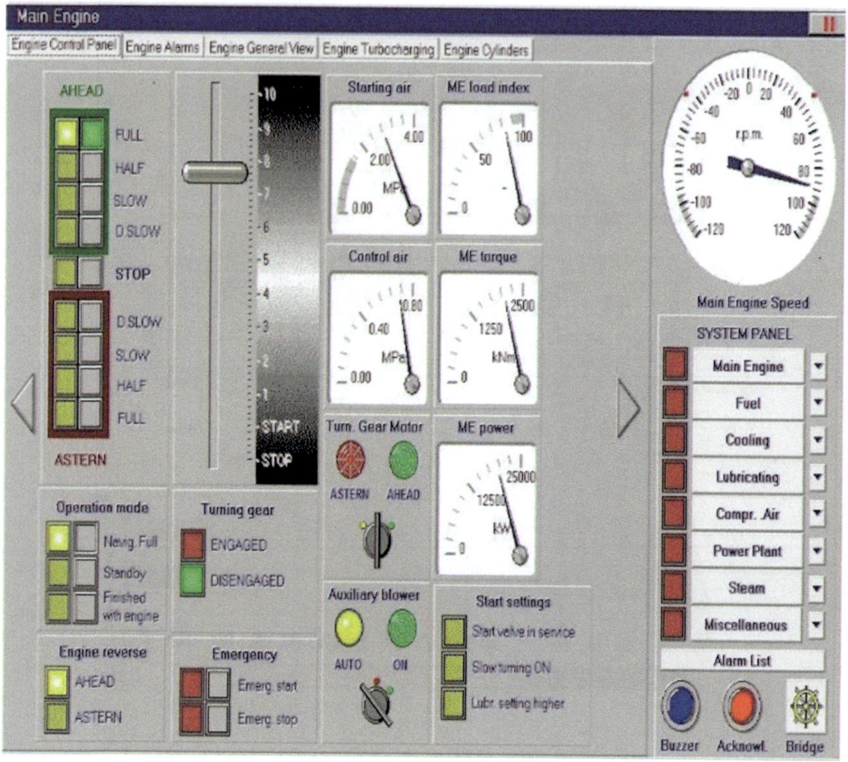

Fig. 5 Screenshot of 'Engine Control Panel' window

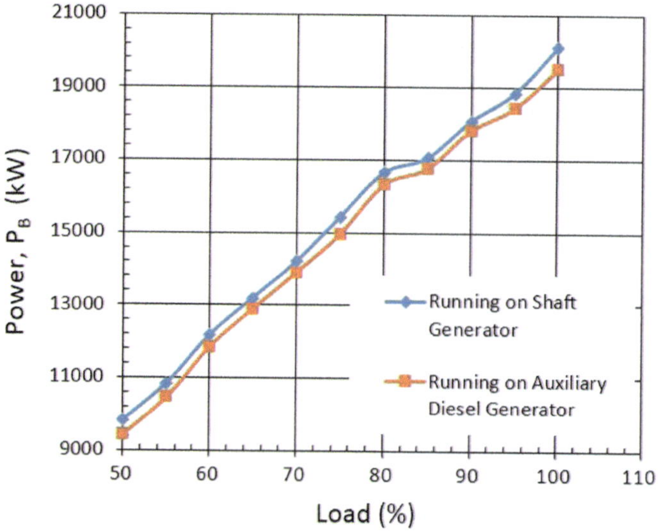

Fig. 6 Main propulsion engine output

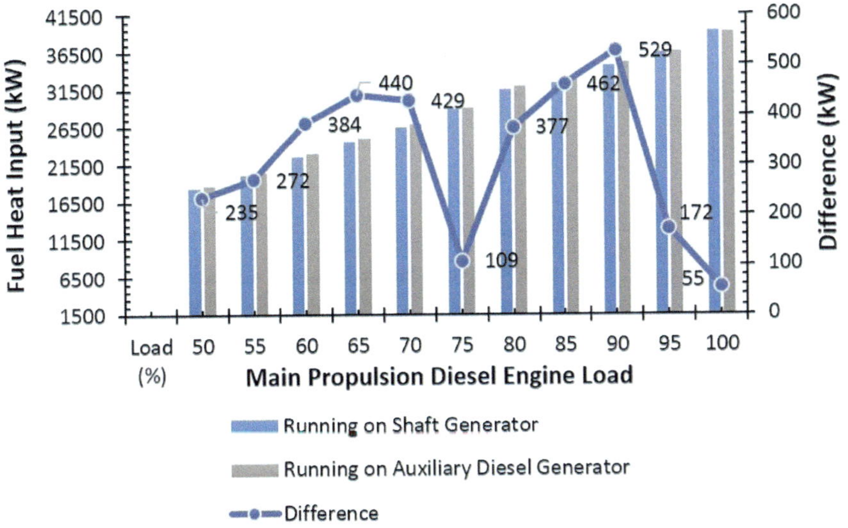

Fig. 7 Combined propulsion and electric power plant fuel consumption

The theoretical fuel heat input to the combined propulsion and electric power plant of the simulated ship throughout the main propulsion diesel engine load range of 50–100% was lesser when running on shaft generator, as shown in Fig. 7. The fuel heat input difference was maximum at 90% load of the main propulsion diesel engine with a difference of 529 kW in favor of the shaft generator as compared to when running an auxiliary diesel generator on the same propeller speed. Likewise, the minimum difference occurs at 100% load with 55 kW lesser fuel heat input, still in favor of the shaft generator.

The overall efficiency of the combined propulsion and electric power plant was calculated using (7) at 5% gradual increase in load on the main propulsion diesel engine while maintaining a steady load on the auxiliary engine. The results were plotted and are shown in Fig. 8. Although consistently higher when running on shaft generator, it can be seen that the difference in overall efficiency was only less than one percent all throughout, ranging from 0.09 to 0.53, with highest difference at 50% and lowest at 100% main propulsion diesel engine load.

5 Conclusion

Based from the results of this investigation, it can be seen that for a ship with such configuration and machinery specification as used in this investigation, the improvement on the theoretical overall efficiency of the combined propulsion and electric power plant when running on a shaft generator was only less than one percent as compared to when a separate auxiliary diesel generator was running to generate and supply

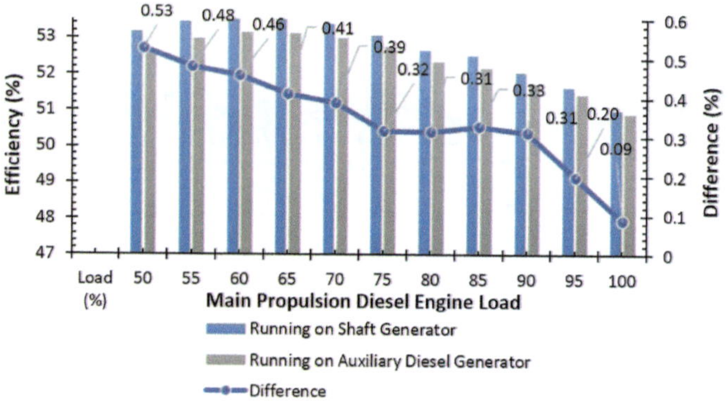

Fig. 8 The overall efficiency of the combined propulsion and electric power plant

electricity onboard. However, although the figure may seem almost insignificant due its very small numerical value, it corresponds to hundreds of kilowatt reduction in fuel heat input, as can be seen in Fig. 7. To ship operators who look for ways and means on how to reduce fuel consumption and improve energy efficiency onboard ships, this is something to consider. In addition, many shipowners and operators still invest on shaft generators due to its acclaimed reliability and reduced installation and maintenance cost compared to auxiliary diesel generators.

It is important to note that the findings and results presented in this investigation are for this particular case only and do not generalize the performance of shaft generators. The benefit in terms of overall efficiency may vary from ship to ship of a different size and configuration.

With the various type of propulsion and electric power plant systems in use today, investigation of a shaft generator's effectiveness onboard a different type of ship is an interesting area for future study. The theoretical test results presented in this paper may also be further validated with other forms of simulation or mathematical modeling.

References

1. Woud HK, Stapersma D (2012) Design of propulsion and electric power generation systems. IMarEST Publication, pp 103–130
2. Hall DT (1999) Practical marine electrical knowledge, 2nd edn. Witherby & Co Ltd., Edinburgh, UK, pp 60–61
3. Woodyard D (2004) Pounder's marine diesel engines and gas turbines, 8th edn. Elsevier, Butterworth Heinemann, Oxford, UK
4. MAN Diesel & Turbo (2018) Basic principles of ship propulsion, p 17. https://marine.mandieselturbo.com/docs/librariesprovider6/propeller-aftship/basic-principles-of-propulsion.pdf?sfvrsn=0. Last accessed on 20 Jun 2018
5. Kluj S (2011) Virtual engine room—free student version user guide. http://beta.drkluj.com/simulators/free-student-version/. Last accessed on 19 Mar 2018

6. ABS (2018) Ship energy efficiency measures—status and guidance, pp 49–50. https://ww2.eagle.org/content/dam/eagle/advisories-and-debriefs/ABS_Energy_Efficiency_Advisory.pdf. Last accessed on 20 Jun 2018
7. Swaboda R (2016) Shaft generators: propelling vessels toward leaner, greener power generation. Wartsila Techn J
8. MAN Diesel & Turbo (2018) Shaft generators for low speed main engines, p 5. https://marine.mandieselturbo.com/docs/librariesprovider6/technical-papers/shaft-generators-for-mc-and-me-engines.pdf?sfvrsn=40. Last accessed on 20 Jun 2018
9. Domonkos T (2010) Computer simulation as a tool for analyzing and optimizing real-life processes. Manage Inf Syst 5(1):13–18
10. Hui C et al (2010) The application and development of marine engine room simulator in the field of maritime training, research notes in information science (RNIS), vol 11, pp 29–39. https://pdfs.semanticscholar.org/27aa/3e073d440966d6f97e084d6089dea6746737.pdf. Accessed on 20 Jun 2018
11. Patel MR (2012) Shipboard electrical power systems. CRC Press, Florida, pp 51–52
12. Eastop TD, McConkey A (1993) Applied thermodynamics for engineering technologists, 5th edn. Pearson Education Limited, England, p 431
13. Nag PK (2014) Power plant engineering, 4th edn. Mc-Graw Hill Education (India), New Delhi, p 362 (2014)
14. IMO Homepage (2018). http://www.imo.org/en/KnowledgeCentre/IndexofIMOResolutions/Marine-Environment-Protection-Committee-%28MEPC%29/Documents/MEPC.281%2870%29.pdf. Last accessed on 24 Oct 2018
15. El-Sharkawi MA (2009) Electric energy: an introduction. 2nd edn. CRC Press, Boca Raton, Florida, p 107
16. Adnanes AK (2003) Maritime electrical installations and diesel electric propulsion. ABB Marine. http://img1.eworldship.com/2012/0913/20120913041849123.pdf. Last accessed on 16 Jan 2017

The Impact of Industrial Revolution 4.0 on Shipbuilding and Ship Repair Activities in Malaysia

Mohd Naim Awang, Effi Helmy Ariffin and Arman Ariffin

Abstract The Industrial Revolution 4.0 has been discussed and debated all around the world. Every country is preparing their strategies to address the Industrial Revolution 4.0. The impact would be more significant than previous industrial revolution. Global economics issue was impacting the marine industry. Unstable oil price creates a substantial impact on country and industries that depend on oil price. In last decade, the marine industry scaled down due to the oil crisis and the oil and gas industry also facing the same problem. In this paper, a brief preparation of existing marine industry status was discussed. A recommendation for the way ahead in Shipbuilding and Ship Repair (SBSR) to confront the revolution was proposed.

Keywords Shipbuilding and ship repair · Industrial revolution 4.0

1 Introduction

In past few years, the Industrial Revolution 4.0 or known as IR 4.0 has become the hot topic in the discussion of every domain related to the industries. In Malaysia, some initiative and organisation have been established at various level (national, state, and regional level) for both government and industry with the objective to keep the industry remain relevant in future. Therefore, significant initiatives to support and drive the identified element should be created to support and stay focus on this industrial revolution.

M. N. Awang (✉)
Politeknik Ungku Omar, Perak, Malaysia
e-mail: mnaim@puo.edu.my

E. H. Ariffin
Institute of Oceanography and Environment, Universiti Malaysia Terengganu, Kuala Nerus, Malaysia
e-mail: effihelmy@umt.edu.my

A. Ariffin
Royal Malaysian Navy, Kuala Lumpur, Malaysia
e-mail: arman.ariffin@ensta-bretagne.org

C. L. Saw et al. (eds.), *Advancement in Emerging Technologies and Engineering Applications*, Lecture Notes in Mechanical Engineering,
https://doi.org/10.1007/978-981-15-0002-2_27

The Third Industrial Master Plan (IMP3), 2006–2020 serves as an important planning instrument in expanding the sources of growth and positioning Malaysia's long-term industrial competitiveness. While the Government will centime to make the investment environment conducive, through enhancing the institutional and infrastructure support, in particular, the regulatory regime and the delivery system, domestic firms and private sectors institutions will need to be more committed, in undertaking measures to enhance productivity, efficiency and competitiveness [1]. Transport Equipment Industry is under the Chap. 13 of the IMP3; the industry comprises of three sub-sectors: 1. Automotive, 2. Marine Transport, 3. Aerospace.

In IMP2 (1996–2005), the achievement of the sub-sector is development of capabilities in marine engineering design and technology upgrading; construction of small ships and boats, including leisure craft and yachts, as well as undertaking of ship repairing activities, such s maintenance, and overhauling of vessel; and increase in exports, comprising mainly small leisure and recreational vessels, as well as tugboats and pusher craft. During IMP2, the total exports of marine transport products grew from RM882 million in 1996 to RM2.2 billion in 2005 (see Fig. 1).

The major export destination was Nigeria Singapore and Myanmar. On the export point of view, it was an increment by an average annual rate of 3.6 per cent, from RM2.3 billion in 1996 to RM3.2 billion in 2005. The significant import items were a cruise ship, ferry boats, cargo ships and barges, tugs, and pusher craft. The sources of import were Japan, Republic of Korea and the Netherlands (see Fig. 2).

Based on the presented data, six identified challenges were identified which are: 1. Increasing global and regional competition, 2. Rising costs of raw material, 3. Excess capacity, 4. A shortage of qualified personnel, 5. Inadequate technology and 6. Lack of infra-structure and support facilities. To overcome these challenges, five strategies and policies have been set for the long-term viability of the marine transport sub-sectors which are 1. Enhancing domestics capabilities in the building of

Products	1996		2000		2005		1996-2005
	Value (RM mil)	Share (%)	Value (RM mil)	Share (%)	Value (RM mil)	Share (%)	Average Annual Growth (%)
Total	**882.0**	**100.0**	**162.9**	**100.0**	**2,214.6**	**100.0**	**10.1**
Major Exports							
Production platforms	847.5	96.1	44.7	27.4	1,913.6	86.4	9.8
Yachts and other vessels for pleasure or sports	21.3	2.4	101.3	62.2	140.1	6.3	18.7
Tugs and pusher craft	3.1	0.4	9.2	5.7	125.6	5.7	33.0

Fig. 1 Exports of marine transport products. *Source* Department of Statistics

Products	1996		2000		2005		1996-2005
	Value (RM mil)	Share (%)	Value (RM mil)	Share (%)	Value (RM mil)	Share (%)	Average Annual Growth (%)
Total	**2,345.7**	**100.0**	**783.8**	**100.0**	**3,237.2**	**100.0**	**3.6**
Major Imports:							
Cruise ships, ferry boats, cargo ships and barges	1,836.4	78.2	750.0	95.6	2,207.2	68.2	3.9
Production platforms	439.5	18.7	6.5	0.8	857.2	26.5	2.4
Tugs and pusher craft	12.0	0.5	2.5	0.3	64.4	2.0	33.3
Yachts and other vessels for pleasure or sports	39.8	1.6	7.3	0.9	16.2	0.5	-6.8

Fig. 2 Imports of marine transport products. *Source* Department of Statistics

smaller vessel, ship repairing and maintenance activities; 2. Intensifying the upgrading of skills and engineering capabilities; 3. Strengthening infrastructure and support facilities; 4. Strengthening the institutional support; and 5. Expanding activities in the fabrication of offshore structures.

The Malaysian Industry-Government Group for High Technology (MIGHT) was launched on 22nd February 1993. MIGHT is a market-driven technological partnership think-tank with emphasis on market intelligence initiative using foresight practices and methodology to identify the technology and business opportunities to be explored. In the report of Malaysian Ship Building/Ship Repair Industrial Report 2015/2016, the former Prime Minister Dato' Seri Mohd Najib Tun Haji Abd Razak reminded the challenge that oil-producing country such as Malaysia once again facing the impact of a significant drop of oil price. It negatively impacted the local oil and gas as well as Shipbuilding and Ship Repair (SBSR) industries. The primary action taken by giant oil companies by cutting down their expenditure inevitably brings the market to a low peak [2].

The term "Industry 4.0" was initially coined by the German government. It describes and encapsulates a set of technological changes in manufacturing and sets out priorities of a coherent policy framework with the aim of maintaining the global competitiveness of German industry. It is conceptual in that it sets out a way of understanding an observed phenomenon and institutional in that it provides the framework for a range of policy initiatives identified and supported by government and business representatives that drive a research and development programme [3]. Since it was debated in European Parliament, other countries around the are preparing their industry to be to date and competitive with this new revolution.

2 Background of Industry Revolution 4.0 and Malaysian Shipbuilding/Ship Repair (SBSR)

The “Industry 4.0” is based on the technology. It describes the organisation of production processes based on technology and devices autonomously communicating with each other along the value chain. The Industry 4.0 concept takes account of the increased computerisation of the manufacturing industries where physical objects are seamlessly integrated into the information network. As a result, “manufacturing systems are vertically networked with business processes within factories and enterprises and horizontally connected to spatially dispersed value networks that can be managed in real time—from the moment an order is placed right through to outbound logistics.” These developments make the distinction between industry and services less relevant as digital technologies are connected with industrial products and services into hybrid products which are neither goods nor services exclusively. Indeed, both the terms ‘Internet of Things’ and ‘Internet of Services’ are considered elements of Industry 4.0.

In Europe, the Industry, Research and Energy Committees under European Parliament is preparing the EU industrial policy, shows a fuzzy picture permanently striving for balance between the three main approaches, which are regulatory; financial and distributed policy [4]. The EU commission has identified some challenges and associates must be confronted. There are five main challenges identified which are: resource efficiency, access to digitisation, access to global value chains, skill development and supportive regulation.

The past few years had become challenging years for the Shipbuilding and Ship Repair (SBSR) industry in Malaysia. It is the impact of the continuing downturn in the offshore market and low oil prices. The SBSR industry annual revenue dropped 33% from the high of RM8.35 billion in 2013 [5]. The Malaysia SBSR industry is well clustered on both Peninsular and East Malaysia.

The SBSR in Malaysia has been around more than 100 years starting from the establishment of Brooke Dockyard in Sarawak. Since then, Malaysia is capable of exporting the Malaysian made vessel to another country worldwide. Unluckily, the ability of SBSR industry was underrated and remain developing in a slow pace. IN copping with the effect of downturn oil prices and scale down of the oil industry, many giant shipbuilders making massive restructuring plan in a struggle to survive. In 2008, only nine shipbuilders were survived compared to 28 shipbuilders in Korea. Figure 3 shows the industry outlook during the low and high peak in SBSR industry around the globe.

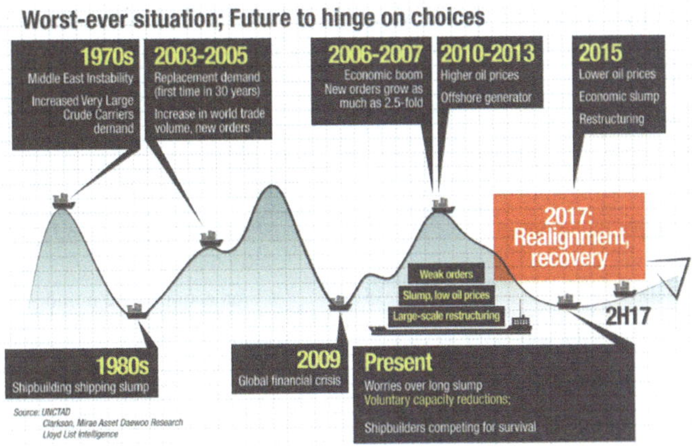

Fig. 3 The industry outlook during the low and high peak in SBSR industry around the globe

3 Industry Revolution 4.0 in Shipbuilding and Ship Repair Industry

The MIGHT as an agency under Prime Minister Office has presented their recommendation for SBSR towards the Industrial Revolution 4.0. There is nine technology trend should be addressed by each key player in SBSR to keep their organisation relevant.

The nine technology trends are 1. Big data analytic; 2. Sensors; 3. Autonomous System; 4. Robotics; 5. Advance materials; 6. Advance manufacturing; 7. Human Augmentation; 8. Clean energy and 9. Blockchain. All technology trends should be addressed in every element in SBSR which are started with Design, Engineering, Workshop setup, Fabrication and assembly, Outfitting and furnishing, Testing and last, is the Operation. In China, the Chinese Government already establish their long-term plan to be shipbuilding superpower in 2020 as shown in Table 1. This plan had registered in the State Commission of Science, Technology and Industry for National Defense [6].

4 Recommendation

With the objective of keeping the SBSR in Malaysia be relevant, authors proposed five main elements to be address related to Industry 4.0.

Table 1 Long-term plan of the Chinese Government (State Commission of Science, Technology and Industry for National Defense) (units: 10,000 dwt; per cent) [6]

	Construction quality	Share	Remarks
2004	800	14	
2005	1000	16	2005 actual performance: 12 mil dwt, 18%
2010	Shipbuilding power	25	Sharing the markets in Japan, S. Korea and China
2015	2400	35	China acquires the number one share
2020	Shipbuilding superpower		China achieves world standards in shipbuilding technology as well and becomes the world leader in both, name and reality

4.1 Development of Comprehensive Blueprint for SBSR

The comprehensive blueprint should consist of a contribution from government, shipbuilder, a local company, and user. Local universities must be a member of each level and process for determining the possible improvement could be implemented for the upgrading or solving existing issues related to SBSR domain.

4.2 LNG Fuelled Engine

Liquefied Natural Gas (LNG) surge in popularity as an alternative fuel for ships are visible these days because of its environmental friendliness. It has a very CO_2 emission. This is aligned with the IMO Tier 3 requirement related to polluted elements contained in the exhaust emitted by a vessel.

4.3 Solar and Wind Powered Ships

Development technologies to utilise renewable energy such as sun and wind. In ship propulsion development, the use of kite is under the development, and it was identified as a potential solution to optimise the clean and renewable energy.

4.4 Integrated Electric Propulsion

The system uses electric transmission instead of mechanical transmission which are bulky and massive. The use of electric propulsion will eliminate the need for clutches and reduce or eliminates the use of a gearbox. Integrated Electric Propulsion offers an excellent operational envelope and may increase the availability of the vessel.

4.5 *Shipbuilding Robotics*

Robotics as a driver of efficiency along with a method to prevent workers from doing dangerous tasks such as welding. The limited resource of high skill welder is also one of the reasons t look upon robotics for carrying out welding, blasting, painting, heavy lifting and another task in shipyards for high production rate.

5 Conclusion

Malaysian should be updated with the Industry 4.0. The move forward plan for SBSR must be taken care of every contributor in this industry. The link in every network is important to make sure each element in SBSR are well-addressed, and the mission and vision of SBSR blueprint could be achieved.

References

1. MITI (2006) Third industrial master plan, 2006–2020: Malaysia, towards global competitiveness: IMP3. Kuala Lumpur
2. Zainal K, Nor AM, Intan SWTAR, Mahfar N, Jalil DSA (2015) Malaysian shipbuilding/ship repair industry report 2015/2016
3. Smit J, Kreutzer S, Moeller C, Carlberg M (2016) Policy department a: economic and scientific policy—industry 4.0. Brussels, Belgium
4. Gyorff M (2018) Policy department a: economy and scientific policy. Brussels
5. Sulaiman MY, Mohamad Z, Rahim RA, Ani Suzila A, Mohd Jalil IH, Ishak M, Mohd Nudri MNA, Sullivan N, Mohd Emran N, Norsam NT, Mohd Razali NT, Mustafar NS, Annabel Anak Bisop P, Hasan R, Raja Intan SWTA (2017) Malaysian shipbuilding/ship repair industry report 2017/2018
6. Chryssolouris G, Makris S, Xanthakis V, Mourtzis D, Ludwig T, Tholen J (2006) Towards the internet-based supply chain management for the ship repair industry. Eur Met 17(1):1–49

BMI Monitoring System Using LabVIEW

Norhayati Mohd Zainee, Md Salim Kamil, Norazimah Mazlan and Ismila Che Ishak

Abstract Unhealthy eating habits and less physical activity had led to the growing number of obesity population and becoming a health problem in Malaysia. The objective of this study is to monitor the BMI index by using LabVIEW and to create a proper filing system. This monitoring system will use LabVIEW as the data processing and result display while Microsoft Excel will be used as the storing medium. This BMI monitoring system by using LabVIEW is user-friendly and provides real-time data storing. This system will the user height and weight based on the BMI categories and automatically stored the information such as time, date, name, age, BMI index, and category accordingly. Continuous BMI monitoring system will help the user to maintain and monitor their fitness by controlling their diet and increased the physical activity and it also can save government costs in health problems.

Keywords Body mass index · LabVIEW · Monitoring system

1 Introduction

At least 2.8 million people die each year as a result of being overweight or obese [1]. Over-weight and obesity are defined as abnormal or excessive fat accumulation [1]. It had increased the risks of coronary heart disease, ischemic stroke and type 2 diabetes mellitus [1]. A Body Mass Index (BMI) that was reinvented in 1950 can be used as a tool to calculate the weight range category according to the height [2, 3]. There are four BMI categories; underweight, normal, underweight and obese. The underweight index range is from 15 to 19.9, normal index range is from 20 to 24.9, underweight index range is from 25 to 29.9 and lastly, the obesity index is 30 and above [4].

Malaysia has been rated as the highest population of obesity among Asian countries according to WHO with the percentages of obese men and women is 43.8%

N. M. Zainee (✉) · M. S. Kamil · N. Mazlan · I. C. Ishak
Universiti Kuala Lumpur Malaysia Institute of Marine Engineering Technology (UniKL MIMET), Lumut, Malaysia
e-mail: norhayati.mz@unikl.edu.my

C. L. Saw et al. (eds.), *Advancement in Emerging Technologies and Engineering Applications*, Lecture Notes in Mechanical Engineering,
https://doi.org/10.1007/978-981-15-0002-2_28

and 48.6%, respectively [5, 6]. Overweight and obesity in Malaysia have been linked to various factors, including physical inactivity, unhealthy dietary habits, alcohol intake, socioeconomic conditions and genetic factors [5]. Based on National Health and Morbidity Surveys (NHMSs) carried out in 2006, 2011 and 2015, an increasing trend of overweight and obesity prevalence was observed among Malaysian adults aged 18 years and older [5]. Malaysian General Hospital had recorded 73% death cases from Non-communicable Diseases (NCD) involved obese patients [7]. A similar phenomenon occurs in Southeast Asia in which overweight or obese has caused 350,000 deaths and led to 2.8 million people died [8]. In term of employment, the obesity affects the performance and productivity at work, increase absenteeism and also reduce the personal income of an individual with 9–10 % lower compared to others with ideal weight [9]. In addition, obesity is not merely a personal problem, but, the government will also indirectly affected by a high number of obesity by having lack number of productive human resources and high allocation for healthcare to cater the increased needs in the medication of citizens [10].

2 Objective

Obesity leads to various negative impacts if it is not monitored and controlled. Thus, the objective of this study is to develop a BMI monitoring system by using LabVIEW. The system will use LabVIEW software to process the information and stored the result in Microsoft Excel. LabVIEW is a graphical programming language that uses icons instead of lines of text to create applications [11]. Comparing to text-based programming languages, the instructions will determine program execution [11]. The user interface in LabVIEW is known as the front panel [11]. Block diagram which contains the graphical source code will be used to control the front panel objects [11].

3 Methodology and Design

For the preliminary data recording, a random data was used. To collect the data, there are procedures need to be followed to make sure all the data are valid. The scale (height and weight) must be calibrated before conducting the measurement. BMI participant needs to remove any outdoor clothing and footwear i.e. coats, jackets, heavy outerwear, shoes or boots and socks or tights. The collected data is measured in kilogram (kg) for weight while the height is measured in centimeter (cm). After the relevant information was obtained, LabVIEW software will be used for data entry and analysis to categorize the data according to the BMI categories.

The system execution process will be based on the flowchart shown in Fig. 1. This system does not only categorize the BMI but also suggest an action plan message to the user. The system inputs are height and weight. Once the BMI index is calculated,

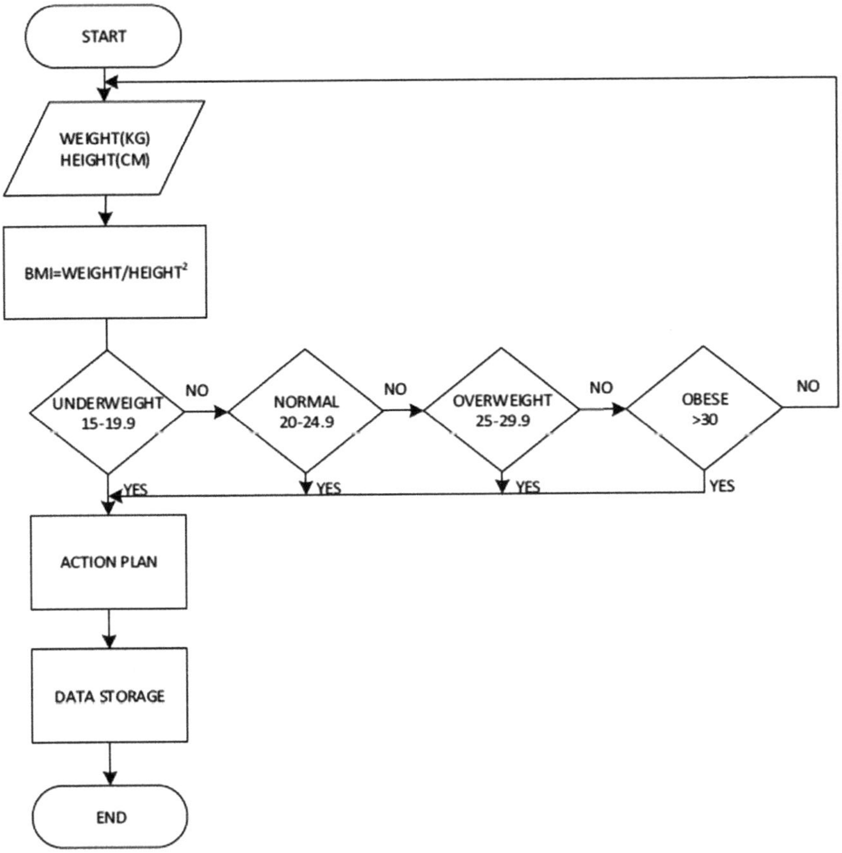

Fig. 1 The system execution flow chart

the system will sort the index based on the BMI categories. The action plan will then be displayed according to their BMI categories.

The BMI monitoring system is designed into two parts; design at the front panel and at the block diagram. The front panel and block diagram window are shown in Fig. 2. At the front panel or the user interface, the boxes are decorated by using decorations functions. The string control function used to insert the name and display the BMI results. While the numeric indicator function is used to insert the age, gender, weight, and height.

In the block diagram, the case structure is used as it can execute one sub-diagram depending on the input value passed to the structure [11]. The case structure is also similar to case statements or and if, then, else statements in text-based programming languages [11]. By using the case structure function, the four categories in the BMI can be executed once the input is being inserted. It will then display the message based on the output of the BMI formula. The formula to calculate the BMI, is weight (kg) divided by square of height (cm^2) i.e. BMI = Weight (kg)/Height^2 (cm^2). Various arithmetic functions like add and multiply were used in the block diagram to compute

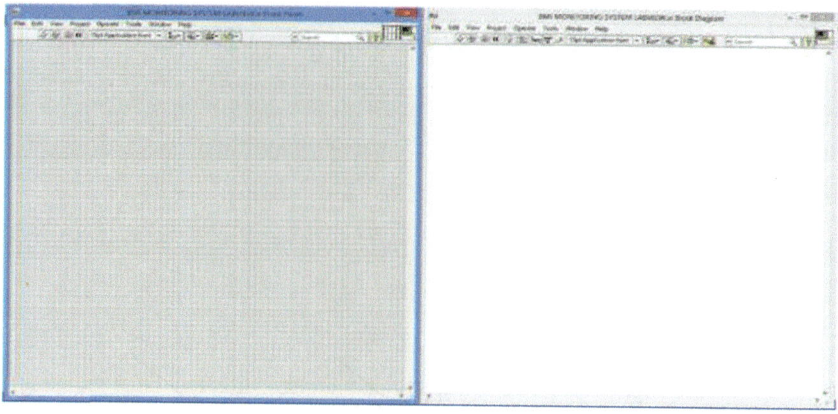

Fig. 2 Left: front panel, Right: block diagram

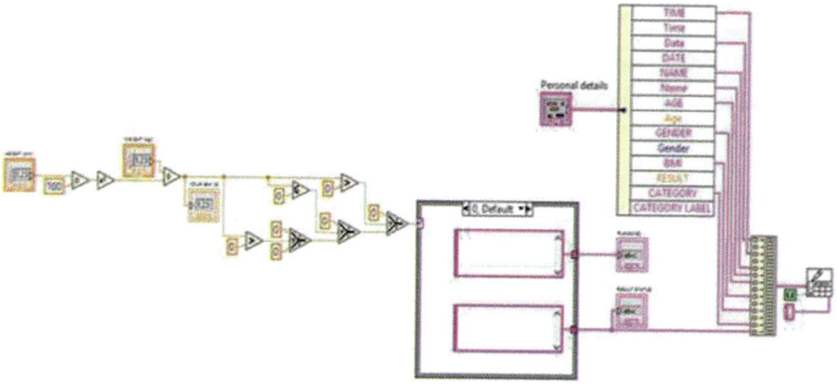

Fig. 3 The system block diagram

the equation. Then the output of the equation will be connected to the case structure and the results will be produced based on if and else statements. The information on personal details, such as name, age and weight were grouped by using bundle function. The bundle function will bundle the arrays into clusters [11]. Lastly, the results of the BMI are stored in the Microsoft Excel by using the write to spreadsheet file function in the LabVIEW.

4 Result

The icons used to develop the system are shown in Fig. 3. Figure 4 (left diagram) shows the front panel before it being simulated while the diagram in the right side shows the result of the BMI categories. Figures 5 and 6 show the user interface panel for all the BMI categories. All the information and results were stored in excel sheet as shown in Fig. 7.

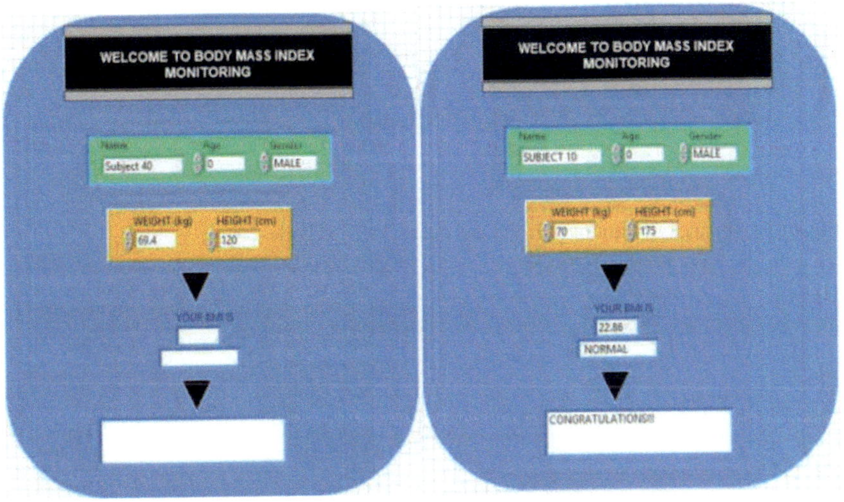

Fig. 4 Front panel view; left: before simulation, right: after simulation

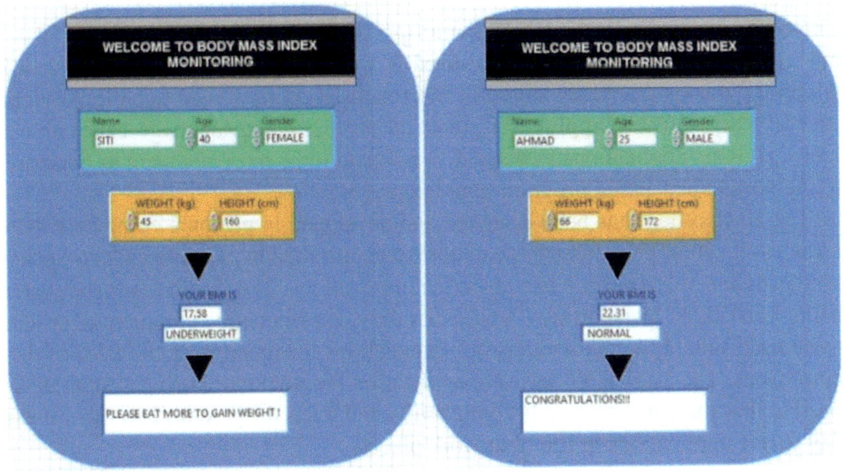

Fig. 5 Front panel view for underweight and normal BMI categories

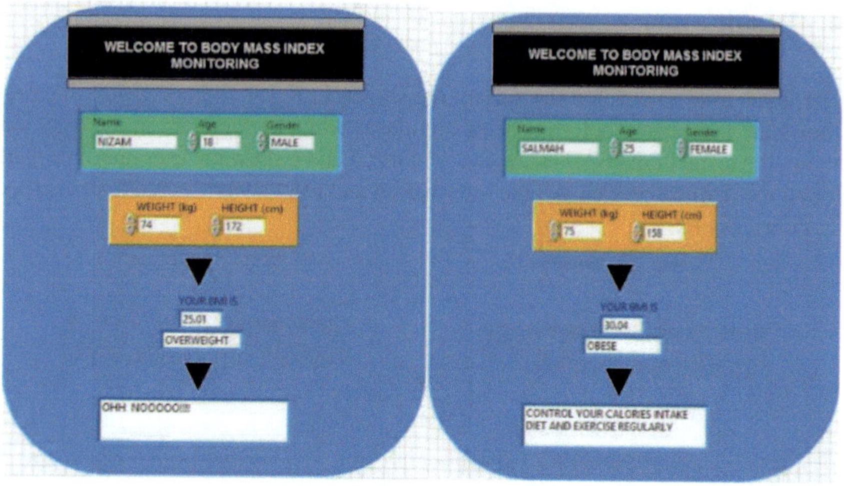

Fig. 6 Front panel view for overweight and obese BMI categories

1	DATE:	7/4/2018	TIME:	4:03 PM	NAME:	SITI	AGE:	40	GENDER:	FEMALE	BMI:	17.58	CATEGORY:	UNDERWEIGHT	
2	DATE:	7/4/2018	TIME:	4:03 PM	NAME:	AHMAD	AGE:	25	GENDER:	MALE	BMI:	22.31	CATEGORY:	NORMAL	
3	DATE:	7/4/2018	TIME:	4:04 PM	NAME:	NIZAM	AGE:	18	GENDER:	MALE	BMI:	25.01	CATEGORY:	OVERWEIGHT	
4	DATE:	7/4/2018	TIME:	4:05 PM	NAME:	SALMAH	AGE:	25	GENDER:	FEMALE	BMI:	30.04	CATEGORY:	OBESE	
5															
6															
7															
8															

Fig. 7 The data stored in excel sheet

5 Conclusion

The objective to develop a BMI monitoring system has been very successful and rewarding. The LabVIEW software used is workable to process the BMI data in which the results are easily stored in Microsoft Excel files. The LabVIEW adopts user-friendly software applying a graphical programming language executable in icons to create applications as desired. It is found to be more practical as compared to the normal text-based programming languages.

The BMI system has been designed and developed in personal computers. Nevertheless it can be conveniently adaptable to be installed in portable gadgets such as note pads and smart phones like any other applications. The viabilities of the system using LabVIEW open software would certainly contribute to a commercial product to be used suitably by individuals or organizations for monitoring the BMI of their employees. The LabVIEW software application for BMI monitoring system developed in this study has proven its applicability and it has the potential for other useful day to day applications in life or at work places.

References

1. World Health Organization (2018) Obesity: situation and trends. Global Health Observatory (GHO) data
2. Ismail MN, Chee SS, Nawawi H, Yusoff K, Lim TO, James WPT (2002) Obesity in Malaysia. Obes Rev 3(3):203–208
3. Hall DMB, Cole TJ (2006) What use is the BMI? Arch Dis Child 91(4):283–286
4. Nuttall FQ (2015) Body mass index: obesity, BMI, and health: a critical review. Nutr Today 50(3):117–128
5. Chan YY et al (2017) Physical activity and overweight/obesity among Malaysian adults: findings from the 2015 National Health and Morbidity Survey (NHMS). BMC Public Health 17(1):1–12
6. https://www.thestar.com.my/news/nation/2014/06/16/obesity-malaysia-highest-in-asia-says-pm-science-advisor (2014)
7. BERNAMA (2013) Obesiti antara punca utama kematian di hospital kerajaan, My Metro
8. World Health Organisation (2011) Jakarta call for action on noncommunicable diseases, no. March
9. Finkelstein EA, daCosta DiBonaventura M, Burgess SM, Hale BC (2010) The costs of obesity in the workplace. J Occup Environ Med 52:971–981
10. Folmann NB, Bossen KS, Willaing I, Sørensen J, Andersen JS, Ladelund S (2007) Obesity, hospital services use and costs. Adv Health Econ Health Serv Res 17:319–332
11. Literature MW (2003) LabVIEW User Manual, no. 320999

Application of Contactless Inductive Charging for Ship—A Review

Mohd Zaifulrizal Zainol, Ismail Zainol, Alif Akmal Abu Bakar, Mohd Faizal Abd Razak and Redzuan Zoolfakar

Abstract The research review the application of contactless inductive charging for marine applications. Contactless power transfer (CPT) is the method to transfer power between two separately coil that known as primary and secondary winding. The CPT system is built from main component of winding that separated from primary and secondary winding through a large air gap for battery charging application. The principle of resonance is used to increase the efficiency of power transfer through large air gap between the winding. In addition, the compensation network is applied in primary and secondary winding of the CPT transformer to reduce VA rating and increased power transfer capability of the CPT system. Next, marine electrical propulsion and energy storage system such as battery, super capacitor, flywheel and super conducting magnetic energy storage will be discussed. The advancement of electrical propulsion ships make it reliable to the ship power system in term of fuel economy, and reduced emission and at the same time meet the increasing of power demands.

Keywords Contactless power transfer (CPT) · Magnetic induction · Marine application · Battery charging · Compensation network

1 Introduction

Contactless power transfer (CPT) system practice two concepts of ampere law and faraday law to generate the operation of each magnetic induction. The electric current flow through conductor with an energy proportional to the current will produce magnetic field which are apply ampere law concept. Besides that, when two separately winding near each other, magnetic field will induce an electromotive force (EMF) in a conductor that is proportional to the magnetic field's strength and its rate of change. There is known as faraday law. Figure 1 illustrates how those legal guidelines may be implemented together to transfer power wirelessly.

M. Z. Zainol (✉) · I. Zainol · A. A. A. Bakar · M. F. A. Razak · R. Zoolfakar
Universiti Kuala Lumpur Malaysian Institute of Marine Engineering Technology, Jalan Pantai Remis, 32200 Lumut, Perak, Malaysia
e-mail: mzaifulrizal@unikl.edu.my

C. L. Saw et al. (eds.), *Advancement in Emerging Technologies and Engineering Applications*, Lecture Notes in Mechanical Engineering,
https://doi.org/10.1007/978-981-15-0002-2_29

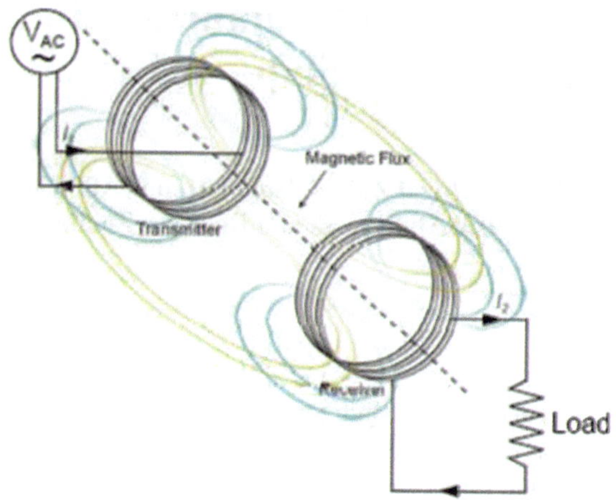

Fig. 1 Principle of operation an Inductive Power Transfer (IPT) system

Contactless power transfer is occurred from the primary coil to the secondary coil. A source direct current will supply and convert to alternating current is through to primary coil and produce a magnetic field. When secondary coil is place in close with primary coil, magnetic field between two coils will result in an electromotive force in secondary coil and current will flow through the load. Technology advancement in the field of CPT, has made many changes in recent years. Among them are electric vehicles, medical and mobile phones. With CPT technology it has a positive impact to further develop by making it more reliable, easy, powerful and comprehensive. In the marine applications, CPT contributes substantially to reducing fuel consumption and reduce emissions. Additionally, for security from physical contact between docking time for charging the batteries. It completely gives a better usage and completely automated operations. Extension and deduction of charging equipment is no longer needed. With CPT, time can be saved for transferring electricity. Finally, at some point of contactless inductive charging will be able to be applied in battery powered ship. The energy can be transfer wirelessly for a few range with a few efficiency using specific design include variety of electronic component that need to be design cautiously. The design of those components is depending on the range of load and source and amount of power to be transferred.

2 Inductive Power Transfer IPT System Description

In Fig. 2, a normal schematic of an Inductive power transfer system is proven. The DC-DC level is highlighted in the figure. For battery charging applications, the electrical power flows from the DC-link to the battery.

Figure 2 shown normal schematic of the proposed IPT system. The inductive power transfer happens among magnetically coupled coils. Their self-inductances are L1 and L2, the mutual inductance is M. L1 and L2 correspond respectively to

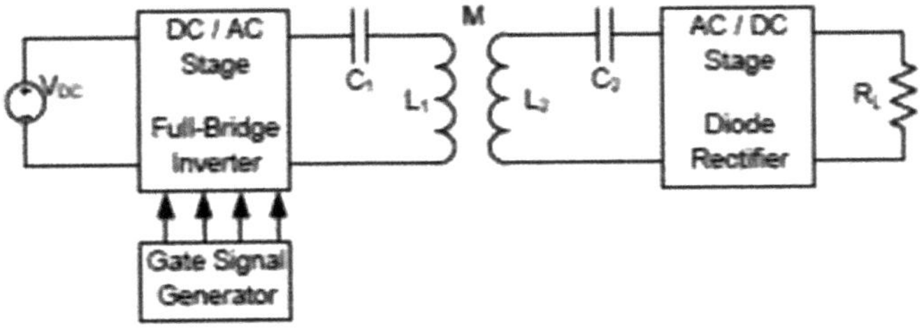

Fig. 2 Normal schematic of the proposed IPT System

the primary and the secondary coil. The primary-side DC voltage supply is attached to the electrical grid, the secondary-side DC section is the load representing the battery to be charged. Because the power transfer among the coupled coils is in AC, intermediate levels are needed, a DC-AC inside the primary side and an AC-DC in the secondary side. Since coils are loosely coupled, a reactive network is wanted to be able to maximize the power transfer efficiency and to optimize the power factor, if the system works on the resonance. This reactive network is known as compensation circuit and consists of capacitors, one for each side. In the instance of the figure, each the compensation capacitors C1 and C2 are connected in series with the primary and the secondary coils. The principle parts of the IPT systems can be defined in the following.

3 Compensation Topology

A reactive network is needed which will maximize the power transfer efficiency closer to the load and the power factor towards the source. Since reactive elements needing to be compensated are the coupled inductors, the compensation elements are capacitors. Referring Fig. 2, the capacitor connected to the primary coil is the primary capacitor C1, while the capacitor connected to the secondary coil is the secondary capacitor C2. In step with the type of connection among the coils and their compensation capacitors, four extraordinary compensation topologies are feasible: series-series (SS), series-parallel (SP), parallel-series (PS) and parallel-parallel (PP). For every of those four solutions, the first word (letter) refers to the primary side, while the second word (letter) refers back to the secondary side, Selecting a specific topology in preference to another one depends on the specific application. The SS preference allows to select the compensation capacitances depending only on the self-inductances, regardless of what the load and the magnetic coupling are. Consequently, in case of misalignments among the coil, the system maintains working under resonance regardless of the mutual inductance variations. For these reasons, the topology seems to be convenient for battery charging. Certainly, in IPT charging of vehicles, the correct alignment among the coils simply constitute an ideal case and even small misalignments are unavoidable.

4 Power Converter

Two power converters are required, the primary and the secondary. In the primary side a double stage is normally hired, because of the cascade of an AC-DC and a DC-AC (this DC-AC is the one highlighted in Fig. 2). The purpose of this double stage is to increase the power frequency from 50 Hz (or 60 Hz) of the grid to tens of kHz of the IPT. In the secondary side an AC-DC stage is wanted to be able to supply the DC battery. Because the AC-DC stage within the primary side is quite standard, the scientific research generally focuses on the stage among the primary DC section and the secondary DC section. Different solutions have been investigated, regarding the primary DC-AC stage and the secondary AC-DC stage. As far as the secondary side is involved, two opportunities have been especially exploited to attach the secondary AC side to the DC battery: either a passive rectifier plus a DC-DC converter or an active AC-DC stage.

5 Inductive Coupling

Both primary and secondary sections can be selected coil shapes. Circular, squared or elliptical coils as shown in Fig. 3. It is form of isolated Litz wire that has very small surface and closeness effect within the frequency range considered [1]. Primary and secondary coil must be designed follow suitable parameter coil, to achieve the comprehensive performance of the system. Mutual inductance M can be enhanced by using the shield. At the same time, it will increase the magnetic flux among the coils. A shield use non-conducting magnetic material and sandwiched primary and secondary coil between two shielding layers as shown in Fig. 4a [2]. Even for the lowest cost materials, ferrites are usually used because they hardly lose the frequency up to several hundred kHz. The ferrites also prevent other parts getting heating because of inductive carrier when run closeness. Moreover, induction create by magnetic circuit that help in enhancing coupling between two coils [3, 4].

To reduce leakage flux, and to act as an additional shield, some designers add other materials (such as Aluminum) as in Fig. 4b which includes ferrite. Because of aluminum resistance, it can produce additional losses at high frequencies and

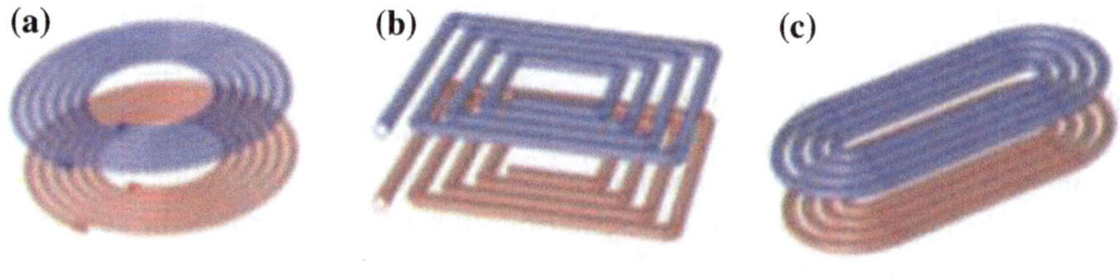

Fig. 3 Coils shapes: **a** circular, **b** square and **c** elliptical

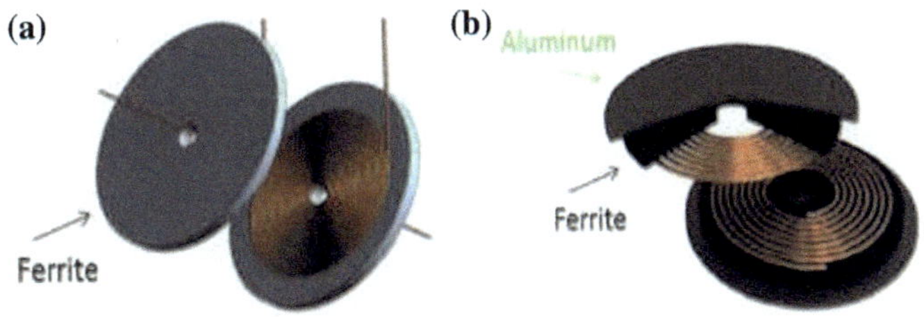

Fig. 4 Planar coils with shielding; **a** 1-layer, **b** 2 layers

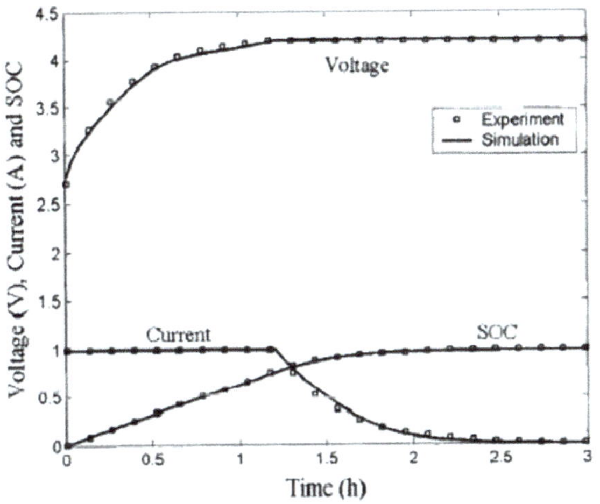

Fig. 5 Typical charging characteristics of a Li-ion cell [6]

ferrite will increases weight in EV [4, 5]. Nonetheless, additional shielding can be considered due to people or devices inside the car if in a real application (Fig. 5).

6 Rechargeable Battery

The most widely used battery typology is lithium ion. This is due to the high energy density and long-life span. Among the applications that suit lithium ion batteries are portable electronics, electric vehicles, spaces and aircraft power systems and stationary power storage. Lithium iron phosphate (LifePO4) is a growing choice in the market, especially for high power consumption. Battery LifePO4 has a high capacity and good stability against chemical and thermal characteristics. Detailed battery models are often needed to optimize battery usage. A complete dynamic model of Li-ion batteries, including nonlinear equilibrium potentials, temperature-dependencies, thermal effects and response to transient power demand, is proposed.

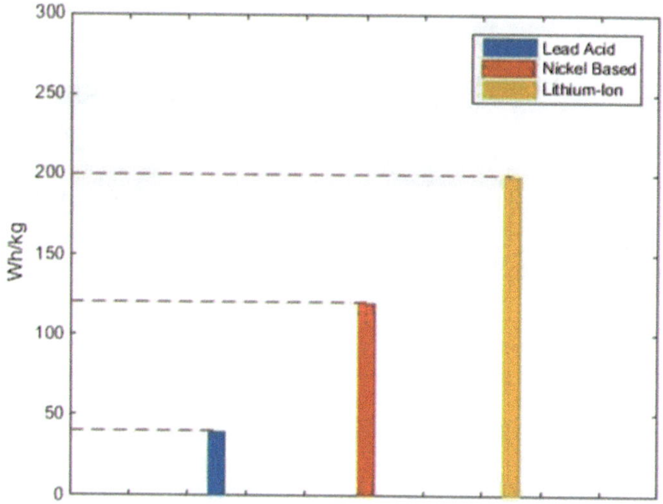

Fig. 6 Energy of battery systems

Experimental tests have been carried out on lithium ion to optimize battery runtime. With particular consideration of the effects of non-linear capacities in, the proposed dynamic LiFePO4 battery model [6–8]. In addition, accurate knowledge of the current state of the batch (SOC) is required for the proper battery charging system. Estimation, according to the nonlinear relationship between the state of charge and the open-circuit voltage of the battery, in [9, 10], the propose an algorithm for the SOC estimation. Shown in Fig. 6 [6], is a typical charge profile of a Li-ion battery cell. Lithium-ion have two charging modes that known as constant current mode and constant voltage mode. During the constant current mode, the current is maintain at a reference value, while the voltage increases to a maximum value. Then, the constant voltage mode starts, and the voltage is maintain at the maximum value, while the current decrease to zero.

Traditional batteries usually used on ships such as lead acids and nickel-based batteries. Nowadays, lithium-ion give more efficiency on shipboard because have eight times more energy per weight unit, as shown in Fig. 6. Normally, to design a battery system for maritime applications it analyses in term of safe and secure operation, combination of high energy density and the usage of flammable electrolytes. Where it is as important to secure the quality of the total integrated system as it is for each individual component.

7 Marine Propulsion System

In the 19th century almost all ships use water as a working liquid for steam engines. External combustion engines work in the Rankine cycle. Initially with a recurrent piston and the first in 1897 with a turbine steamer. It is used in several very large vessels, and in nuclear submarines. The name of the steam vessel is often made with SS. Another that, in the 1930s, most ships used diesel engines. Internal combustion engine (ICE), working in the Diesel cycle, using marine diesel or heavy fuel oil. Also known as motor boats. At present there is no limit on practice, diesel engines have been in power for decades. The name of a motor boat is often preceded by MS. Furthermore, gas turbines come from an aviation turbine that can burn marine diesel, kerosene or jet fuel, used in several fast ships (hydrofoils), warships (for fast action), and large cruisers. The internal combustion engine (ICE), which works in the Brayton cycle. To reduce fuel costs, diesel engines have been added and became the first vessel with diesel-electric and gas (CODAG). Finally, dual fuel engines, using liquefied natural gas (LNG) as the main fuel. Internal combustion engines that function in diesel cycles such as LNG engines. Sometimes work in dual-fuel mode with partial marine-diesel injection (Table 1).

8 Marine Hybrid Propulsion System

Systems which have multiple type of power source for propellers or systems that may operate the propeller electrically or mechanically depending at the speed and power required are known as marine hybrid propulsion systems. To harness the best power source with the aid of optimizing the propulsion performance and operating factor of power in their area of greatest efficiency it focuses on the hybrid propulsion system. A few of the power sources that can be used are diesel engines, gas turbines, steam turbines, capacitors, batteries, fuel cells, and so on. Since various power sources can be used, hybrid propulsion systems are more relevant than conventional schemes. For example, depending on operating loads that produce a greater economical and reliable system, the propeller may be driven mechanically or electrically. More redundancy and less noise on hybrid system. The growth of energy management plans to optimally optimize hybrid propulsion systems and in some cases it involves an extra area or weight. Due to that, hybrid propulsion systems can be greater complex for the same purpose for multiple power sources. Since diesel engines dominate the marine propulsion system, an overview of the hybrid diesel engine propulsion system was first introduced.

Table 1 Analysis type of Battery powered ship

No.	Title	Type of Ship	Type of battery	Power	Voltage	Air-gap	Capacity
1	Wireless Charging for Ships: High-Power Inductive Charging for Battery Electric and Plug-In Hybrid Vessels (2017)	Norled MF Ampere	Lithium-ion	1 MW above	690-Vac or 1000-Vdc	20 and 50 cm	–
2	Feasibility Study of Integrated Power System with Battery Energy Storage System for Naval Ships (2012)	Korean new naval destroyer class (KDDX)	Lithium-ion	8 MW	6.6 kV/450 V	–	2 MWh
3	Stability Improvement of DC Power Systems in an All-Electric Ship Using Hybrid SMES/Battery (2018)	AES	Lithium-ion	14 MW	6.6 kV	–	13.88 kWh
4	Battery-Powered Ships (2015)	Car ferry, the Ampere	Lithium-ion	–	–	–	1–4.2 MWh

8.1 Hybrid Diesel Propulsion Systems

Different architectures and different modes can be used in hybrid propulsion systems by diesel engines (DE). DE using the Combined Diesel Electric or Diesel (CODLOD) or using both in the Combined Diesel Electric and Diesel (CODLAD) so that electrical or mechanical power can drive the propeller. As shown in Fig. 7. The ships of anchor handling suppliers have used this system, to minimize the use of fuel and pollution [11].

Table 2 shows variety of power sources such as steam turbines, gas turbines, and gasoline and sterling engines [12]. As can see, diesel engines offer benefits such as inexpensive heavy fuel oil, lower fuel consumption and higher reliability, higher efficiency, system power density and lifetime. Consequently, diesel engines play an important role in hybrid propulsion systems as easily as conventional propulsion systems. More than 99% of large commercial vessels use the diesel engine for propulsion system [13].

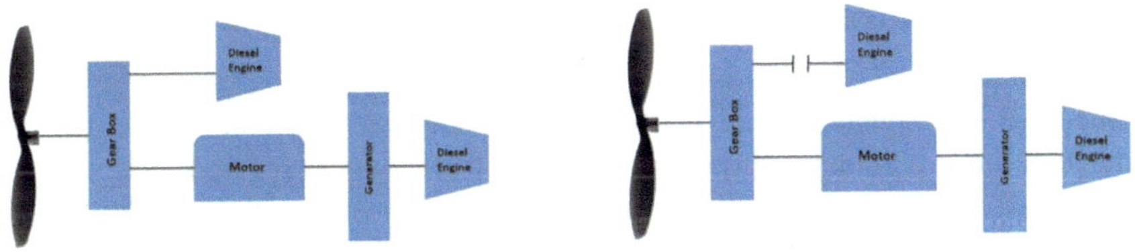

Fig. 7 Different hybrid diesel propulsion systems CODLAD versus CODLOD

Table 2 Comparison of different power source [14]

	Diesel engine	Turbine generator	Photo voltaic	Wind turbines	Fuel cells
Efficiency (%)	35 48–51	29–42	6–19	25	40–60 25–70
Capital cost ($/kW)	200–350 500	450–870	6600	1000	50–10,000
Operation & ($/kW) Maintenance cost	0.005–0.015	0.005–0.0065	0.001–0.004	0.01	0.0019–0.0153
Lifetime (years)	30	20–35	20–30	20	0.5–5 5–10

8.2 Hybrid Fuel Cell Systems

Combined fuel cells and energy storage systems contribute to higher efficiency and reduce CO_2 emissions [15]. Hybrid fuel cell propulsion systems have been applied successfully in several applications such as the automotive industry. Which leads to fuel consumption, higher efficiency and low fuel emissions [16]. It is a combination of the high energy density of fuel cells and the high power density of storage systems. The electricity generated is from fuel cell systems such as methanol or hydrogen depending on fuel cell type. Generated electricity is used in propulsion and used to charge the energy storage system which can be a battery or a capacitor. Hybridization and control methods used [17] which are considered to be the main focus of this research to provide the system's efficiency and behavior. There are two main categories of hybrid fuel cell propulsion system; series hybrid and parallel hybrid as shown in Fig. 8.

9 Energy Storage Devices

Due to its slow dynamics, fuel cells have a time-delayed reaction. In order to meet dynamic and rapid changes in power requirements, energy storage devices are usually combined with fuel cells [18]. Chemical, electrical, heat or kinetic media [19], is energy storage, including energy, technology and scale and can be stored in different forms. For marine applications, super-capacitors, hydrogen, batteries and flywheels are being considered [20]. Fuel cell output is electricity. Fuel cell power plants used energy storage such as electrochemical capacitors and batteries, often referred as super-capacitors or ultra-capacitor. As shown in Fig. 9 [21], fuel cells have lower dynamically but batteries and capacitor are higher dynamically. By the way, batteries have slower charging rates and response times, slower discharge cycles and lower power density, compare batteries to electrochemical capacitors. However, in marine applications the important is batteries have higher power range and more energy density. That is why for fuel cell applications, batteries will considered to be the main energy storage [22].Chemical processes happen at the electrode/electrolyte interface's boundaries when ions and electrons are sent separately, also both include

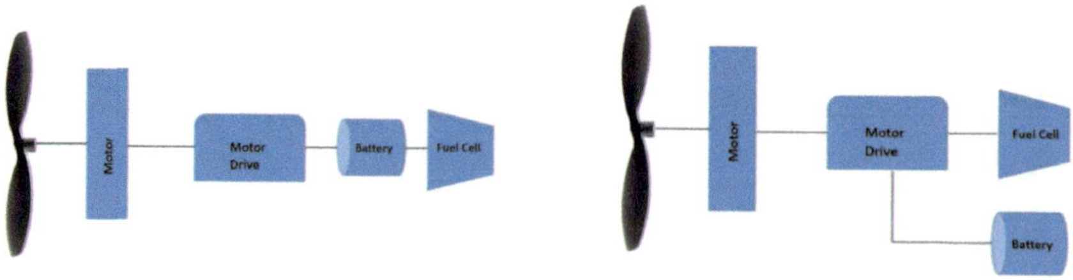

Fig. 8 Hybrid fuel cell propulsion systems block diagram Series versus Parallel

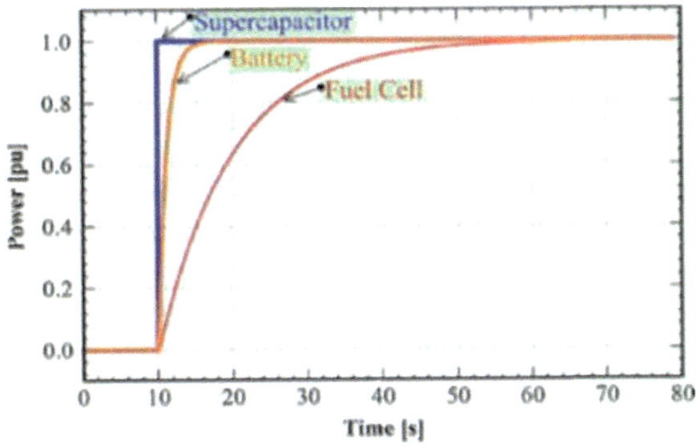

Fig. 9 Dynamic classification of fuel cells, batteries, and capacitors

of electrodes in contact with electrolyte. That is electrochemical similarities in fuel cells and batteries. However, the system of batteries and fuel cells are differences. Batteries are closed systems where conversion and energy storage take place in the same cell. Another that, advantage for fuel cells as it doesn't need to be recharged because an open system where the fuel and oxidant come from outside the cell. It will produce electricity, as long as fuel and oxidant are supplied. As shown in Table 2, is the comparison between fuel cells and batteries.

The use of batteries combined with fuel cells in transportation applications has been investigate. That is showing increase in fuel efficiency and consequently reduction in emissions such as trams [18], locomotives [24], vehicles [25], trucks [26] and buses [27]. Greenhouses gas reduction percentage will depend on the cycle efficiency of the batteries in hybrid applications [28]. Because of space limitation and weight of naval applications, there are different types of batteries for lighter and higher energy density types have to be identified. Differences among battery types shown in Table 3.

Table 3 Differences between fuel cells and batteries [23]

	Fuel cells	Batteries
Efficiency (%)	25–70	70–90
System specific power (W/kg)	100–500	1–1000
System power density (W/l)	10–600	200–300
Lifetime of electrochemical stack (year)	0.5–5	0.1–10
Capital cost ($)	50–10,000/kW	10–1000/kWh
Operating and maintenance cost ($/kWh)	0.1–1	≈0

10 Potential Energy Storages for Shipboard Power System

The choice of proper energy storage has become very important to minimize size, weight, and improve the efficiency of the ship's power system. Ship power systems play an important role in determining the type of energy storage, load profile and generation profile. They have two modes energy storage of operation for ship power system. During normal operation, energy storage is required to maintain a load balance, and while their transient mode reduces the effect of load variations. To achieve efficiently, cannot perform these tasks in a single type of energy storage. Energy storages with high energy density are required for energy balance. Another that, high power density energy storages are required for mitigating the load variation. Battery can be used for energy storages in normal operation due to their high energy density characteristic. Super capacitor, FESS, and SMES can be used for energy storage in transient operation due to their high power density characteristic [29]. AES have a limited size, they will be big issue of size and weights of the energy storages. Major characteristics of the energy storages for example energy storage capacity, specific energy, energy density, power density, life time, round-trip efficiency, capacity degradation, complexity, cost, and safety issue are required to be considered, in order to feed the shipboard power system with the energy storages. Shipboard power system application is given with description about the potential energy storages.

10.1 Battery

For batteries, chemical reactions arise in cells in which two electrodes are set to electrolytes. A battery is an electrochemical device that convert chemical energy immediately into electrical energy via oxidation-reduction reactions [31]. It's far the most familiar technology to store electrical energy. Load create a reaction while transfer of electrons from one electrode to the other. To get the required voltage and current, battery modules are connected in series or in parallel. High energy density, high round trip efficiency, cycling capability is some of benefits of the battery. Initial cost of installation is low, and the life span of the battery is also reliable. The issues with battery because of deep discharging and overcharging that reduce the lifetime of the battery [32]. Lead-acid battery have disadvantages comparing to lithium-ion battery that shows several superior capabilities in case of power density, energy density, life time and round-trip efficiency. The comparison of lead acid and lithium ion battery is shown in Table 4 in terms of different characteristics.

Table 4 Differences of battery types [30]

Type	Wh/kg	Wh/l	Durability (years)	Cost ($/kWh)
Lead-acid	30–50	50–80	5–15	200–400
Nickel-cadmium	50–75	60–150	5–20	800–1500
Nickel-metal hydride	30–110	140–435	3–15	350–500
Lithium ion	75–250	200–600	5–20	600–2500
Sodium-sulphur	150–240	150–240	15	300–500
ZEBRA (Sodium-nickel chloride)	100–140	150–280	8–14	100–200
Zinc-bromine	60–85	30–60	5–20	150–1000

10.2 Supercapacitor

Capacitor can process several thousand charging/discharging cycles without material degradation as compared to batteries. Its additionally can be direct storage of electricity and deliver electricity at a totally fast rate. With the characteristics fast charging and discharging rate made it appropriate for transient voltage stability applications [33]. Parallel plates separated by a dielectric material can store capacitor energy, both of electrostatic and electrolytic. Dielectric is used for the separation of the plates for electrostatic capacitor. Another that, liquid electrolyte (aluminum oxide) is used for electrolytic. Instead of dielectric, electrolyte solution (propylene carbonate) is used among plates for electrochemical double-layer capacitor (super capacitor) [34, 35]. Super capacitor have higher energy and power density than the conventional capacitor but still energy density is lower than the battery [36].

10.3 Superconducting Magnetic Energy Storage (SMES)

SMES shows the advantageous characteristics like super capacitor (high power density, high energy storage efficiency, long application lifetime and few environmental pollution). To store electrical energy, SMES uses superconducting coil within the form of magnetic power with the aid of developing magnetic field with the flow of direct current (DC). SMES system have three important parts efficiency, long application lifetime and few environmental pollution). To store electrical energy, SMES uses superconducting coil within the form of magnetic power with the aid of developing magnetic field with the flow of direct current (DC). SMES system have three important parts [37]. Though it is advantageous like a super capacitor. Superconducting easy to get the higher temperature, the coil must be placed in liquid helium to get the temperature below 260 °C [38]. Besides that, additional equipment needed for cooling. It is very costly to maintain this kind of very low temperature and has significant auxiliary equipment. It is still not suitable for use in ship power systems as it has no resistance to unpredictable environment.

10.4 Flywheel Energy Storage (FES)

A FES devices have five essential components. That is flywheel, bearings, electrical rotating machine, power electronics converter, and a vacuum chamber [39]. Flywheel is an electromechanical energy storage device. The energy stores inside the rotation of a cylindrical rotor. Flywheel is connected to the rotating machine. The rotating machine have function which could work as a generator or motor depending on the power demand. When rotating machine works as a motor, electrical energy is used to accelerate the flywheel and it stores energy in the form of kinetic energy. That is call charging process. When rotating machine works as a generator, the flywheel decelerates and delivers electrical energy to the connected load. That is call discharging process. The energy storing capability of flywheel energy storage system relies upon at the rotating speed of flywheel and its inertia. Figure 14 shows the simple schematic diagram of a flywheel energy storage system [38]. FES have an advantages for using power system application as it has high cycle life and power density. Besides that, it can meet high power demand for only a short time. The disadvantages of FES are low energy density, high self-discharge and additional auxiliary equipment. It cannot be used as standalone backup power because of low energy density. It has to be used with different high energy density storage device like a battery. Another that, not suitable for the usage of longer storage time due to high self-discharging. It is able to store energy for a very short time. Moreover, FES in the shipboard power system is that it has large auxiliary equipment. For reduction energy loss because of air resistance, the flywheel had to installed high vacuum environment.

11 Conclusion

This work was devoted to have preliminary review on the possibility of contactless power technology for the ship. Next, marine electrical propulsion and energy storage system which is related to this technology had been discussed. Further simulation works could be carried out to study the reliability of this system for the ship power system.

References

1. Budhia M, Covic GA, Boys JT (2011) Design and optimization of circular magnetic structures for lumped inductive power transfer systems. IEEE Trans Power Electron 26(11):3096–3108
2. Valtchev S, Borges B, Brandisky K, Ben Klaassens J (2009) Resonant contactless energy transfer with improved efficiency. IEEE Trans Power Electron 24(3):685–699
3. Su YP, Liu X, Hui SYR (2009) Mutual inductance calculation of movable planar coils on parallel surfaces. IEEE Trans Power Electron 24(4):1115–1123
4. Acero J, Carretero C, Lope I, Alonso R, Lucia Ó, Burdio JM (2013) Analysis of the mutual inductance of planar-lumped inductive power transfer systems. IEEE Trans Ind Electron 60(1):410–420
5. Onar OC, Miller JM, Campbell SL, Coomer C, White CP, Seiber LE (2013) A novel wireless power transfer for in-motion EV/PHEV charging. In: Conf. Proc. - IEEE Applied Power Electronics Conference and Exposition—APEC, pp 3073–3080
6. Gao ADL, Liu S (2002) Dynamic lithium-ion battery model for system simulation. IEEE Trans Components Packag Technol 25(3):495–505
7. Chen M, Rincón-Mora GA (2006) Accurate electrical battery model capable of predicting runtime and I-V performance. IEEE Trans Energy Convers 21(2):504–511
8. Yao LW, Aziz JA (2012) High capacity LiFePo 4 battery model with consideration of nonlinear capacity effects. In: Proceedings of The 7th International Power Electronics and Motion Control Conference—ECCE Asia, IPEMC 2012, vol 1, pp 182–187
9. Kamal E, Hajjaji AE, Mabwe AM (2015) State of charge estimation based on extened Kalman Filter algorithm for lithium-ion battery. In: 2015 23rd Mediterranean conference on control and automation, pp 734–739
10. Zhu Z, Sun J, Liu D (2012) Online state of charge EKF estimation for LiFePO4 battery management systems. In: ISPACS 2012—IEEE International Symposium on Intelligent Signal Processing and Communications Systems, pp 609–614
11. Xiao N, Zhou R, Lin X (2016) Type selection and design of hybrid propulsion system of ship. IOP Conf Ser Mater Sci Eng 157:12034
12. Benvenuto G, Campora U (2007) Performance prediction of a faulty Marine Diesel Engine under different governor settings. In: 2nd International conference on marine research and transportation, pp 35–44
13. Deepan G (2013) Experimental investigation of performance and emission characteristics in a variable compression ratio diesel engine using partly cold exhaust gas recirculation (EGR) method. In: 2013 International Conference on Energy Efficient Technologies for Sustainability, ICEETS 2013, pp 727–739
14. Kirubakaran A, Jain S, Nema RK (2011) DSP-controlled power electronic interface for fuel-cell-based distributed generation. IEEE Trans Power Electron 26(12):3853–3864
15. Kamiya Y, Sumida Y, Daisho Y (2014) Analysis of adverse effects on vehicle performance due to battery deterioration installed in BEV and HEV. Energy Power Eng 8(1):183–189
16. Saadi R. et al., “Energy management of fuel cell/supercapacitor hybrid power sources based on the flatness control. In: International conference on power engineering, energy and electrical drives, pp 128–133
17. Cho J et al (2011) Analysis of transient response of a unit proton-exchange membrane fuel cell with a degraded gas diffusion layer. Int J Hydrogen Energy 36(10):6090–6098
18. Garcia P, Fernandez LM, Garcia CA, Jurado F (2009) Fuel cell-battery hybrid system for transport applications. In: Proc - 12th International Conference on Electrical Machines and Systems ICEMS 2009
19. Vazquez S, Lukic S, Galvan E, Franquelo L, Carrasco J, Leon J (2011) Recent advances on energy storage systems. In: IECON 2011-37th Annual Conference of the IEEE Industrial Electronics Society, pp 4636–4640
20. Naish C, McCubbin I, Edberg O, Harfoot M (2007) Outlook of energy storage technologies. In: Eur. Parliam. - Policy Dep. - Econ. Sci. Policy, January 2004, pp 1–57

21. Thounthong P, Sethakul P, Raël S, Davat B (2009) Performance evaluation of fuel cell/battery/supercapacitor hybrid power source for vehicle applications. IEEE Ind Appl Soc Annu Meet IAS 2009:1–8
22. Scherson DA, Palencsár A (2006) Batteries and electrochemical capacitors. Electrochem Soc Interface
23. Shahrukh Shamim JA, Sudhakar K, Choudhary B (2016) A review on recent advances in proton exchange membrane fuel cells : materials, technology and applications. Adv Appl Sci Res 6:89–100
24. Final Technical Report U.S. Department of Energy Office of Energy Efficiency and Renewable Energy Golden Field Office Hydrogen, Fuel Cells, and Infrastructure Technologies Program (2007)
25. Burke A (2007) Batteries and ultracapacitors for electric, hybrid, and fuel cell vehicles. Proc IEEE 95(4):806–820
26. Tazelaar E, Shen Y, Veenhuizen PA, Hofman T, van den Bosch PPJ (2012) Sizing stack and battery of a fuel cell hybrid distribution truck. In: Oil Gas Sci Technol – Rev d'IFP Energies Nouv, vol 67, no 4, pp 563–573
27. Zaetta BMR (2011) Hydrogen fuel cell bus technology state of the art review, no. 245133
28. Samaras C, Hauffe R (2009) Impact of battery weight and charging patterns on the economic and environmental benefits of plug-in hybrid vehicles, vol 7
29. Nourai A (2002) Large-scale electricity storage technologies for energy management. In: Power engineering society summer meeting, 2002 IEEE, p 6
30. Mccluer S, Christin J (2011) Comparing data center batteries, flywheels, and ultracapacitors. In: Schneider Electr – Data Cent Sci Center - White Pap, 65 Rev 2, pp 1–16
31. Kloess M (2012) Electric storage technologies for the future power system—an economic assessment. In: 2012 9th international conference on the European energy market, pp 1–8
32. Albright G, Edie J, Al-Hallaj S (202) A comparison of lead acid to lithium-ion in stationary storage applications. AllCell Technol. LLC, March, pp 1–14
33. Denholm P, Ela E, Kirby B, Milligan M (2010) The role of energy storage with renewable electricity generation the role of energy storage with renewable electricity generation. In: Contract, vol NREL/, no January, pp 1–53
34. Bizuayehu AW, Medina P, Catalão JPS, Rodrigues EMG, Contreras J (2014) Analysis of electrical energy storage technologies' state-of-the-art and applications on islanded grid systems
35. Vazquez S, Lukic SM, Galvan E, Franquelo LG, Carrasco JM (2010) Energy storage systems for transport and grid applications. IEEE Trans Ind Electron 57(12):3881–3895
36. Whittingham BMS (2012) History, evolution, and future status of energy storage, vol 100
37. Chen L, Liu Y, Arsoy AB, Ribeiro PF, Steurer M, Iravani MR (2006) Detailed modeling of superconducting magnetic energy storage (SMES) system. IEEE Trans Power Deliv 21(2):699–710
38. IEC (2011) Electrical energy storage—White Paper. In: Int Electrotech Comm, pp 1–78
39. Amiryar M, Pullen K (2017) A review of flywheel energy storage system technologies and their applications. Appl Sci 7(3):286

Ro-Ro Port Facilities Toward Customer Satisfaction: Evidence from Kuala Perlis Terminal, Perlis, Malaysia

Amayrol Zakaria, Aminuddin Md Arof, Ismila Che Ishak and Abdul Qayum Mukti

Abstract Door-to-door logistic chain is structured affording to the principle of short sea service. Despite to encourage the use of the environmental friendly mode of transport, the formation of this principle is to enhance the facilities quality in order to capture customer value. Therefore, facilities quality towards customer satisfaction already become superficial. Objective of this study to examine the current status of customer satisfaction on facilities quality of Ro-Ro operations in Kuala Perlis Terminal and to determine the affiliation of port facilities quality and customer satisfactions in Kuala Perlis Terminal. Data has been collected from 150 respondents from the Ro-Ro service customers in Kuala Perlis, Malaysia. The data were analyzed by SPSS statistical method analysis. Based on this approach, the link between facilities quality and customer satisfaction has been investigated. The results indicate that facilities quality influences customer satisfaction. Results moreover indicate that the more facilities quality is provided further customer satisfaction will be achieved.

Keywords Door-to-door · Ro-Ro · Customer satisfaction

1 Introduction

1.1 Ro-Ro Operation

Roll on/roll off (Ro-Ro) vessels are the most suitable to deal with Short Sea Shipping (SSS) services since they do not incur in long times in the terminals, because the freight can be quickly loaded and unloaded into the Ro-Ro vessel [1]. In order to prevent the unnecessary activities occur in Ro-Ro operations, some variables such as inadequate number of terminal gates and customs control units, number of vehicles arrived to a terminal, ship capacity, distance between terminals, and bunkering should be considered before operate the system [2].

A. Zakaria (✉) · A. M. Arof · I. C. Ishak · A. Q. Mukti
Universiti Kuala Lumpur Malaysian Institute of Marine Engineering Technology, Kuala Lumpur, Malaysia
e-mail: amayrol@unikl.edu.my

C. L. Saw et al. (eds.), *Advancement in Emerging Technologies and Engineering Applications*, Lecture Notes in Mechanical Engineering,
https://doi.org/10.1007/978-981-15-0002-2_30

Problem Statement. Based on the preliminary finding, most of the customer dissatisfied due to unorganized facilities provided at the terminal such as parking and vehicle compilation. It is the intention of the researcher to investigate on the other determinants that could contributed to the customer dissatisfied for the Ro-Ro operation in Kuala Perlis.
Research questions. These question are made to achieve the main objective of this research (a) what are the current situation on the customer satisfaction toward the port facilities at Kuala Perlis Terminal? and (b) What are the relationship between the port facilities and customer satisfactions at Kuala Perlis Terminal?

2 Literature Review

2.1 *Short Sea Shipping*

Short Sea Shipping can also being stood to cover all the maritime transport. It is which the service does not involve the ocean crossing [3]. It is also included the maritime transportation along the coasts between the mainland coast, and island of the European Union to protect purely the national transport and cross-border service. And also as a sea-river transportation by coastal vessel and from port in the hinterland [4]. In the ASEAN nation are sharp into investigating such chances to fortify their hold as a major worldwide assembling giant and these country are positioned on the international highways, the South China Sea and Malaccan Straits [5].

2.2 *Ro-Ro Operation at Kuala Perlis, Perlis, Malaysia*

Kuala Perlis Terminal is the main port that is located in their second largest town in Perlis. It is in the northwest of Peninsular Malaysia, near the Thailand's border. Kuala Perlis Terminal is operated under Maritime Department that acts as major departure point for Langkawi Island. There are two companies that operates the Ro-Ro operations in Kuala Perlis which are Langkawi Ferry Sdn Bhd and Langkawi Auto Sdn Bhd. Both companies are focusing on the movement of passenger and cargo to Langkawi Island

Customer satisfaction. Zairi [6] revealed that customer is the purpose what seller do and rather than them depending on us, we very much on them. They are not the source of a difficult but we must not perhaps make a demand that customer 'should go away' because our future and our safety will be put in risk [7]. This is one of main point or reason why organization today focus on the customer satisfaction, loyalty and retention [7]. Hansemark and Albinsson [8] stated that "satisfaction is a general customer attitude to a service provide, or an emotional feedback to the dissimilar between what customers anticipate and what they receive, regarding the completion of some need, goal or desire". Hoyer et al. [9] also indicate that customer retentive is the practice of working to content customers with the purpose of emergent lasting relationships with them. In order to succeed customer satisfaction, organizations must be intelligent to satisfy their customers' needs and wants [10].

3 Methodology

About 150 respondents selected from the population through a purposive sampling technique. Data is analyzed after they have been collected. All the answer from respondents analysed using SSPS on the method of Multiple Regression Analysis.

4 Results of Reliability Test

The result of reliability test shows that the result obtained through SPSS for the research question, which is to identify factors of project success in shipping company. The number of questions that had been asked are 30 questions within the total number of respondents which are 150 respondents. Based on the result showed in the Table 1, Cronbach alpha for the reliability test is 0.721. Which is more than 0.7 and it is an acceptable consistency.

Correlation analysis show the relationship between Ro-Ro port facilities toward customer satisfaction from Kuala Perlis terminal as listed in Table 2. There is a strong relationship between customer satisfaction and administrative facilities which is 0.648 close to 1 and the lowers is port facilities 0.420 it a bit far to 1.

Table 1 Result of reliability test

Cronbach's alpha	No of items
0.721	30

Table 2 Pearson Correlation analysis

Variables	Administration facilities	Port facilities	Customer_satisfaction
Administration facilities	1	0.520	0.648**
Port facilities	0.520	1	0.402
Customer_satisfaction	0.648**	0.402	1

**There is a strong relationship between two variables

5 Conclusion

For the conclusion, this study revealed that there are 3 important thing that Ro-Ro port Kuala Perlis must focus on its administration facilities. For this part, port administrative necessity to ensure that service provided at the greatest level. In order to increase the level of customers satisfaction. They are 6 activities in Ro-Ro operation in Kuala Perlis terminal. The activities are listed below:

1. Ticketing
2. Check-in
3. Entering the terminal gateways
4. Waiting at loading site
5. Check-in process that can be done while customer entering the terminal gateways
6. Online ticketing can reduce the customer waiting time.

On the other hand, improvement of port facilities at Kuala Perlis terminal also one of the key success factor to capture customer value and to increase the level of customer satisfaction. Adequacy is the aspect which refer to the port facilities' availability. It is related in total, size or quantity that suit with number of customer [11]. Facilities are also have been prepared also always placed with the right ways, so it's coincide with numbers of customer expected [11]. According to Abdullah [11] lack of the facilities will make the unpredictable situation and also make customer not satisfied. This part are important because to ensure all the thing at the gate are smooth and easy for the customer. As an example, the total of ramp at Ro-Ro vessel, the ramps is large steel construction consisting of longitudinal beam plated over to provide a vehicle roadways. External ramps are used to allow wheeled vehicle to travel between a quay and Ro-Ro vessel. Fixed or movable internal ramps provide the access from deck to deck.

Gegeckaitė [12] revealed that customer satisfaction is important because it is the way or the key for a successful business. Organizations search for approaches to discover clients craves however in the meantime need to guarantee their fulfillment with administrations [12]. Along these lines it is essential to clear up not just the most imperative determinants of consumer satisfaction on administrations additionally investigate which of them are the best [12].

6 Recommendation

In the recommendation part, it is suggested that to concentrate more towards consumer satisfaction, because it will add more consumer to use Ro-Ro vessel. Once this situation is totally overcome, no accident will occurred and returnable customer will reach at the optimum level. Consequently, this study suggests that the company to take a look on the customer point of view in order to lead the port activities to become faster and time efficient. Ro-Ro Company must improve port facilities because it will improve the quality of operations at the same time they can increase the level of customer's satisfaction. On the other hand, the other determinants that could have contributed to the dissatisfied customer were not elaborated. Hence, this study suggested that the other scholars or academia will discern the other key determinants that must be addressed to ensure the success of their future Ro-Ro port operation.

References

1. Fusco PM, Sauri S (2009) Performance indicators for Ro-Ro terminal: a planning assessment tool. Transp Res Rec J Trans Res Board 2100:38–46
2. Özkan ED, Nas S, Güler N (2016) Capacity analysis of RO-RO terminals by using simulation modeling method. Asian J Shipp Logist 32(3):139–147
3. Arof AM (2015) Determinants for a feasible short sea shipping: lessons from Europe for ASEAN. Asian Soc Sci 11(15):229–238
4. Arof AM, Nair R (2017) The identification of key success factors for interstate Ro-Ro short sea shipping in Brunei-Indonesia-Malaysia-Philippines: a Delphi approach. Int J Shipp Transp Logist 9(3):261–279
5. Adnan M, Fatima B (2016) China-Pakistan economic corridor: a road to development and its challenges. South Asian Stud 31(2):225
6. Zairi M (2000) Managing customer satisfaction: a best practice perspective. TQM Mag 12(6):389–394
7. Singh H (2006) The importance of customer satisfaction in relation to customer loyalty and retention. Acad Mark Sci 60:193–225
8. Hansemark OC, Albinsson M (2004) Customer satisfaction and retention: the experiences of individual employees. Manag Serv Qual: Int J 14(1):40–57
9. Hoyer WD, Macinnis DJ, Pieters R (2001) Customer behavior Houghton Mifflin Company. South-Western Cengage Learning, Ohio, USA
10. LaBarbera PA, Mazursky D (1983) A longitudinal assessment of consumer satisfaction/dissatisfaction: the dynamic aspect of the cognitive process. J Mark Res 20(4):393–404
11. Abdullah AA, Wan HL (2013) Relationships of non-monetary incentives, job satisfaction and employee job performance. Int Rev Manag Bus Res 2(4):1085
12. Gegeckaitė (2011) Factors of customer satisfaction and its determinants. http://scholararticles.net/factors-of-customer-satisfaction-on-services/. Accessed 08 May 2017

Finite Element Modeling and Updating for Dynamic Study of Exhaust Structure

M. S. M. Fouzi and M. S. M. Sani

Abstract Modal analysis is an approach to determine the dynamic behavior of a structure through numerical analysis or experimental approach. Nowadays, numerical analysis has been preferred by most researchers and engineers compared to experimental analysis to analyze the structure due to expenditure and time consuming factors of the research. Recently, numerical analysis is done computationally or well known as Finite Element Analysis (FEA) by using finite element (FE) model to replicate the real test structure. Instead of advantage of FE analysis, the trustworthiness of FE model has been questioned since simplification has been made during design stage for complex structure and inaccurate of geometry input of the model. Hence, this paper is carried out to reduce the discrepancies between numerical prediction results with measured test data on FE model of exhaust structure by implementing FE model updating using SOL200 algorithm. Initially, the FE model is running on normal mode analysis SOL103 to extract the modal parameters such as natural frequency and mode shape. Then, the numerical pre-diction result has been correlated and validated with measured test data obtained through Experimental Modal Analysis (EMA). Due to disagreement between numerical result and experimental data in correlation process, the FE model has been updated using SOL200 algorithm. The percentage error between numerical prediction result and measured test data is minimized from 5.67 to 3.23% after been updated. It's shown the capability SOL200 algorithm in preparing the reliable FE model before been used for further analysis such as structural dynamic modification.

M. S. M. Fouzi (✉) · M. S. M. Sani
Advanced Structural Integrity and Vibration Research (ASiVR), Faculty of Mechanical Engineering Universiti Malaysia Pahang, 26600 Pekan, Pahang, Malaysia
e-mail: msahril@puo.edu.my

M. S. M. Sani
Automotive Engineering Centre, Universiti Malaysia Pahang, 26600 Pekan, Pahang, Malaysia

M. S. M. Fouzi
Department of Mechanical Engineering, Politeknik Ungku Omar, Jalan Raja Musa Mahadi, 31400 Ipoh, Perak, Malaysia

C. L. Saw et al. (eds.), *Advancement in Emerging Technologies and Engineering Applications*, Lecture Notes in Mechanical Engineering,
https://doi.org/10.1007/978-981-15-0002-2_31

Keywords Dynamic study · Joints strategy · Finite element analysis · Experimental modal analysis · FE model updating

1 Introduction

Since introduction of internal combustion engine (ICE) in automotive field in early 19th century, it was designed with exhaust system. The paramount function of exhaust system to filter the hazardous gaseous and lowering the noise formed from combustion process in ICE before entering the surrounding. For light weight vehicle such as car, exhaust structure normally attached under body of vehicle's chassis via the hangers. There are numbers of study [1–6] reporting the source of noise and vibration is excited from engine driveline and uneven road condition spread along the exhaust structure via the hangers and brackets.

These noise and vibration were affected the comfort level of passenger and influenced the durability of the exhaust structure. Hence, engineers and researchers involved in noise, vibration and harshness (NVH) area are giving intense of this issue as it has become major criteria of customer during purchasing a vehicle. Since past few years, several active researchers [7–9] were conducted to study the dynamic behavior of exhaust structure via modal analysis. Through modal analysis, the researchers are able to identify dynamic characteristics of the test structure via modal properties such as natural frequency, mode shape and damping ratio. Recently modal analysis is conducting through experimental and computational depend on the capital and purpose of the study or research.

In addition, the FE model updating process is adopted in this study since it has been reported by [10] able to reduce the discrepancies between FEA and EMA result for dynamic behavior of body-in-white (BIW) structure.

2 Finite Element Analysis (FEA)

In recent years, numerical prediction analysis is well known as finite element analysis (FEA). Parallel with rapid evolution of computer performance and abilities, a numbers of available FEA software packages nowadays are able to operate using even personal computer or laptop compared to past few decades which been limited to super-computer only own by giant company such as aeronautics company. As reported in their work [11], frequency analysis is done using FEA on the FE model of the go-kart structure to obtain the dynamic properties of the structure.

2.1 *Execution of FE Model for Exhaust Structure*

Preliminary, the FE model has been sketched using CAD software, SolidWork as demonstrated in Fig. 1 due to the complexity of the structure to be modeled directly in FEA package MSC. Patran. Referred to Fig. 2, the FE model was imported into MSC Patran and has been meshed using CTETRA 4 element for solid geometry. The meshing process has produced 137043 numbers of elements and 46548 numbers of nodes. Once the meshing process is completed, the input geometry is assigned into FE model. The material properties used in this study is Stainless Steel SUB304 (Young Modulus, E is 195 GPa; Density, ρ is 8000 kg/m^3; Poisson's Ratio, ν is 0.29).

3 Experimental Modal Analysis (EMA)

Experimental approach to study the dynamic behavior of structural constantly offered a major contribution to our endeavors to understand and to control most of vibration phenomena encountered in practice. One of the most applicable methods which been

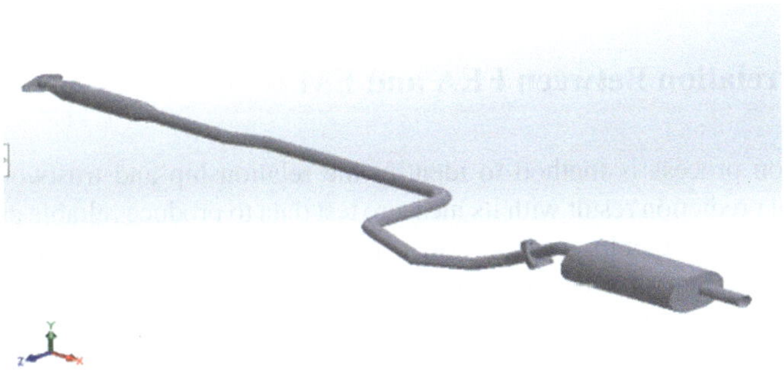

Fig. 1 Free-free boundary condition of exhaust structure for modal testing

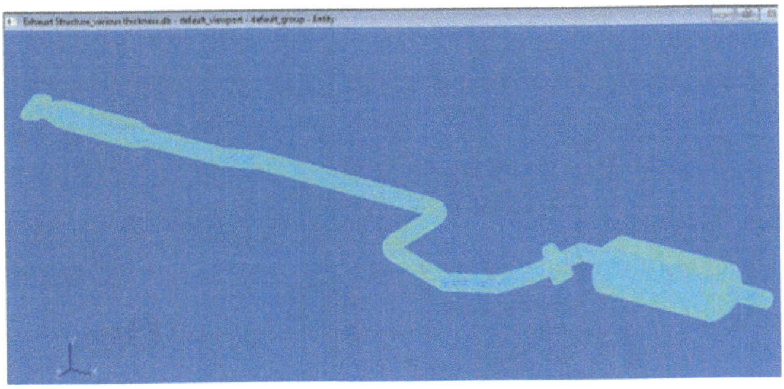

Fig. 2 Equipments for modal testing execution

used until today to describe the dynamic behavior of structural is called modal testing, recently called as experimental modal analysis (EMA). Another researcher [12] has explained experimental observation have been made for determining the nature and extent of vibration response level, and verifying the numerical prediction.

The test structure will be excited with initial load and let the energy spreading along the structure till it decays within time. The force given on the test structure is measured through force transducer as input source while the response is assessed using accelerometer as output function.

3.1 *Description of Modal Testing for Exhaust Structure*

In the way to replicate the free-free boundary condition for the test structure, the exhaust has been hanged using bungee cords as illustrated in Fig. 3. The modal testing is carried out with the assistance of equipments as portrayed in Fig. 4. The equipments used such as; (a) USB cable, (b) NI-DAQ's rack, (c) NI-DAQ, (d) tri-axial accelerometer's cable, (e) impact hammer, (f) tri-axial accelerometer, and (g) BNC cable.

4 Correlation Between FEA and EMA

Correlation process is method to identify the relationship and trustworthiness of numerical prediction result with its measure test data to produce reliable and verified FE model before employing for further analysis. Normally correlation process is done through comparison of modal parameters; natural frequency (NF) and mode shape (MS) between EMA and FEA such as tabulated in Table 1.

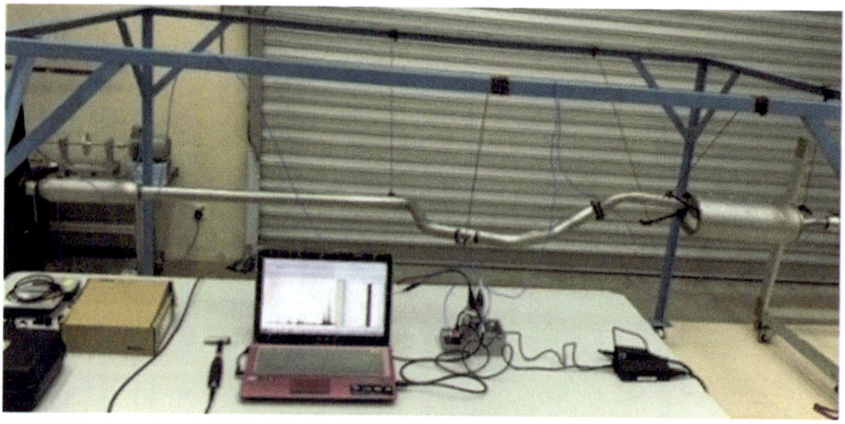

Fig. 3 Free-free boundary condition of exhaust structure for modal testing

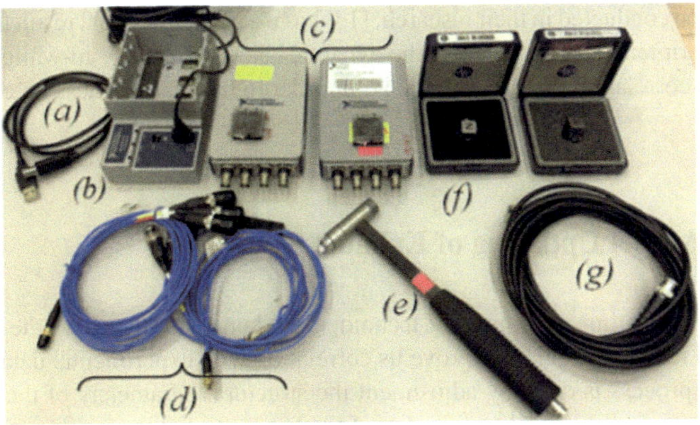

Fig. 4 Equipments for modal testing execution

Table 1 Comparison of modal properties between modal testing result and numerical predition

Mode	Experimental Modal Analysis (EMA)		Finite Element Analysis (FEA)		Error (%)
	MS	NF (Hz)	MS	NF (Hz)	
1		21.1		19.22	8.93
2	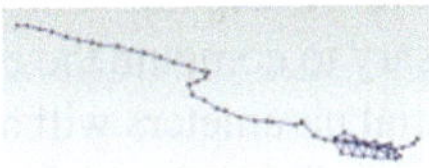	46.9		46.76	0.29
3	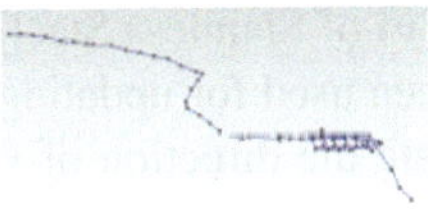	65.1		61.87	4.97
4	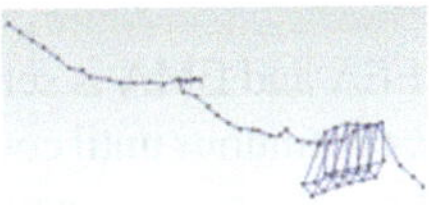	93	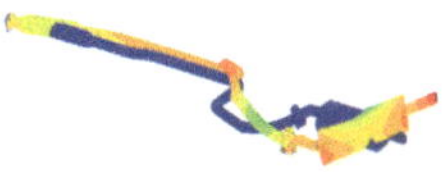	85.12	8.47
	Total average error				5.67

As been conducted in their research, [13] has used the numerical result to correlate with experimental data to analyze the dynamic behavior of body-in-white (BIW). In addition, correlation of data is essential in order to has accurateness on existing FE model.

5 FE Model Updating of Exhaust Structure

FE model updating is commonly technique implemented to update the numerical predicted result in order to improve its correlation with experimental data [14]. The updating process is done by adjustment the structural parameters of the initial FE model by using the modal testing data. Structural parameters such as stiffness and mass are identified often by minimizing a residual error between FE model and test data [15].

5.1 Sensitivity Analysis

Ahead of updating process, the sensitivity analysis using SOL200 in MSC Nastran is adopted to choose the suitable parameter to been updated [16]. Reported by [17], selecting the updating parameters is an important aspect of the FE model updating process. The parameters selected should be justified by engineering understanding of the structure and the number of parameters should be kept to a minimum to avoid ill-conditioning problem. Therefore, it is necessary to compute the eigenvalue sensitivities beforehand, so that only the most influential parameters will be chosen. Since it showed only two parameters is most sensitive in Table 2 which are Young Modulus (E) and density (ρ) except Poisson's ratio (ν) of Stainless Steel SUB304, for that reason only two sensitive parameters have been used for updating purpose; (E) and (ρ). The negative sign in Table 2 just indicate the direction of vector and not represented the value of parameter since it is sensitivity matrix coefficient to distinguish the sensitiveness of updating parameter.

An objective function based on residuals between FEA and EMA is set for minimization in the updating procedure [17]. The procedure continuous until convergence is accomplished when the difference between values of the objectives function J from

Table 2 Sensitiviy matrix coefficient for three parameters

NF	Young modulus (E)	Density (ρ)	Poisson's ratio (ν)
NF 1	9.6502	−10.893	−0.4984
NF 2	23.394	−26.407	0.0739
NF 3	31.114	−35.120	−2.2663
NF 4	42.602	−48.087	−0.3798

Table 3 Correlation of natural frequencies (Hz) between EMA and Updated FEA

Mode	Natural frequencies (Hz)				
	EMA	Initial FEA	Error %	Updated FEA	Error (%)
1	21.1	19.22	8.93	20.45	3.10
2	46.9	46.76	0.29	49.76	6.09
3	65.1	61.87	4.97	65.83	1.12
4	93	85.12	8.47	90.57	2.61
Total average error			5.67		3.23

consecutive iterations is sufficiently small. The objective function of eigenvalue residuals is given by (1);

$$J = \sum_{j=1}^{n}\left(\frac{\lambda_j}{\lambda_j^{exp}} - 1\right)^2 \tag{1}$$

where λ_j^{exp} is the j-th experimental eigenvalue and λ_j is the j-th eigenvalue predicted by the FE model. It is important to note that (1) only holds if the measured eigenvalue and its predicted counterpart are paired correctly, and therefore it is essential to ensure that the experimental and numerical data relate the same mode.

After treated with the FE model updating, the updated FE model is correlated with modal testing result such as Table 3. From the updating process it is showed the updated numerical prediction has better agreement with the reduction of percentage of error from 5.67 to 3.25%. Hence, it was certified the SOL200 algorithm is capable to shrink the discrepancies between FEA and EMA as proved in the predecessor researchers in their studies [11, 16].

The design variables after updating treatment of FE model are; 1.06 for Young Modulus (E) and 0.94 for density (ρ). Deviation between initial value (i.v) and updated (u.v.) is calculated; |[(u.v.)/(i.v)]/i.v.|. The updated value of parameters in this study is summarized in Table 4.

Table 4 Updated value of parameters for exhaust structure

Parameters	Initial value (i.v.)	Updated value (u.v)	S.I. Unit	Deviation
(E)	195	206.7	GPa	0.06
(ρ)	8000	7520	kg/m^3	0.06

6 Conclusion

The reliable FE model of exhaust structure is capable to be designed and used for further analysis. It can be verified using measured test data from experimental using impact hammer method. Then, the algorithm of SOL200 is adopted to update the discrepancies between numerical predictions results with its measure test data to be close as much as possible. Throughout FE model updating process, the sensitivity analysis is made in advance to identify the most sensitive parameters should be picked to be updated. As a result, the initial discrepancies between FEA and EMA were managed to be reduced from 5.67 to 3.23%. Hence, it has been proved that updating process for FE model using SOL200 is proficient to prepare reliable FE model for further analysis.

References

1. Rajadurai MS, Kavin R, Rejinjose, Prabhakaran, Rajeshraman (2016) A system approach to dynamic characteristics of hanger rod in exhaust system. Int J Innov Sci Eng Technol 3(5)
2. Lupea I (2016) An exhaust system lumped model–identification and simulation. ACTA Technica Napocensis-Series: Appl Math Mech Eng 59(4)
3. Bötke A, Yazgaç D (2016) Exhaust mount placement optimisation with comparable methods. In: 23rd International congress on sound & vibration (ICSV23)
4. Nyein O, Abu AB, Moe AL (2015) Numerical analysis of vehicle exhaust system to determine hanger location using root mean square value. ARPN J Eng Appl Sci 10
5. Maruthi BH, Rudresh M, Vikram BV (2015) Evaluation of structural integrity of passenger car exhaust system. Int J Res Advent Technol 3
6. Gaonkar CD (2015) Modal analysis of exhaust system to optimize mounting hanger location. Int J Eng Res Technol (IJERT) 4(03)
7. Balaji DJ, Srihari PV, Sheelvanth VB (2014) Model frequency analysis of automotive exhaust system. Int J Mech Eng Robot Res 3
8. Gadre KR, Jadhav TA (2015) Vibration analysis of an automotive silencer. Int J Innov Eng Res Technol 2(6)
9. Noorazizi MS, Aminudin BA, Faraziah H (2014) Technique for hanger location of vehicle exhaust system using finite element method. Appl Mech Mater 663
10. Izham MHN, Abdullah NAZ, Zahari SN, Sani MSM (2017) Structural dynamic investigation of frame structure with bolted joints. MATEC Web Conf 90:01043
11. Abdullah NAZ, Sani MSM, Husain NA, Rahman MM, Zaman I (2017) Dynamics properties of a Go-kart chassis structure and its prediction improvement using model updating approach. Int J Automot Mech Eng (IJAME) 14(1):3887–3897
12. Ewins DJ (1984) Modal testing: theory and practice, vol 15. Research Studies Press, Letchworth
13. Abdullah NAZ, Sani MSM, Rahman MM, Zaman I (2017) Correlation of numerical and experimental analysis for dynamic behaviour of a body-in-white (BIW) structure. MATEC Web Conf 90:01020
14. Nehete DV, Modak SV, Gupta K (2015) Structural FE model updating of cavity systems incorporating vibro–acoustic coupling. Mech Syst Signal Process 50:362–379

15. Chen H-P, Maung TS (2014) Regularised finite element model updating using measured incomplete modal data. J Sound Vib 333(21):5566–5582
16. Zahari SN, Sani MSM, Ishak M (2017) Finite element modelling and updating of friction stir welding (FSW) joint for vibration analysis. MATEC Web of Conf 90
17. Husain NA, Khodaparast HH, Snaylam A, James S, Dearden G, Ouyang H (2010) Finite-element modelling and updating of laser spot weld joints in a top-hat structure for dynamic analysis. Proc Inst Mech Eng Part C: J Mech Eng Sci 224(4):851–861

Natural Mode Solar Evacuated Tubes Desalination

Chun Lin Saw, Jeefferie Abd. Razak, Azmi Naroh, Mohd Nazrin Nahar and M. Bakhyaraj

Abstract Solar desalination using solar still is the simplest method to produce clean distilled water but it yield low potable water. Hence, evacuated tube solar distiller is used in this study and modified with triangle shape glazing. Waste water from restaurant in Ipoh, Perak Malaysia was collected for distillation process for duration of 1 month. The average yield of the distilled water output is about 18.2% from 40 Litres of daily waste water experiment and comply with Ministry of Health Malaysia raw water requirements. The average efficiency of solar evacuated tube desalination performance is 60.5%.

Keywords Waste water · Desalination · Clean water · Evacuated tubes

1 Introduction

Clean water demand increases with increase population growth in the world. Polluted water from industrial waste and human activities such as waste water from restaurant, uses of excess chemical fertilizer in the farm and improper plastic dispose lead to contamination to water source then increase the risk of clean water production. There are numerous methods used for water purification including filtration, disinfection, sedimentation and distillation [1]. Solar distillation is the cheapest way to produce clean water using solar radiation. Solar still device is need to convert waste water to become clean water either for human consumption or draining back to the environment. UNICEF reported in 2015 that 5.2 billion people globally that accounted of 71% of world population mange to use safe drinking water services. On top of that, 2.1 billion people still lack safe water at home where 263 million people spend more than 30 min per round trip to collect water, while 844 million people do not have

C. L. Saw (✉) · A. Naroh · M. N. Nahar · M. Bakhyaraj
Politeknik Ungku Omar, Jalan Raja Musa Mahadi, 31400 Ipoh, Perak, Malaysia
e-mail: clsaw78@gmail.com

J. Abd. Razak
Fakulti Kejuruteraan Pembuatan, Universiti Teknikal Malaysia Melaka, Hang Tuah Jaya, 76100 Durian Tunggal, Melaka, Malaysia

C. L. Saw et al. (eds.), *Advancement in Emerging Technologies and Engineering Applications*, Lecture Notes in Mechanical Engineering,
https://doi.org/10.1007/978-981-15-0002-2_32

basic drinking water services and 159 million people still drink water directly from surface sources such as streams of river or lakes [2]. Diarrhea is the most widely known disease due to contaminated food and water. Some 842,000 people are estimated to die each year from diarrhea as a result of unsafe drinking-water, sanitation, and hand hygiene [2]. Yet diarrhea is largely preventable, and the deaths of 361,000 children aged under 5 years could be avoided each year if these risk factors were addressed. When water is not readily available, people tend to decide handwashing is not a priority, thereby adding to the likelihood of diarrhoea and other diseases.

At the end of 2015, there were approximately 18,000 desalination plants or about 150 countries worldwide, with a total installed production capacity of 86.55 million m^3/day or 22,870 million gallons per day (MGD) [3]. Around 300 million of people are relying on desalinated water for daily usage that accounted a total of 44% of production capacity (37.32 million m^3/day or 9860 MGD) is located in the Middle East and North Africa. While desalination in that region is projected to grow continuously at a rate of 7–9% per year, the "hot spots" for accelerated desalination development over the next decade are expected to be Asia, the US and Latin America.

Solar evacuated tube collector has a efficiency is 16–20% much higher than solar still collector [4]. Solar evacuated tube collector efficiently absorbed more heat than solar still and work well in the medium to high temperature operation [5]. In this paper, an evacuated tubes solar collector is used in the experiment to distil water from restaurant waste water and river by mean of natural mode. The experiment has been carried out for one month at Solar Research Site, Politeknik Ungku Omar, Ipoh, Perak, Malaysia. The purpose is to analyse the performance of solar evacuated tubes collector and to analyse the treated water or distilled water contents produced whether it comply with Ministry of Health Malaysia raw water standard requirements.

2 Methodology

Figure 1 shows the test rig of solar evacuated tube desalination. The system consists of eight (8) vacuum tubes, 40 litres tank, triangle cover and reflector. Vacuum tubes is used as a heat transfer enhancement technique to improve thermal energy harness of the conventional solar still. The bottom of each vacuum tubes is integrated with a reflector, the function of the reflector is to reflect back the solar radiation at the two sides of vacuum tubes to the vacuum tubes.

Figure 2 shows the experimental method of desalination under direct sunlight. Firstly, 40 litres of waste water is poured into the storage tank to occupied storage tank 80% capacity. The test rig is placed facing south with inclination angle of 30°. Two type of water sample was used in this study that is river water in front of Politeknik Ungku Omar and waste water from a restaurant. Every morning, the storage tank will be assured to be filled with 40 litres of waste water from restaurant before beginning of experimentation. The distilled water will be collected to be analyzed. Natural mode of the solar desalination was applied to the desalination system.

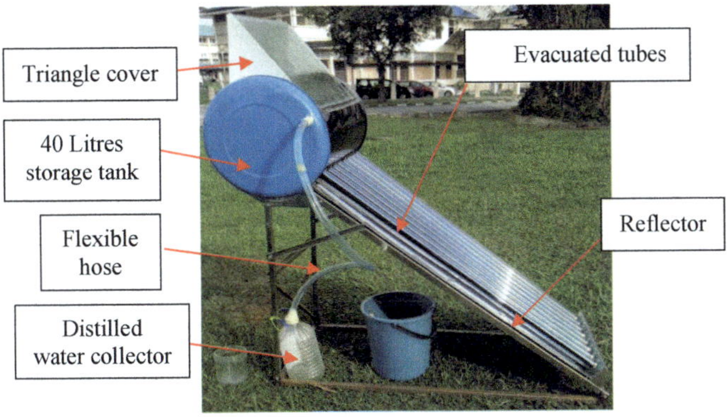

Fig. 1 Solar evacuated tube desalination

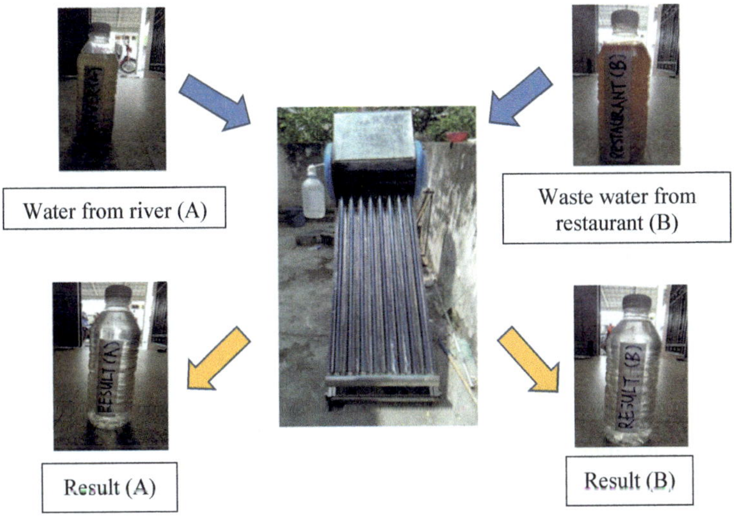

Fig. 2 Desalination method

Three (3) thermocouples type-k is installed where two thermocouple is placed inside the storage tank to measure the heating waste water, one more thermocouple is placed inside distilled water collector to measure distil water temperature. The experimentation was carried out for one month starting 1st of October till 31st of October 2017 on desalinate the waste water from restaurant. Temperature of heating water and distil water was measured using the portable temperature measurement device PCE-T 1200 with accuracy of ±1 °C. Weight of distilled water is measured by using standard portable weighting machine with accuracy ±1 kg. The temperature data and weight of distilled water is measured hourly from 7 a.m. until 7 p.m. daily.

However, the contaminations of water are analyzed by using Pro DSS multiparameter device.

The equation that is used to calculate the performance of solar evacuated tube collector as below [6]:

$$Instantaneous\ efficiency, \eta_i = \frac{M_{di} \times L}{I_t x A_c \times 3600} \quad (1)$$

where M_{di} = hourly amount of distilled water produced, L = latent heat of vaporization of water, I_t = hourly solar radiation, A_c = vacuum tubes surface area. Instantaneous efficiency is the hourly efficiency of solar evacuated tube collector, however the average efficiency is the total average of 12 h of solar evacuated tube collector efficiency. Percentage of distilled water production equation as below [6]:

$$Percentage\ of\ distilled\ water\ production\ (\%) = \frac{M_{daily}}{M_{waste}} \times 100\% \quad (2)$$

3 Results and Discussions

There are two data analysis was done in this study. The firstly is the experimental work to measure the productivity of distilled water and performance of solar evacuated tube collector, secondly is the laboratory analysis at Lembaga Air Perak (LAP) Malaysia to analyse the contents of waste water and distilled water. Tables 1 and 2 shows the analyzed restaurant waste water and water from river. A simple of these water were brought to Lembaga Air Perak Malaysia to analyze the parameters on colour, pH, Turbidity and total dissolved solid (TDS). The device used for this analysis is the Pro DSS multiparameter.

The results show that the water source from river and restaurant waste water is alkaline, after distillation the water back to pH neutral level. Both the turbidity of the waste water improved after distillation. Turbidity is a measurement to show the

Table 1 Distilled water of restaurant waste water

Parameters	Type of water	
	Waste water from restaurant	Waste water after distillation
Temperature (°C)	29.7	28.6
TDS (mg/L)	1000	10
Ammonia	–	–
Colour (TCU)	0.42	0.00
pH value (pH)	10.87	7.08
Turbidity (NTU)	209.3	6.0

Table 2 Distilled water of river

Parameters	Type of water	
	River	Waste water after distillation
Temperature (°C)	29.6	28.7
TDS (mg/L)	500	60
Ammonia	–	–
Colour (TCU)	0.03	0.00
pH value (pH)	8.25	7.26
Turbidity (NTU)	8.0	6.2

Table 3 Ministry of Health Malaysia (MoH) guidelines [7]

Parameters	Type of water
	MoH guidelines (Raw water)
Temperature	–
TDS	1500 mg/L
Ammonia	1.5 mg/L
Colour	300 TCU
pH value	5.5–9.0 pH
Turbidity	1000 NTU

amount of suspended solid in the water or cloudiness of water, hence less cloudiness in the water found. Figure 2 clearly shows that the colour of distilled water is transparent that eliminate the colour in the waste water. Total dissolved solid (TDS) of both waste water also reduced below the WHO guidelines of treated water for save water consumption as listed in Table 3.

Performance of solar evacuated tube collector was analyzed based on productivity of distilled water from waste water of restaurant due to the objective of this study. Daily distilled water yield as depicted in Fig. 3 shows the experimental data on 6th October 2017, a total of 7.2 litres distilled water was collected. For data from 1st October until 31st October 2017, the daily distilled water yield between 5 and 9.6 litres can be produced from a total of 40 litres of restaurant waste water during desalination process. A total of 800 litres of waste water has been poured into the solar evacuated tube but only 145 litres of distilled water has been collected due to efficiency of the collector. Hourly instantaneous efficiency of solar evacuated tube was calculated on 6th of October 2017 too. Figure 4 shows that the hourly instantaneous efficiency of solar evacuated tube desalination with overall efficiency of 63% and heat loss of 1.343 W/m^2 °C. The lower the value of heat loss, the solar evacuated tube collector efficiency will increase.

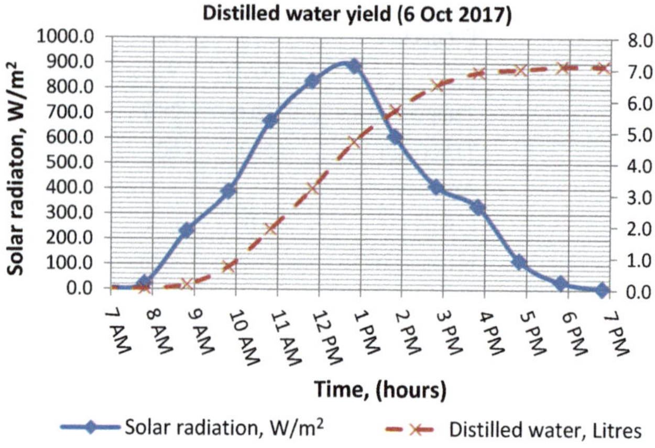

Fig. 3 Daily distilled water yield

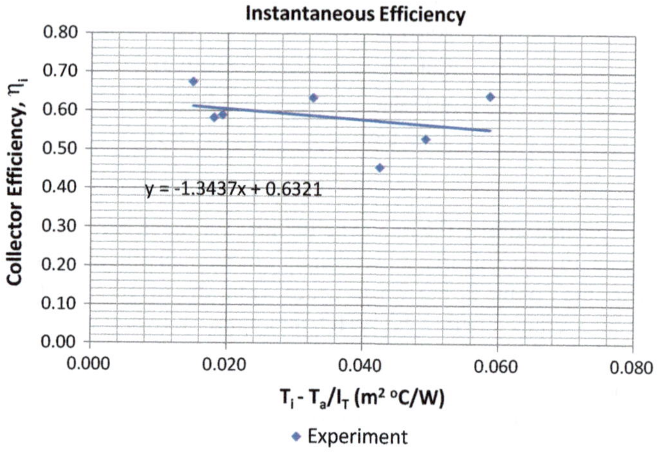

Fig. 4 Instantaneous efficiency of solar evacuated tube collector

4 Conclusions

Overall solar evacuated tube solar collector efficiency is 60.5% and able to produce 145 litres of distilled water. This accounted 18.1% of 800 litres of waste water distilled to become clean water. Investigation on the angle of inclination, different shape of cover and reduction on leakages are recommended for further improvement of the desalination system.

References

1. Al-harahsheh M, Abu-Arabi M, Mousa H, Alzghoul Z (2018) Solar desalination using solar still enhanced by external solar collector and PCM. Appl Therm Eng 128:1030–1040
2. Al-Hamadani A, Shukla S (2014) Modelling of solar distillation system with phase change material (PCM) storage medium. Therm Sci 18:347–362
3. Al-Zghoul Z (2016) Solar desalination with solar still having phase change material and connected to a solar collector. Jordan University of Science and Technology, Master Thesis
4. Badran OO (2007) Experimental study of the enhancement parameters on a single slope solar still productivity. Desalination 209(1–3):136–143
5. Kabeel AE, Arunkumar T, Denkenberger DC, Sathyamurthy R (2017) Performance enhancement of solar still through efficient heat exchange mechanism—a review. Appl Therm Eng 114:815–836
6. Al-Kayiem HH, Saw CL (2014) Performance evaluation of solar water heater integrated with PCM nanocomposite TES at various inclinations. Sol Energy 109(1):82–92
7. Ministry of Health Malaysia. Drinking water quality standard. http://kmam.moh.gov.my/public-user/drinking-water-quality-standard.html. Last Accessed 26 July 2018

Optimal Power Generation in Microgrid System Using Particle Swarm Optimization

M. N. Abdullah, N. F. A. Mohd Azlan, W. M. Dahalan, N. F. Naswan, R. Hamdan and M. N. Ismail

Abstract This paper presents an application of Particle Swarm Optimization (PSO) algorithm for minimizing the total generation cost in microgrid system within 24 h. The microgrid system consists of conventional and renewable energy power plants are considered in this project. The main objective is to minimize the generation cost while satisfied the load demand and system constraints. This case study consists of three fuel cells, two diesel generators and two wind turbines. The proposed PSO algorithm has been simulated in Matlab software to determine optimal generation cost. The results are compared with other existing algorithms to validate performances of PSO in term of minimizing generation cost in microgrid. It found that the PSO algorithm gives the lower optimal cost compared to other methods.

Keywords Generation scheduling · Microgrid · Particle swarm optimization · Renewable energy

1 Introduction

Nowadays, the development is going up in all around the world. At the same time, the demand of electricity has increased greatly. Therefore, the microgrid play as a vital role in the electricity transmission [1]. In addition, microgrid is known as electrical network of various size bounded from main grid by the power transformer or the electrical interface [2]. Microgrid are small-scale networks to supply several types of power including electricity, heating and cooling for a small community, rural area and island [3]. Likewise, microgrid are able to become a backup supply when the main

M. N. Abdullah (✉) · N. F. A. Mohd Azlan · N. F. Naswan · R. Hamdan · M. N. Ismail
Green and Sustainable Energy (GSEnergy) Focus Group, Faculty of Electrical and Electronic Engineering, Universiti Tun Hussein Onn Malaysia, 86400 Parit Raja, Batu Pahat, Johor, Malaysia
e-mail: mnoor@uthm.edu.my

W. M. Dahalan
Department of Marine Electrical and Electronic Technology, Universiti Kuala Lumpur, Malaysian Institute of Marine Engineering Technology, 32200 Lumut, Perak, Malaysia

C. L. Saw et al. (eds.), *Advancement in Emerging Technologies and Engineering Applications*, Lecture Notes in Mechanical Engineering,
https://doi.org/10.1007/978-981-15-0002-2_33

grid have a fault or failure to operate. For example, microgrid are used for important application such as hospitals, airports and railways that necessary to connect with the electricity supply [1]. Microgrid can operate either in grid-connected mode or stand-alone mode. The energy sources in microgrid including photovoltaic, wind, diesel engines, and fuel-cell [4]. Moreover, microgrid is suitable as renewable energy sources for distribution application and provide good impact on the environment.

The optimization technique has been applied to solve the optimal power generation scheduling in microgrid system in order to minimize the generation cost while fulfill the load demand and system constraints. Artificial Fish Swarm Algorithm (AFS) is applied in [5] solve the problem of optimal power generation. The aims of this article are reduce the cost and approve the applicability of AFS algorithm in solving economic generation scheduling in microgrid. Kamboj and Chanana [6] applied mixed integer non-linear programming (MINLP) method in their research to minimize the total operational cost and emission of microgrid to obtain optimal power schedule in a day. In [7], Hongbin Wu, Xingyue Lin and Ming Ding presents economic dispatch in microgrid which consider lowest cost in a scheduling cycle and coordinates between different distribution generation over a period. Next, Yang and Deb was developed cuckoo search algorithm (CSA) for solving complex, non-linear and non-convex optimization problem. The aims of this algorithm is to solve multimodal optimization problem and minimize the generation cost in microgrid within 24 h [8]. Binary gravitational search algorithm (BGSA) was developed to achieve their objective which is obtain optimal operation at minimum operating cost while satisfied the load demand [9]. The new algorithm well known as iterated-based algorithm is designed to solve economic dispatch problem in microgrid. The main objectives of this paper is to minimize total generation and solve the lossless economic dispatch problem in microgrid system [8].

In this paper, particle swarm optimization (PSO) is implemented to optimize operation scheduling in microgrid system in order to minimize total generation cost within 24 h with consider system constraints and meet the power demand. The rest of paper is organized as follows. Section 2 described the problem formulation. Section 3 present the proposed PSO algorithm and it implementation steps. The simulation results are explained in Sect. 4. Lastly, the conclusion described in Sect. 5.

2 Problem Formulation

The problem of this project is to minimize the generation cost of the microgrid within 24 h. It is considered as a single objective optimization problem which are include operation constraints of generator and system.

2.1 Cost Modelling

In this section, cost modelling for microgrid system is described including wind generator, fuel-cell plant and diesel generator are described in details [3].

Cost of Diesel Generator. Diesel generator is known as conventional power plant and the equation of cost function is form in quadratic function. The equation of (1) expresses the cost function of diesel generator.

$$F_{(diesel,i)}(t) = a_i + b_i P_{(diesel,i)}(t) + c_i P^2_{(diesel,i)}(t) \tag{1}$$

where, $F_{diesel,i}\ (t)$ is the generation cost of i-th diesel unit in the scheduled period, (t), $P_{diesel,i}\ (t)$ is the output power of i-th diesel unit in the scheduled period, (t) and a_i, b_i, c_i is the cost coefficient of i-th, ($).

Cost of Wind Power Plant. Equation (2) is modelled for cost function of wind generator. The output power of wind generator is reliant on wind speed at the time. The relationship between wind power and the strength of wind are presented in Eq. (3) [2].

$$F_{(wt,i)}(t) = b_i P_{(wi,t)}(t) \tag{2}$$

where, $F_{wt,i}\ (t)$ is generation cost of i-th wind power plant in the scheduled period, (t), $P_{wt,i}\ (t)$ is the power of i-th wind power plant in the scheduled period, (t) and b_i is the cost coefficient.

$$P_{(wi,t)}(t) = \begin{cases} 0, & v < v_{cut-in} \\ P^r_{(wi,t)} \times \frac{v - v_{cut-in}}{v_r - v_{cut-in}}, & v_{cut-in} \le v < v_r \\ P^r_{(wi,t)}, & v_r \le v < v_{cut-out} \\ 0, & v \ge v_{cut-out} \end{cases} \tag{3}$$

where, $P^r_{(wt,i)}$ is the rated of power wind turbine number i, V is the wind speed in (m/s), V_{cut-in} is cut-in wind speed, $V_{cut-out}$ is cut-out wind speed and v_r is nominal wind speed.

Cost for Fuel-cell Plant. For fuel-cell system, the Eq. (4) presents the generation cost by considering the energy conversion efficiency [3].

$$F_{(fc,i)}(t) = \frac{b_i P_{(fc,i)}(t)}{\eta_{(fc,i)}} \tag{4}$$

where, $F_{(fc,i)}$ is the generation cost of i-th fuel-cell plant at time, (t), $P_{(fc,i)}$ is the output power of i-th fuel-cell plant in the scheduled period, (t), b_i is cost of natural gas in ($/kg) and $\eta_{(fc,i)}$ is efficiency of the i-th fuel-cell plant.

2.2 Objective Function

The objective function is to minimize the total generation cost of microgrid operation. Equation (5) presents the cost function of microgrid and the generation scheduling consider for a whole day [3].

$$F_{total} = \sum_{t=1}^{T} \left(\sum_{i=1}^{N_D} F_{(diesel,i)}(t) + \sum_{i=1}^{N_{wt}} F_{(wt,i)}(t) + \sum_{i=1}^{N_{fc}} F_{(fc,i)}(t) \right) \tag{5}$$

where, N_D is the numbers of diesel units, N_{wt} is the numbers of wind turbine, N_{ft} is the number of fuel-cell plant, T is the parameter of scheduling period, and F_{total} is the total generation cost with period T.

Constraints. Two types of constraints that are considered in problem of microgrid optimization which are equality constraints and boundary limits.

Power Balance Constraints. The power generated for each distribution generation (DG) must fulfill the power demand of each scheduled period [3].

$$\sum_{i=1}^{N_D} P_{(diesel,i)}(t) + \sum_{i=1}^{N_{wt}} P_{wt}(t) + \sum_{i=1}^{N_{fc}} P_{fc,i}(t) = P_D(t) \tag{6}$$

where, P_D is the power demand at time (t), $P_{diesel,i}(t)$ is the power of i-th diesel at time (t), $P_{wt,i}(t)$ is the power of i-th wind turbine at time (t) and $P_{fc,i}(t)$ is the power of i-th fuel-cell at time (t).

Boundary Limits. The output power for every generator have lower and upper limits. Equations (8)–(9) demonstrates the generator boundary limits for different types of generators [3].

$$P_{diesel,i}^{\min} \leq P_{diesel,i}(t) \leq P_{diesel,i}^{\max} \tag{7}$$

$$P_{wt,i}^{\min} \leq P_{wt,i}(t) \leq P_{wt,i}^{\max} \ldots t = \{1, 2, \ldots T\} \tag{8}$$

$$P_{fc,i}^{\min} \leq P_{fc,i}(t) \leq P_{fc,i}^{\max} \tag{9}$$

where, $P\frac{\min}{diesel,i}$ is the minimum power of i-th diesel power, $P\frac{\max}{diesel,i}$ is maximum power of i-th diesel power, $P\frac{\min}{wt,i}$ is the minimum power of i-th wind turbine, $P\frac{\max}{wt,i}$ is maximum power of i-th wind turbine, $P\frac{\min}{fc,i}$ is minimum power of i-th fuel-cell and $P\frac{\max}{fc,i}$ is maximum power of i-th fuel-cell.

3 Particle Swarm Optimization

The algorithm was first introduced by J. Kennedy and R. Eberhart in 1995. The behavior of PSO algorithm based on a herd of insects such as an ant, termites, bees, birds and fish [10]. Each particle make a decision based on its own experiences and its neighbor's experiences. The particles achieve the optimum result by its previous experiences, current velocity and its neighbor's best experience [11]. The techniques of PSO can produce high quality solution in less of calculation time and perform good convergences characteristic [8].

In PSO, there are two types of vector which is V_i and x_i described for each particles. Each particles will updated two value for each movement step of particle population. P_{best} is known as first value of best previous position of particle that calculate using fitness function. Next, the value of best particle amongst all P_{best} that known as G_{best} [12, 13].

All particles will update their new velocity and position using Eqs. (10)–(12) [14].

$$V_i^{k+1} = wV_i^k + c_1r_1(P_{best,i} - x_i^k) + c_2r_2(G_{best,i} - x_i^k) \tag{10}$$

$$x_i^{k+1} = x_i^k + V_i^{k+1} \tag{11}$$

$$w = w\text{max} - \frac{(w\text{max} - w_{\text{min}})}{k\text{max}} \times k \tag{12}$$

where, c_1, c_2 is acceleration constant in the range [0, 2], r_1, r_2 is uniform random value in the range [0, 1], $P_{best,i}$ is best previous position of particle *I*, $G_{best,i}$ is best particle among all $P_{best,I}$, V_i, x_i is velocity and position vectors of particle *i*, *k* is number of iteration, *w* is inertia weight factor, k_{max} is maximum iteration number and w_{min}, w_{max} is minimum and maximum inertia weight factor.

3.1 *Implementation of PSO for Optimal Power Generation in Microgrid System*

This section discussed the implementation procedures of Particle Swarm Optimization (PSO) with consider constraints handling for optimizing generator cost in microgrid [7, 15]. The implementation steps are as follows:

Step 1: Data Collection

Collect data for generation cost coefficient and boundary limits of distribution generation.

Step 2: Initializing

Initialize randomly number of particles, maximum iteration, acceleration constant and inertia weight factor. Then, each particle generated in range of power limit constraints using Eqs. (7)–(9).

Step 3: Fitness Evaluation

Evaluate fitness of each individual particles using the objective function in Eq. (5) and power constraint in Eq. (6).

Step 4: Memory Update

P_{best} and G_{best} are selected. P_{best} is personal best position obtain by particle while G_{best} is global best position among all the best. For first iteration, P_{best} same as randomly initialized. In next iteration, the position of particle having best fitness value defined as G_{best}.

Step 5: Update the Velocity and Position

All particle is updated their velocity and position vector using Eqs. (10)–(12).

Step 6: Constraints Handling

MIP constraints handling techniques are applied as discussed in [13].

Step 7: Update P_{best} and G_{best}

Repeat step 3 to evaluate the fitness. P_{best} are stored as a P_{best} if the current value is better than the previous value. Else, the previous value remained as P_{best}. The G_{best} value is defined when the best value for current P_{best}.

Step 8: Termination Process

Step 4–7 are repeated until maximum iteration is reached. Then, the final results are recorded.

4 Result and Discussion

This case study considered seven unit generators including two diesels, two wind turbines and three fuel-cell plants as illustrated in Fig. 1. The system and the data is taken from [2].

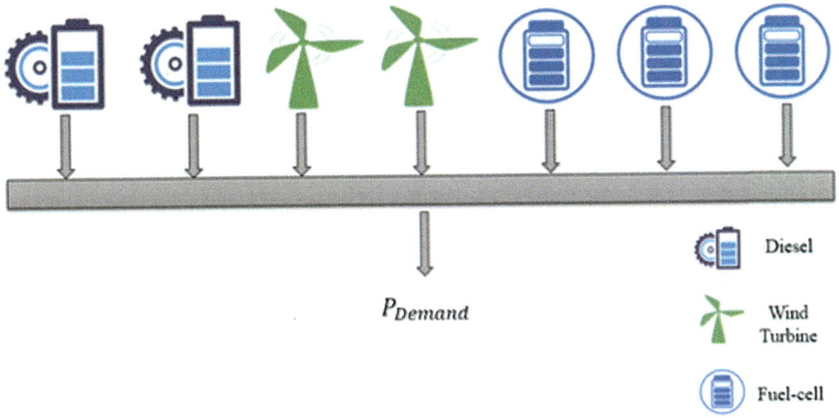

Fig. 1 7 Unit test system

Table 1 Statistical results for different number of particles

No. of particles		10	30	50	70
Total generation cost ($)	Minimum	31357.84	30849.70	30774.68	30726.34
	Average	32192.46	31585.46	31455.35	31352.29
	Maximum	33412.22	32696.13	32426.89	32194.56
Standard deviation		468.16	412.69	425.37	366.60
Computation time (s)		6.51	20.67	34.72	67.21

4.1 *Effect of Number of Particles*

The performance of PSO has been tested based on different number of particle as shown in Table 1. From the table, it can be seen that the higher the number of particles will be generated the lowest optimal cost. However, the simulation time is proportionally increased to the number of particle.

4.2 *Optimal Power Output and Comparison*

Based on Table 1, it shows that number of particle 70 can the lowest optimal cost ($30726.34) and satisfied all the constraints in (6)–(9). The optimal cost obtained by PSO has been compared with the results of other algorithms as presented in Table 2. It found that the PSO obtained the lower generation cost among the other methods. Therefore, PSO provided high of cost saving of $3204.6 compared to DE [15]. While CSA [15] has the lowest cost saving of $106.84. It revealed the effectiveness of the PSO for solving optimal microgrid.

Table 2 Comparison of the optimal cost and the total generation within 24 h

Method		DE [15]	CSA [15]	CSA [2]	FA [16]	PSO
Total generation cost within 24 h (kWh)	Diesel	14664	14482	14402	13763	13396
	Wind	6144	8860	6144	6144	6144
	Fuel-cell	7012	4477	7274	7913	8280
Total cost ($)		33930	33824	33440	31223	**30726.3**

5 Conclusion

This paper investigated the optimal operation scheduling in microgrid system in order to minimize the total generation cost within 24 h by PSO algorithm. The PSO algorithm is simulated in Matlab software to determine the optimal generation cost. Based on this study, PSO has produced the lower optimal cost of $30726.34 compared to other methods such as FA, DE and CSA. It can be concluded that PSO algorithm is capable to provide a good performances in terms of minimizing the total generation cost in microgrid system as well as higher cost saving of $3213.6 compared to DE.

Acknowledgements The authors gratefully appreciate the Universiti Tun Hussein Onn Malaysia (UTHM) under Incentive Grant Scheme for Publication (U684) and Department of Marine Electrical and Electronic Technology, Universiti Kuala Lumpur, Malaysian Institute of Marine Engineering Technology for supporting this research work.

References

1. Toma L, Tristiu I, Bulac C, Neagoe-Stefana AG (2016) Optimal generation scheduling strategy in a microgrid. In: 2016 IEEE Transportation Electrification Conference and Expo, Asia-Pacific, ITEC Asia-Pacific 2016
2. Modiri-Delsha M, Rahim NA, Taheri SS, Seyed-Shenava SJ (2014) Optimal generation scheduling in microgrids by Cuckoo search algorithm. In: 3rd IET Int Conf Clean Energy Technol 10(5)
3. Wei D, Zheng D (2014) Optimal energy management strategy for an isolated industrial microgrid using a modified particle swarm optimization, pp 494–498
4. Karthikeyan A, Manikandan K, Somasundaram P (2016) Economic dispatch of microgrid with smart energy storage systems using particle swarm optimization. In: 2016 international conference on computation of power, energy information and communication
5. Kumar KP, Saravanan B, Swarup KS (2016) Optimization of renewable energy sources in a microgrid using artificial fish swarm algorithm. Energy Procedia
6. Kamboj A, Chanana S (2017) Optimization of cost and emission in a renewable energy microgrid. In: 1st International Conference on Power Electronics, Intelligent Control and Energy Systems (ICPEICES)
7. Wu H, Liu X, Ding M (2014) Dynamic economic dispatch of a microgrid: mathematical models and solution algorithm. Int J Electr Power Energy Syst
8. Modiri-delshad M, Koohi-kamali S, Taslimi E, Aghay SH (2013) Economic dispatch in a microgrid through an iterated-based algorithm

9. Li P, Zhou Z, Lin X, Yang X, Niu X (2014) Dynamic optimal operation scheduling of microgrid using binary gravitational search algorithm
10. Lee K, Park J (2006) Application of particle swarm optimization to economic dispatch problem: advantages and disadvantages. In: 2006 IEEE PES Power Systems Conference and Exposition
11. Gaing Z (2003) Particle swarm optimization to solving the economic dispatch considering the generator constraints. IEEE Trans Power Syst, 1187–1195
12. Khan NA, Awan AB, Mahmood A, Razzaq S, Zafar A, Sidhu GAS (2015) Combined emission economic dispatch of power system including solar photo voltaic generation. Energy Convers Manag
13. Abdullah MN, Tawai R, Yousof MF (2017) Comparison of constraints handling methods for economic load dispatch problem using particle swarm optimization algorithm. Int J Adv Sci Eng Inf Technol 7:1322–1327
14. Madi M, Markovi D, Radovanovi M (2013) Comparison of meta-heuristic algorithms for solving machining optimization problems. Facta Universitatis 11:29–44
15. Basu M, Chowdhury A (2013) Cuckoo search algorithm for economic dispatch. Energy 60:99–108
16. Abdullah MN, Abdullah NL, Jamian JJ (2017) Optimal power generation in microgrid system based on firefly algorithm. In: 2017 6th International Conference on Electrical Engineering and Informatics (ICEEI), pp 1–6

Tensile Properties of Diffusion Bonded Duplex Stainless Steel to Low Carbon Steel

B. A. Baharudin, P. Hussain, M. Mustapha, F. Ayob, A. Ismail, F. Ab Rahman, P. Z. M. Khalid, D. A. Hamid and M. A. Rojan

Abstract The diffusion bonding is one of the methods used to join dissimilar metals. Specimens of duplex stainless steel and low carbon steel were joined by diffusion bonding under varied temperature and holding time. The specimens were clamped using jigs and heated in a furnace. Tensile test was performed on the joined samples. The effect of bonding time and holding temperature on tensile strength and quality of the diffusion bonds were observed. Tensile strength of the joints was examined and the highest tensile strength of 116.12 MPa are obtained at temperatures of 900 °C and holding time of 180 min.

Keywords Diffusion bonding · Duplex stainless steel · Dissimilar metals

1 Introduction

The development of fusion/arc welding technologies, to a certain degree, still pose problems in joining common materials such as high tensile steel, stainless steel, aluminium alloy, cast iron, cast steel, titanium, nickel alloy and others. With the development of advanced materials such as the super alloy metals and composites

B. A. Baharudin (✉) · F. Ayob · A. Ismail · F. Ab Rahman · P. Z. M. Khalid
Universiti Kuala Lumpur, Malaysian Institute of Marine Engineering Technology,
Jalan Pantai Remis, 32200 Lumut, Perak, Malaysia
e-mail: bakhtiarab@unikl.edu.my

P. Hussain · M. Mustapha
Department of Mechanical Engineering, Universiti Teknologi PETRONAS,
32610 Seri Iskandar, Perak, Malaysia

D. A. Hamid
Malaysia Italy Design Institute, Universiti Kuala Lumpur, 119, Jalan 7/91,
Taman Shamelin Perkasa, 56100 Kuala Lumpur, Malaysia

M. A. Rojan
School of Mechatronic Engineering, Universiti Malaysia Perlis, Kampus Putra,
02600 Arau, Perlis, Malaysia

C. L. Saw et al. (eds.), *Advancement in Emerging Technologies and Engineering Applications*, Lecture Notes in Mechanical Engineering,
https://doi.org/10.1007/978-981-15-0002-2_34

such as metal matrix composites, the challenges faced by the wide range of industries to produce quality joints are even much greater.

In this study, diffusion bonding which is relatively new as compared to the fusion/arc welding aim to explore a potential solution to some of the problems as mentioned above. The materials being chosen in this study are duplex stainless steel (DSS) and marine grade low carbon steel. DSS can be fusion-welded with steels of any grade. However, the high heat input during fusion welding causes the formation of brittle phases effects, which reduces the corrosion resistance properties especially in the heat affected zone area. Diffusion bonding method could prevent microstructural problems caused by fusion welding method, while providing a joint that is significantly stronger than that produced by other non-fusion processes.

Diffusion bonding is a very attractive technology for the advanced materials and the process is dependent of a series of variables: time, applied pressure, bonding temperature, atmosphere, material characteristics, roughness and contamination degree of mating surfaces [1]. Both similar and dissimilar materials can be joined by diffusion bonding [2, 3]. Vigraman et al. [4] performed diffusion bonding of DSS to medium-carbon steel by applying heavy pressure at elevated temperature and their strength and characterization have been already reported in the literatures [5, 6]. The issues that have been identified to be addressed in this research are the capability of the joints produced by the diffusion bonding method and the suitable conditions to join dissimilar metals. The objective of this research is to determine the relationship of the diffusion bonding process variables such as temperature and holding time on the mechanical properties of the joints produced.

2 Methodology

Further information on the experimental setup are presented.

2.1 *Equipment for Diffusion Bonding*

The main equipment used for diffusion bonding is the furnace which provided heating temperature. The samples are clamped together by using jigs. As shown in Fig. 1, a plate thickness of 20 mm was cut into a rectangular shape with the length of 2.5 inches and 1.0 inches width. Two holes of 13 mm diameter were drilled onto the plate at 15 mm from the side of the jigs to accommodate bolts and nuts.

Fig. 1 Jig used to clamp the specimen

2.2 *Specimen Preparation for Diffusion Welding*

Materials used in this study as parent metals were commercial duplex stainless steel (UNS S31803) and marine grade low carbon steel (ASTM A36). The materials were cut into 2.5 inch length each as required for the diffusion welding. The samples were turned using lathe machine to reduce their diameter from 1 to 0.5 inch as shown in Fig. 2. After the specimens had been prepared, inspection has been conducted to make sure the specimens follow the dimension required for the diffusion bonding process. The specimen ends that are to be joined were polished on 240 grit emery paper and cleaned with alcohol to remove any loose grit or dirt and grease or other contaminants. The specimen is then blown dry. Specimen of DSS and ASTM A36 were aligned together and joined as required by ASTM E8 as shown in Fig. 3.

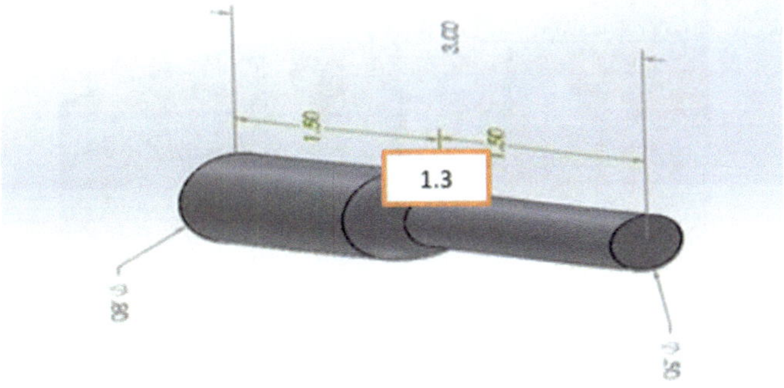

Fig. 2 Specimen Dimension

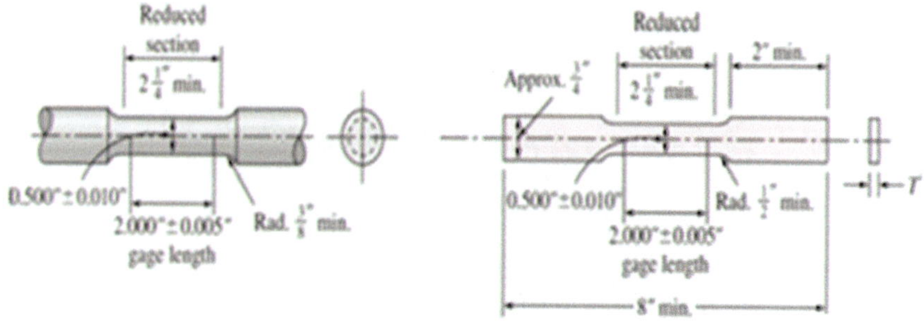

Fig. 3 Dog Bone Specimen According to ASTM E8 Standard [7]

Next, a mixture of boron nitride and acetone were evenly coated on to the jig's surfaces and specimen's surfaces to avoid the from getting attached during the diffusion bonding process. The DSS specimen and ASTM A36 specimen was placed in joining position between two bolts and nuts. Force were applied onto the bolts and nuts using a torque wrench at the reading of 70 Nm. The two-mating surface of specimen were aligned as shown in Fig. 4.

Diffusion Bonding Procedure. The specimens were positioned in the furnace as shown in Fig. 5. The furnace was heated up at the rate of 36.7 °/min until it reaches the required holding temperature. The holding temperature was maintained as the required time for example for specimen S1, the holding time was 180 min. When the required time was reached, the furnace was turned off. The specimen was left to cool

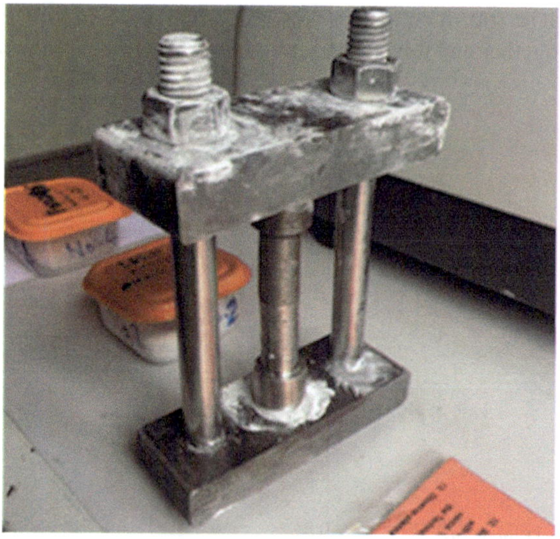

Fig. 4 Specimen aligned and clamped using jigs

Fig. 5 Clamped specimen in the furnace

down at furnace temperature until the temperature reaches room temperature. The specimen was then taken out from the furnace and the joint produced was inspected.

Tensile Test. In this study, tensile testing was used to measure the level of success of diffusion bonded joint. This testing was conducted by referring to ASTM E8 Standard Test Methods for Tension Testing of Metallic Materials [7]. By conducting this mechanical test, the time required for the perfect joining was investigated.

3 Results and Discussion

The results for the tensile strength are shown in Fig. 6. From the test, the highest tensile strength obtained was 116.12 MPa at the process parameters of 900 °C and

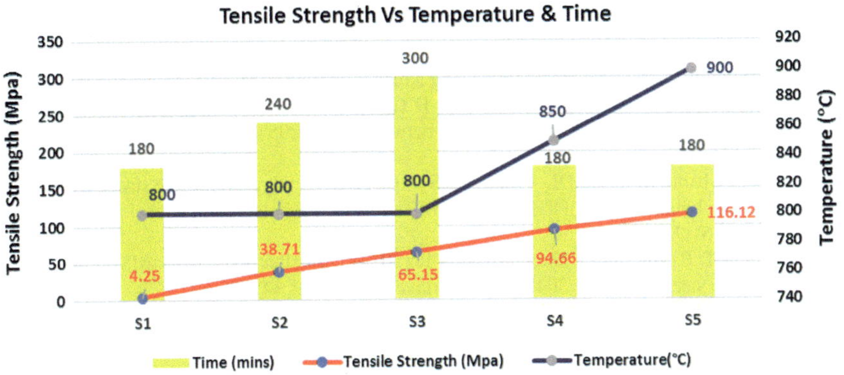

Fig. 6 Graph of Tensile Strength versus Temperature and Time

180 min from sample S5. This condition maybe due to diffusion bonded interface. This can be attributed to higher diffusion efficiency and more complete bonding process for the bonds made at 900 °C. Sample S1, S2, S3 and S4 shows poor tensile strength of below 100 MPa. This could be attributed to the poor contact between the bonding surfaces and lack of diffusion.

4 Conclusion

As a conclusion, the experiment of diffusion bonding of dissimilar materials of ASTM A36 and 2205 DSS mechanical properties values were investigated. The following are the outcomes deduced from the investigation:

a. Dissimilar metals of ASTM A36 and 2205 DSS could be joined by diffusion bonding method at temperatures of 800 °C, with holding time of 180 min and clamped by jigs. However, the tensile strength is quite low due to lack of diffusion
b. The highest tensile strength is obtained at the parameter of 900 °C and 180 min with the value of 116.12 MPa.
c. Further investigation is needed to obtain the optimum tensile strength for the diffusion bonding of DSS and ASTM A36.

Acknowledgements The authors thanked both universities; Universiti Teknologi PETRONAS and Universiti Kuala Lumpur for supporting this research work.

References

1. Lucia L, Georgeta A, Icpeca I (2007) Materials bonding by diffusion welding technology. Conferinţa Internaţională, Innovative Technologies For Joinning Advanced Materials, June, Timisoara, Romania, pp 134–139
2. Kundu S, Ghosh S, Laik A (2005) Diffusion bonding of commercially pure titanium to 304 stainless steel using copper interlayer. Mater Sci Eng A 407:154–160
3. Kundu S, Chartterjee S (2006) Interfacial microstructure and mechanical properties of diffusion bonded titanium–stainless steel joints using a nickel interlayer. Mater Sci Eng A 425:107–113
4. Vigraman T, Narayanasamy R, Ravindran D (2012) Microstructure and mechanical property evaluation of diffusion-bonded joints made between SAE 2205 steel and AISI 1035 steel. Mater Des 35:156–169
5. Kundu S, Mishra B, Olson DL, Chatterjee S (2013) Interfacial reactions and strength properties of diffusion bonded joints of Ti64 alloy and 17-4PH stainless steel using nickel alloy interlayer. Mater Des 51:714–722
6. Vigraman T, Ravindran D, Narayanasamy R (2012) Diffusion bonding of AISI 304L steel to low-carbon steel with AISI 304L steel interlayer. Mater Des 34:594–602
7. ASTM E8/E8M—Standard test methods for tension testing of metallic materials. No. C. ASTM Int., West Conshohocken, Pennsylvania, USA. 1–27 (2016)

Carbon Emissions Measurement Using Portable Emission Device in Coastal Fishing Boats

Ismail Zainol, Z. M. Ahmad, S. N. Ridwan, Z. M. Zaifurizal, M. R. Dahalan, A. Kamal and Zulzamri Salleh

Abstract This paper measure the carbon emission from small scale fishing boats. Carbon emissions were measured from two types Class A Malaysian fishing boats marine engines covering various size of inboard engines and outboard engines. The gas emission analyser MSA Altair 5x used is portable and easy to handle. Selected fishing boats were sampled from a total of 664 fishing boats operates in Manjung, Perak. The results of carbon emission measurement were presented in parts per million (ppm). The outboard powered fishing boats recorded the highest carbon value in ppm about 2000 ppm meanwhile inboard powered fishing boats constitutes slightly less carbon emission with 379 ppm. The results signify that outboard engine fueled by petrol producing more carbon value if compared with inboard diesel engine. These results can be an initial data for carbon emission inventory for local fishing boats.

Keywords Carbon emission · Fishing boats · Altair 5x

1 First Section

Emission issues are always become an important topic over the decades as it is contributing to air quality level. Emission come from various source including marine transport. In general, the marine transport responsible about 2.8% of global carbon

---

I. Zainol (✉) · Z. M. Zaifurizal · M. R. Dahalan · A. Kamal · Z. Salleh
Universiti Kuala Lumpur, Malaysian Institute of Marine Engineering Technology, Dataran Industri Teknologi Kejuruteraan Marin, Bandar Teknologi Maritim, Jalan Pantai Remis, 32200 Lumut, Perak, Malaysia
e-mail: ismailz@unikl.edu.my

Z. M. Ahmad
Faculty of Mechanical Engineering, Universiti Teknologi Malaysia, Skudai, Malaysia

S. N. Ridwan
Department of Marine Engineering, Centre of Technology in Marine Engineering, Politeknik Ungku Omar, 31400 Ipoh, Perak, Malaysia

C. L. Saw et al. (eds.), *Advancement in Emerging Technologies and Engineering Applications*, Lecture Notes in Mechanical Engineering,
https://doi.org/10.1007/978-981-15-0002-2_35

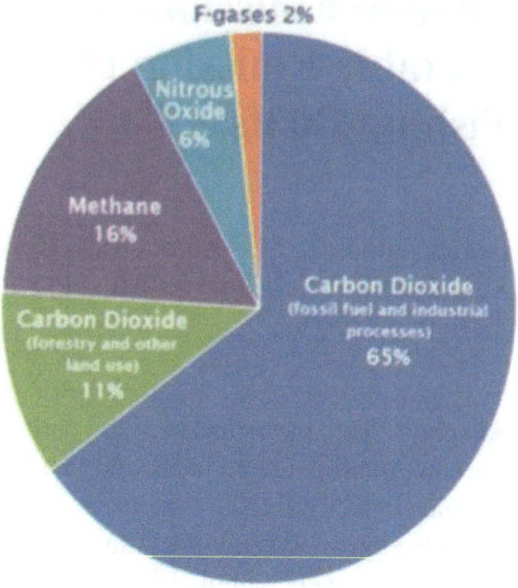

Fig. 1 Global emissions shares

emission and expected to be increase by 50–250% in period to 2050 [1]. The increasing demand for marine transport service has cause the rise of emission as well. In case of fishing industry, the carbon emissions from fishing vessels still make a significant contribution if not much. Fishing vessels are heavily dependent to fossil fuels such as diesel and petrol to generate power. Combustion of fossil fuel however would generate greenhouse gas (GHG) emission and the concomitant impact on the environment.

The main exhaust gas emissions from ships include CO_2, NO_x, SO_x, CO, hydrocarbons, and particulate matter. As there are raised concern on reduction of emissions, greenhouse gases (GHG) measure from fuel uses has become important aspect to be addressed. Among the gas emission, carbon emission is considered as the most important anthropocentric GHG as illustrated in Fig. 1. Carbon emissions are form of CO and CO_2. Numerous papers have been published on determining CO_2 gas [2] and compared to CO gas. Carbon monoxide (CO) is a tasteless, odorless and colorless gas. It is a by-product of partial combustion of organic compounds. This paper focus to use portable emission to measure CO gas. However, others emission gaseous can be measured using the similar method as well depending on the sensor equipped.

Emissions can be estimated using various method and formula. There were many study in emission estimation with various methodologies such highlighted [2–8]. In general, emission can be determined through direct measurement or empirical approach. The most widely used method of estimating emissions from ships is through an empirical approach. A complete methodology was developed in 2006 [9]. Later in 2009, EMEP/EEA proposed the estimation model which categories into 3 methods namely Tier 1, Tier 2 and Tier 3 [10]. Tier 1 method is the simplest method used to estimate emission by using emission factor. Tier 2 can be regarded as updated version of Tier 1 which default emission factors were replaced by country-specific emission factors. Emission factor is a representative value that attempts to relate the

Table 1 Emission factor from Lloyd Register

Substance	Emission factor scientific notation (kg/kWh)
CO	0.0016
CO_2	0.66
HC	0.0005
SO_2	0.0064
NO_x slow	0.017
NO_x medium	0.012
PM fuel oil	0.0015
PM gas oil	0.0002

quantity of a pollutant released to the atmosphere with an activity associated with the release of that pollutant. These factors are usually expressed as the weight of pollutant divided by a unit weight, volume, distance, or duration of the activity emitting the pollutant. Such factors facilitate estimation of emissions from various sources of air pollution. In most cases, these factors are simply averages of all available data of acceptable quality, and are generally assumed to be representative of long-term averages for all facilities in the source category [11]. The Lloyds Register (2015) [12] provide the emission factor (kg/kWh) of different substances for diesel engines as listed in Table 1.

The field measurement of emission will provide more accurate data rather than simply using existing emission factor however costly and time consuming. Hence, the empirical calculation using emission factor is considered more preferably due to simple calculation and cost effective. However, there is an accuracy issue on emissions estimation from local fishing boats when some inputs emission factors for local fishing boats are not exist whereby the recent GHG data are hard to be applied due to localized factor. Furthermore, the lack of local inventory data of emission factors. Several studies have been conducted to determine emission from Malaysian fishing vessel as carried out by Saad [13]. Saad has determined carbon emission from fishing vessel using tier 1 method waterborne navigation method. The author measure at four fisheries districts in Selangor during fishing vessels anchored for operation using emission factor derived from EPA. It is important to mention that emission estimation using non localized emission factor would not reflect to the actual results. The regulations of controlling emission from marine transport was established by IMO in 2005 [14]. The International Maritime Organization (IMO) adopted an international convention protocol to reduce air pollution from ships, in order to achieve sustainable maritime development. This protocol approved by 15 member countries and enforced in May 2005. Thus, a field emission measurements of Malaysian fishing vessels is important steps towards developing a country emission inventory for fishing industry.

2 Methodology

The gas emission analyzer MSA Altair 5x was used for measuring the carbon emission. The device specifications of the exhaust gases were shown in Fig. 2 and Table 2. Altair 5x is a multi-gas detector that able to detect combustible gases and certain combustible vapors, oxygen-deficient or oxygen-rich atmospheres and specific toxic gases for which a sensor is installed. This device measure gas concentration in particle per minute (ppm) In house calibration was made to ensure the accuracy of device

Fig. 2 Altair 5x

Table 2 Altair 5x specifications

Dimensions (L × W × D)	6.69″ H × 3.49″ W × 1.79″ D without belt clip or IR sensor
Weight	1 lb (without IR sensor)
Dropt test	10 feet
Audible alarm	>95 dB typical
Visual alarm	2 ultra-bright LEDs, on top
Battery	Rechargeable Li-ION or AA alkaline
Run time	20 h @ room temperature
Charging time	≤6 h
Operating temperature	−20 °C to +50 °C
Short-period operation	−40 °C to +50 °C
Data log	Adjustable, 200 h minimum

at accepting level according to manufacturer manual. Test preparation is set according to equipment specification as shown in Table 2 before field measurement was done.

2.1 Field Test Location and Boat Particulars

For research area, fishing community in Manjung, Perak has been selected as case study. These locations were selected due to large number of fishing boats operation which constitutes about 664 fishing boats [15]. Exhaust carbon emission tests were made both from inboard fishing boat and outboard powered fishing boat as shown in Figs. 3 and 4 respectively. The activities of fishing operations are divided into four zones fishing operations A, B, C, and C2. For this study class a fishing boat was selected. Class A is classified fishing zone less than 5 or 8 nautical miles from shore, reserved solely for small-scale fishers using traditional fishing gear and owner-operated vessels. Class A fishing boats engines come in two broad categories outboard or inboard. These different engine types have markedly different emissions characteristics. Engine horsepower selected according the Number of Licensed Fishing Vessels by State and Horsepower on 2015 issued by the department of fisheries Malaysia as shown in Table 3 [15]. The most widely used craft was the wooden

Fig. 3 Inboard fishing boats

Fig. 4 Outboard fishing boats

Table 3 Number of licensed fishing vessels by horsepower [15]

Engines (HP)	Inboard	Outboard
20–39	820	1370
40–59	65	1001
60–99	209	425
100–149	601	19

Table 4 Engine capacity

Boats	Engine capacity (HP)	Fuel type
Outboard A	60	Petrol RON 95
Outboard B	90	Petrol RON 95
Outboard C	115	Petrol RON 95
Inboard A	24	Diesel
Inboard B	60	Diesel
Inboard C	113	Diesel

vessel. All vessels were motorized and the engine horsepower (HP) ranged from 10 to 149 hp. The average of inboard wooden fishing boat not exceed 24 m in length according to small fishing vessel regulation.

2.2 *Engines and Fuels Data*

The analysis of the estimation of fishing vessel emission is based on observation of fuel oil consumption from fish catching activity and engine power rating. Table 4 shows a statistic of inboard engines and outboard engines for Peninsular Malaysia in 2016. Majority inboard engine in range of 20–39 HP meanwhile majority fishermen equipped their boats with engines in range of 20–39 HP. The emission tests were conducted from both inboard and outboard engines as details out in Table 3.

2.3 *Sampling Procedures*

The emission level increase proportional to combustion level which indicated by engine load. Engine loading vary according to boat activities. For fishing boat activities, it can be divided into 3 phase as suggested by Ismail [16] and shown in Fig. 5. Ismail discovered the maximum speed of fishing boats. However, by considering the fact that maximum emission level emitted when the boat at maximum load. Hence, to simulate the boat activities, it is sufficient to run the engine at idle and full throttle.

Turn on the multi gas detector by pressing the power button. The user should verify that the alarm LEDs flash, horn sounds briefly and vibrating alarm triggers

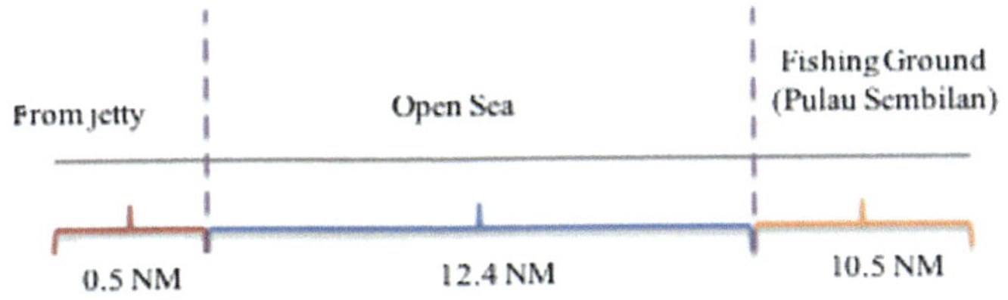

Fig. 5 Fishing route

Fig. 6 Sampling locations at exhaust outlet outboard

briefly. Block the pump inlet to perform the pump test. After pump test passed, all the sensors will warm up before measurement can be taken. After all the sensors have been warmed up, set the fresh air setup before performing the carbon monoxide testing. The sampling location shows at Figs. 6 and 7.

3 Results and Discussions

Results of emission were read directly from multi gas detector as shown in Figs. 8 and 9. Three (3) reading were taken for each engines capacity and presented in Tables 5 and 6.

Many for CO emission reading in 60 HP engine was recorded as 1453 ppm during idle and increased to 2000 ppm at full throttle. Table 5 shows the results of outboard

Fig. 7 Sampling locations at exhaust outlet Inboard

Fig. 8 Snapshot of outboard reading

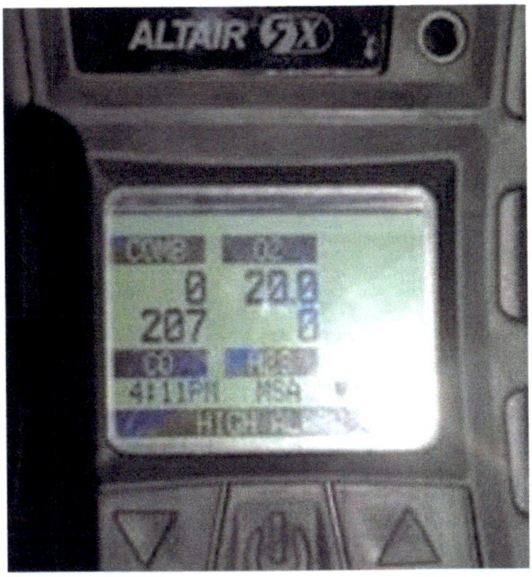

Fig. 9 Snapshot of inboard reading

Table 5 Outboard results

Engine (HP)	Idle (ppm)				Full throttle (ppm)			
	1	2	3	Means	1	2	3	Means
60	1309	1049	2000	1453	2000	2000	2000	2000
90	310	287	286	294	384	382	368	378
115	424	402	382	403	488	507	506	500

Table 6 Inboard Results

Engine (HP)	Idle (ppm)				Full throttle (ppm)			
	1	2	3	Means	1	2	3	Means
24	96	88	98	94	370	385	381	379
50	182	207	208	199	333	351	343	342
115	244	254	251	250	288	342	360	330

engines for 60, 90 and 115 HP. Similar results from inboard engines were presented in Table 6. Based on these table, the graphs are plotted for different engine capacity as shown in Figs. 10 and 11.

As referred from the graph, highest value was recorded by 115 HP engine which about 500 ppm. Meanwhile for inboard engine is about 379 ppm. Comparison between 115 HP outboard engine and 113 HP inboard engine has shown major variance in term of CO concentration despite both engines have in range engine size. This is due to difference fuel and engine system were used. The results show that CO value increase proportionally to the engine capacity as more fuel was burn for larger engines power. The power requirement for one particular engine may vary depending on various condition such hoteling, cruise and fishing operation mode.

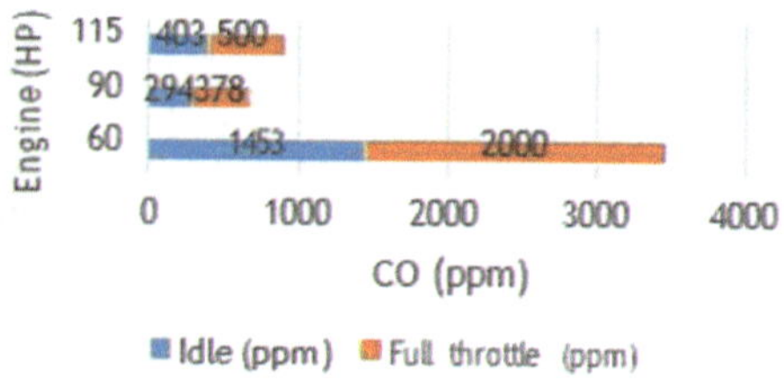

Fig. 10 Outboard engine carbon monoxide results

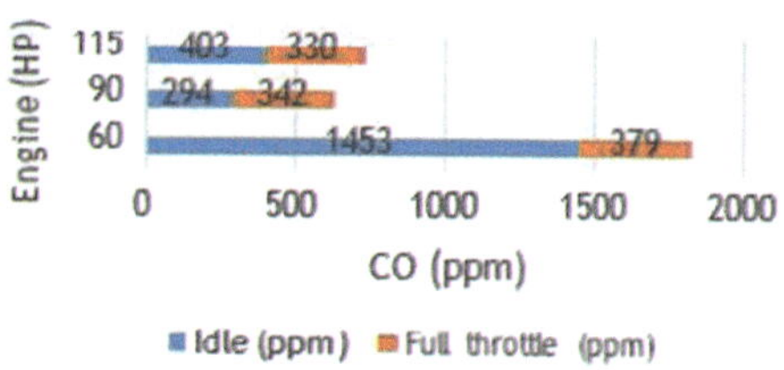

Fig. 11 Inboard engine carbon monoxide results

In addition, the quantity of emission produced are much influenced by combustion efficiency, engine condition and engine lifetime. New purchased engine is mostly to give better emission reading than older engine.

4 Summary

Exhaust emission test has been successfully carried out to determine CO gas in Malaysian fishing boats. Portable emission device has shown comparable results with other gas detection instruments. Additionally, it easy to use, less cost, less weight and user friendly. It can be owned by fishermen and this could help them to self-monitor exhaust emission of their boats. Carbon emission data for Malaysia Class A fishing boats will assist authority bodies to monitoring the contribution level of emissions in Peninsular of Malaysia. Sources of emission can be used as decision support for national rule making and for choice of preventive action. Further works will benefit from the data collected by field measurement to generate the new emission factor that will improve the empirical emission estimation for class A fishing boats.

References

1. Smith TWP, Jalkanen JP, Anderson BA, Corbett JJ, Faber J, Hanayama S, Raucci C et al (2014) Third imo ghg study 2014. International Maritime Organization (IMO), London, http://www.iadc.org/wp-content/uploads/2014/02/MEPC-67-6-INF3-2014-Final-Report-complete.pdf
2. Trozzi C (2010) Emission estimate methodology for maritime navigation. Techne Consulting, Rome

3. Khan MY, Ranganathan S, Agrawal H, Welch WA, Laroo C, Miller JW, Cocker DR III (2013) Measuring in-use ship emissions with international and US federal methods. J Air Waste Manage Assoc 63(3):284–291
4. Mustapa SI, Bekhet HA (2016) Analysis of CO_2 emissions reduction in the Malaysian transportation sector: an optimisation approach
5. Park JA, Gardner C, Chang MI, Kim DH, Jang YS (2015) Fuel use and greenhouse gas emissions from offshore fisheries of the Republic of Korea. PLoS ONE 10(8):e0133778
6. Balzani Lööv JM, Alfoldy B, Gast LF, Hjorth J, Lagler F, Mellqvist J, Swart DP (2014) Field test of available methods to measure remotely SO_x and NO_x emissions from ships. Atmos Measure Tech 7(8):2597–2613
7. Saputra H, Maimun A, Koto J (2013) Estimation and distribution of exhaust ship emission from marine traffic in the straits of Malacca and Singapore using Automatic Identification System (AIS) data. Jurnal Mekanikal 36:86–104
8. Arapatsakos C, Karkanis A, Anastasiadou C (2015) The load and the gas emissions measurement of outboard engine. Int J Heat Technol 33(4):221–228
9. Trozzi C, Vaccaro R (2006) Methodologies for estimating air pollutant emissions from ships: a 2006 update. In: Poster presented at 2nd Environment & Transport Conference (including 15th Transport and Air Pollution conference, pp 12–14
10. EEA E (2016) EEA air pollutant emission inventory guidebook 2016, Technical guidance to prepare national emission inventories. Report, 21, 2016
11. About Emissions Factors (nd) Retrieved from United States Environmental Protection Agency: https://www.epa.gov/air-emissions-factors-and-quantification/basic-information-air-emissions-factors-and-quantification
12. Entec UK Limited (2002) Quantification of emissions from ships associated with ship movements between ports in the European Community. European Commission Final Report
13. Saad N (2013) Contribution fuel consumption of fishing vessel operation to greenhouse gas emission. From Sources to Solution, pp 343–346
14. IMO International Convention for the Prevention of Pollution from Ships MARPOL 73/78. The Regulations for the Prevention of Air Pollution from Ships (Annex VI) London (1997)
15. Perangkaan Perikanan Tahunan 2016 (2016) Retrieved from Jabatan Perikanan Malaysia: https://www.dof.gov.my/dof2/resources/Perangkaan/jadual_nelayan_pdf_2016.pdf
16. Zainol IB, Yaakob OB, Zulkifli SNB, Jimata NS (2014) Feasibility of using surface piercing propellers and diesel engine, science and engineering

Friction Stir Welding on AA5083 Tee Joint in 1F Position

A. Ismail, M. Awang, M. S. M. Zuhir, F. A. Rahman, B. A. Baharudin, M. K. Puteri Zarina, D. A. Hamid, M. A. Rojan and W. M. Dahalan

Abstract Friction stir welding (FSW) is a recent technique that exploits a non-consumable rotating welding tool to generate frictional heat and plastic deformation at the welding location. The principal advantages of FSW, being a solid-state process, are low distortion, absence of melt-related defects and high joint strength, even in those alloys that are considered non-weldable by conventional techniques. Hard materials such as steel and other important engineering alloys can now be welded efficiently using this process. The understanding has been useful in reducing defects and improving uniformity of weld properties and expanding the applicability of FSW to new engineering alloys. This project focuses mainly on the experimental study of FSW on AA5083 T-joint in 1F position. This study investigates the jig fixture and tool pin design with effects of pin rotation speed on the macrostructure of the joint. The present work deals with an experimental campaign aiming on FSW on aluminium alloy 5083 T-joint. Tool pins are fabricated using material high tensile steel H13 with angle shoulder with heat treatment process for surface-hardened tools.

Keywords Friction stir welding · Aluminium alloy · T-joint · 1F position · MILKO 37 conventional milling machine

A. Ismail (✉) · M. S. M. Zuhir · F. A. Rahman · B. A. Baharudin · M. K. Puteri Zarina · W. M. Dahalan
Universiti Kuala Lumpur, Malaysian Institute of Marine Engineering Technology, Jalan Pantai Remis, 32200 Lumut, Perak, Malaysia
e-mail: azman@unikl.edu.my

M. Awang
Department of Mechanical Engineering, Universiti Teknologi PETRONAS, 32610 Seri Iskandar, Perak, Malaysia

D. A. Hamid
Malaysia Italy Design Institute, Universiti Kuala Lumpur, 119, Jalan 7/91, Taman Shamelin Perkasa, 56100 Kuala Lumpur, Malaysia

M. A. Rojan
School of Mechatronic Engineering, Universiti Malaysia Perlis, Kampus Putra, 02600 Arau, Perlis, Malaysia

C. L. Saw et al. (eds.), *Advancement in Emerging Technologies and Engineering Applications*, Lecture Notes in Mechanical Engineering,
https://doi.org/10.1007/978-981-15-0002-2_36

1 Introduction

Joining technologies are in an exciting phase today due to some relevant innovations concerning dissimilar aspects. One innovative process to be considered is that of friction stir welding (FSW) [1]. FSW is a new joining process invented by Wayne Thomas, from The Welding Institute (TWI) in England and patented in 1991. FSW is a solid state joining method based on friction heating and local plastic flow in the joint region by stirring with a rotating tool pin. Basically all wrought and cast aluminum alloys can be joined by FSW process [1].

A substantial benefit of FSW is that it has significantly fewer process fundamentals to control. In a fusion welding, there are many process factors that must be controlled such as purge gas, voltage and amperage, wire feed, travel speed, shield gas, arc gap [2]. In FSW, there are only three process variables to control which are the rotation speed, travel speed and pressure. FSW can be used to produce butt, corner, lap, tee, spot, fillet and hem joints, as well.

This method can produce joints utilizing equipment based on conventional machine tool technologies, and it has been used to weld a range of similar and dissimilar alloys as well as for welding metal matrix composites and the for the repair of current joints [3]. It enables to weld almost all types of aluminum alloys, even the one classified as non-weldable by fusion welding due to hot cracking and poor solidification microstructure in the fusion zone [4]. FSW is a green technology due to its energy efficiency, environment friendliness, and versatility. FSW involves complex material movement and plastic deformation. Welding parameters which are the tool geometry and joint design exert significant effect on the material flow pattern and temperature distribution, thereby influencing the micro structural evolution of material [5, 6].

2 Methodology

Technical specification of the milling machine is shown in Table 1. It is important to understand the capabilities and the limitations of the machine for a smooth process. The setup of the experiment is shown in Fig. 1.

Tables 2 and 3 show the chemical and mechanical properties of AA5083 respectively. Meanwhile, the chemical and mechanical properties of AISI H13 high carbon steel is shown in Tables 4 and 5 respectively.

The jig fixture works as a clamp to hold the work piece during the welding process. The jig are made from the sheet metal of mild steel, angle iron, solid rod of mild steel and pillow block bearing. The design of the jig was based on the specimen size and the position of welding process. The setup of jig on Milko 37 milling machine is shown in Fig. 2. The straight cylindrical tapered tool pin design is used in this setup. The diameter of both shoulders are 20 mm as shown in Fig. 3.

Table 1 MILKO37 milling machine specification

Working surface	(1300 × 300) mm
T slot (5)	16 mm
Both sides Swivel	45°
Traverses	
Longitudinal	970 mm
Cross	315 mm
Vertical	450 mm
Spindle	*ISO-40*
Standard Taper	12
Number of speeds	50–1700 r.p.m.
Range	
Work feeds	
Number of feeds	18
Longitudinal and cross	16–1080 mm/min
Vertical	6–430 mm/min
Rapid feed	
Longitudinal and cross	2000 mm/min
Vertical	800 mm/min

AA5083 was chosen for the experimental study of friction stir welding in 1F position. The AA5083 is the marine grade alloy which is widely used in heavy marine industry due to lightweight and corrosion resistance. The specimens of AA5083 which are sized 200 mm × 200 mm skin and 200 mm × 100 mm stringer with a thickness of 6 mm (width) based on the previous literature [1].

For the welding process, the welding parameters of tool rotational speed and traverse speed on milling machine were set according to the preset speed on the Milko 37 conventional milling machine. The determination of welding parameters was an important part in this experiment because it will determine the acceptance or rejection of the weld quality. The welding parameters used for this experiment setting is shown in Table 6. The setup during the friction welding process is shown in Fig. 4.

The specimens were cut to small samples according to AWS D17.3 by a vertical band saw unit. The cross cut of FSW specimen were then mounted by the mounting press machine. Sanding and polishing on the cross section surface was performed to inspect the joining cross section under optical microscope with the magnification of 10×.

Fig. 1 Milko 37 milling machine

Table 2 Chemical compositions of AA5083 [7]

Element	% Present
Manganese (Mn)	0.40–1.00
Iron (Fe)	0.40 Typical
Copper (Cu)	0.10 Typical
Magnesium (Mg)	4.00–4.90
Silicon (Si)	0.40 Typical
Zinc (Zn)	0.25 Typical
Titanium (Ti)	0.15 Typical
Aluminium (Al)	Balance

Table 3 Mechanical properties of AA5083 at room temperature [7]

Property	Value
Proof strength	145 MPa
Tensile strength	300 MPa
Elongation	23%
Shear strength	175 MPa
Hardness Vickers	75 HV

Table 4 Chemical composition of AISI H13 Carbon Steel [8]

Element	Content (%)
Chromium, Cr	4.75–5.50
Molybdenum, Mo	1.10–1.75
Silicon, Si	0.80–1.20
Vanadium, V	0.80–1.20
Carbon, C	0.32–0.45
Nickel, Ni	0.3
Copper, Cu	0.25
Manganese, Mn	0.20–0.50
Phosphorus, P	0.03
Sulfur, S	0.03

Table 5 Mechanical properties of AISI H13 at room temperature [8]

Properties	Metric
Tensile strength, ultimate (@20 °C/68 °F, varies with heat treatment)	1200–1590 MPa
Tensile strength, yield (@20 °C/68 °F, varies with heat treatment)	1000–1380 MPa
Reduction of area (@20 °C/68 °F)	50.00%
Modulus of elasticity (@20 °C/68 °F)	215 GPa
Poisson's ratio	0.27–0.30

Fig. 2 Welding jig for T-joint setup on Milko 37 milling machine

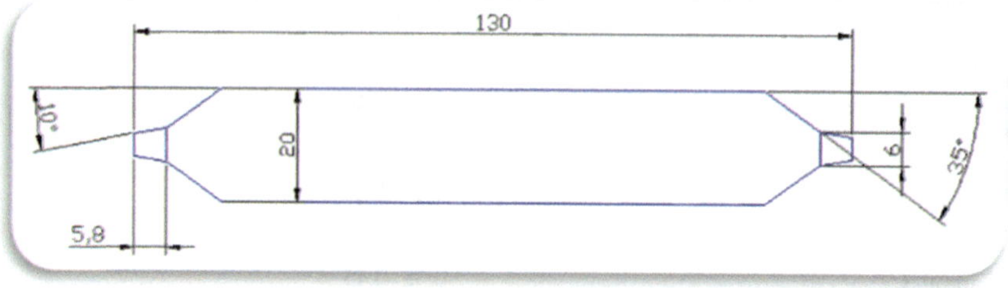

Fig. 3 Surface-hardened H13 welding tool

Table 6 Welding parameters for FSW

Specimen No.	Rotation speed (rpm)	Traverse speed (mm/min)
FSW 1	910	16
FSW 2	910	29
FSW 3	910	44
FSW 4	1280	44
FSW 5	1700	44

Fig. 4 Weld setup during friction stir welding process

3 Results and Discussion

The results for the surface conditions for all FSW specimen are shown in Figs. 5 and 6.

Various designs were analyzed in order to produce this jig. Jig can be moved to the left and right so that both sides of the specimen can be welded easily without additional time to setup. Plates can be joined together in T-joint configuration by using this jig with rough surface and under-filling. Rough surface and underfill weld surface happened due to absence of stationary shoulder to hold the materials intact

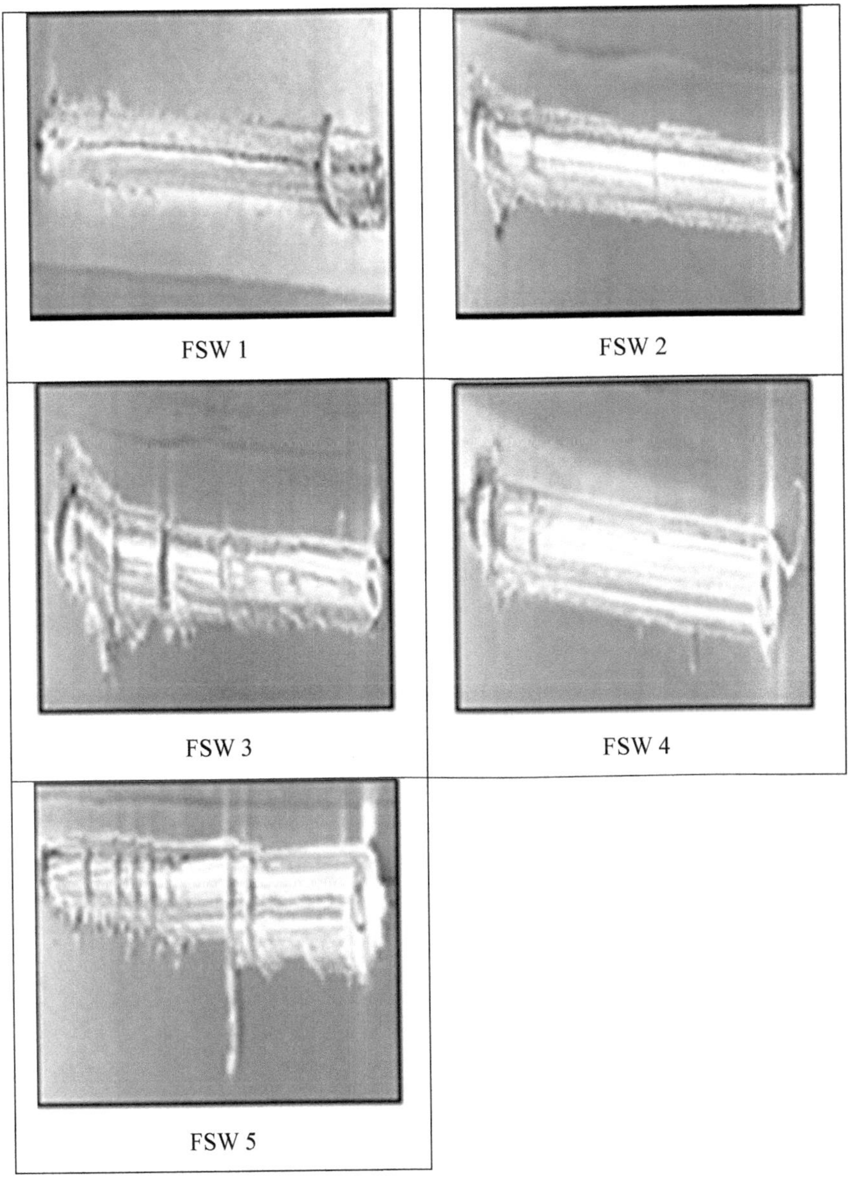

Fig. 5 Visual inspection on weld surface for FSW 1-5

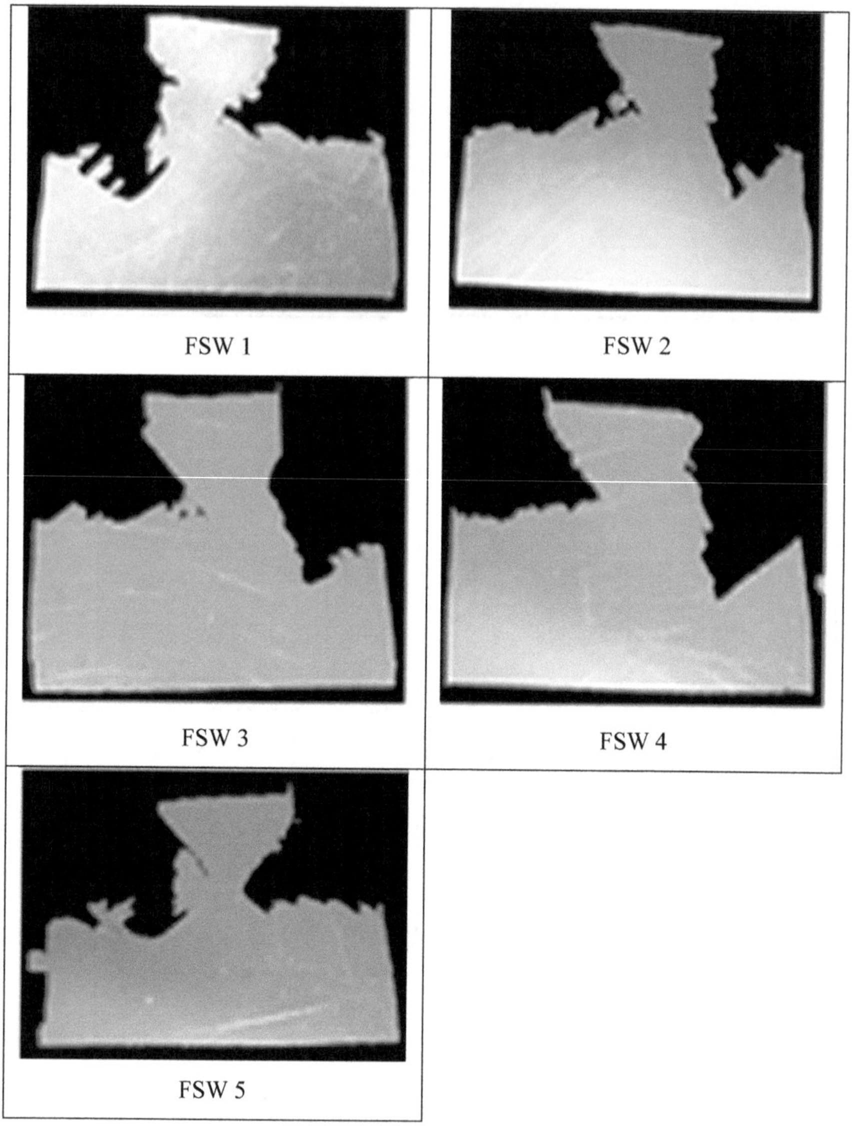

Fig. 6 Cross section view for FSW 1-5

together during the friction process. The visual inspection describes that all the specimens had a low quality of weld seam.

The cross section inspection showed the same result. It happened due to the retracting side was not in contact with the advancing side of the tool pin. With the increase of the traveling speed, the quality of the weld seam increased due to the suitable travel speed at 44 mm/min. With the increment of the rotational speed, it caused excessive caving due to plastic region being spitted out due to excessive rotational speed. The most suitable rotational speed rate was 910 rpm.

4 Conclusion

This experiment has successfully shown the effects of the absence of stationary shoulder friction stir welding (SSFSW) on the weld structure during the welding process. This is the contributing factor to poor quality of the weld seam. This experiment managed to indicate a suitable clamp/jig fixture for T-joint in 1F position friction stir welding process. The plate of 6 mm AA5083 T-joint was joined in 1F position using the conventional MILKO37 milling machine. A macrostructure analysis on weld samples had also been conducted to trace for any defect. The following recommendations should be considered for future improvement:

a. Make improvements or minor modifications to the jig fixture and tools for better results of friction stir welding process in 1F position.
b. Utilize a Computer Numerical Control (CNC) milling machine for more accurate results due to limited speed on Milko 37.
c. Introduce stationary shoulder for better quality of welded joints.

Acknowledgements The authors thanked Universiti Kuala Lumpur for supporting this research work. Grateful acknowledgment to all people who were always around to motivate and assist in this project.

References

1. Fratini L, Buffa G, Filice L, Gagliardi F (2005) Friction stir welding of AA6082-T6 T-joints: process engineering and performance measurement. Proc IMechE Part B: J Eng Manuf 220:669–676
2. Khaled T (2005) An outsider looks at friction stir welding. ANM-112N-05-06, 1–71
3. Threadgill PL, Leonard AJ, Shercliff HR, Withers PJ (2009) Friction stir welding of aluminium alloys. Int Mater Rev 54(2):49–93
4. Blignault C (2002) Design, development and analysis of the friction stir welding process, Nelson Mandela Metropolitan University, Master Thesis
5. Beckman A, Sundstrom M (2014) Development of stationary shoulder for friction stir welding. Chalmers University of Technology, Master's Thesis

6. Martin JP, Stanhope C, Gascoyne S (2011) Novel techniques for corner joints using friction stir welding. Paper presented at TMS 2011 Annual Meeting & Exhibition. San Diego, CA., USA. 27 Feb–3 Mar 2011, TWI Technology Centre (Yorkshire)
7. https://www.thyssenkrupp-materials.co.uk/aluminium-5083.html. Last accessed on 28 Oct 2018 at 10.47pm
8. https://www.azom.com/article.aspx?ArticleID=9107. Last accessed on 28 Oct 2018 at 11.40pm

A Study of a Heat Exchanger Sizing of a Gas Turbine Inlet Air Cooling for Power Enhancement

Didi Asmara Salim, Mohd Amin Abd Majid and Adzuieen Nordin

Abstract A gas turbine operated at the University Technology PETRONAS Gas Turbine District Cooling Plant produces an average of 3258 kWh with 30% of estimated gas turbine efficiency. A previous study done by the writer has shown of 20% increase in the efficiency of a gas turbine with the implementation of the turbine inlet air cooling. The sizing of the Heat Exchanger needed to be determined as the existing air inlet is only a 30 ft^2 flange attached directly to the compressor of the gas turbine. This studies conclude that a 320 ft^2 heat exchanger is needed in order to reduce the air inlet fluctuations (a range of 23–34 °C) to 15 °C. By placing the heat exchanger in the gas turbine, a significant Pa pressure loss of approximately 187.55 was experienced.

Keywords Gas turbine · Heat exchanger · Sizing

1 Introduction

The performance of a Gas turbine (GT) is undeniably highly dependent on the inlet air conditions [1–4]. It can be described as a condition in which the power output of a GT decreases as the inlet air temperature increases. The GT's performance will experience a drawback if the air inlet of the GT is above the International Standardization of Organization (ISO) rate of 15 °C. Depending on the GT model, power input decreases are in the range of 5%–13% when the ambient temperatures are increased to 10 °C [5]. Barigozzi et al. [5] in his research, indicated that a GT characteristic of either an aero-derivative GT or Industrial GT were correlated relatively to the inlet cooling performance. Inlet air cooling (IAC) technology can be classified into Evaporative Cooling and Mechanical/Electric Chiller. The suitability

D. A. Salim (✉) · M. A. A. Majid
Mechanical Engineering Department, Universiti Teknologi PETRONAS, Tronoh, Perak, Malaysia
e-mail: ddasmara@gmail.com

A. Nordin
Politeknik Ungku Omar, Jalan Raja Musa Mahadi, 31450 Ipoh, Perak, Malaysia

C. L. Saw et al. (eds.), *Advancement in Emerging Technologies and Engineering Applications*, Lecture Notes in Mechanical Engineering,
https://doi.org/10.1007/978-981-15-0002-2_37

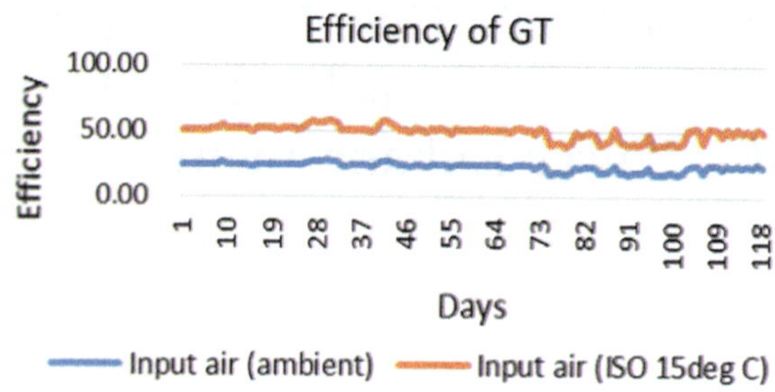

Fig. 1 Efficiency of the GT with the input air according to the ISO rated value

of the IAC is depends on the cost and climatic conditions. Al-Ibrahim and Varnham [6] in their research, concluded that evaporative cooling worked ideally for a hot and dry climate as the research was done in Saudi Arabia. The study recorded that by using Evaporative Cooling the cost of lowering the inlet GT temperature to 20 °C was relatively lower than by using Mechanical/Electrical chiller. Other findings by Farzaneh and Deymi [7], which used the chiller method for the IAC technology, found that it could cool the inlet air of a GT regardless of the ambient conditions.

The adoption of inlet air cooling was a very popular strategy for GT power augmentation mostly in the Middle East as the region experiences a hot climate [2, 4, 8, 9]. Malaysian climate which is warm and humid should practice the strategy to increase the power output of GT operated in the country. As the University Technology of Petronas (UTP) has its own Gas District Cooling (GDC) plant, a study on the effectiveness of implementing the IAC technology to its GT has been done. Retrieving the data of electricity produce in 2015 by UTP GDC and the local data of ambient temperatures in Setiawan, Perak as a reference were used to determine the effectiveness of the implementation of the IAC technology.

As shown in Fig. 1, a promising chance of approximately 20% in the efficiency of the GT can be achieved if the Inlet Air Cooling was implemented with the inlet air of the GT set according to the ISO rated value of 15 °C. The proposed system of the IAC in the GT is shown in Fig. 2 where the chilled water (CHW) is tapped from the Steam Absorption Chiller (SAC) of the GDC plant.

As shown in the Fig. 2, the IAC will require a Heat Exchanger (HEX) to cool down the ambient air flow to the desired temperature (15 °C) as this is the intake of the GT. This paper will discuss the selection and sizing process of the heat exchanger which will be proposed to the IAC system. The selection of the type of heat exchanger used is very important as it is the main component in the IAC system to cool down the ambient temperature. Common types of heat exchangers used in heating and cooling projects are shell-in-tube, plate-fin, plate-and-frame and plate-finned-tube [10]. Adapting from previous studies [2, 4, 5] and according to the situations at the UTP-GDC plant, the plate-fin-tube type of heat exchanger was selected. The selection of this type of HEX was largely encouraged by the limited exchange of the sensible heat during the exchange heat process and the finned tubes which were specifically employed in water-to-air coils [11]. The logical explanation was the use of fin in the HEX can increase the face area of HEX without increasing the coil for

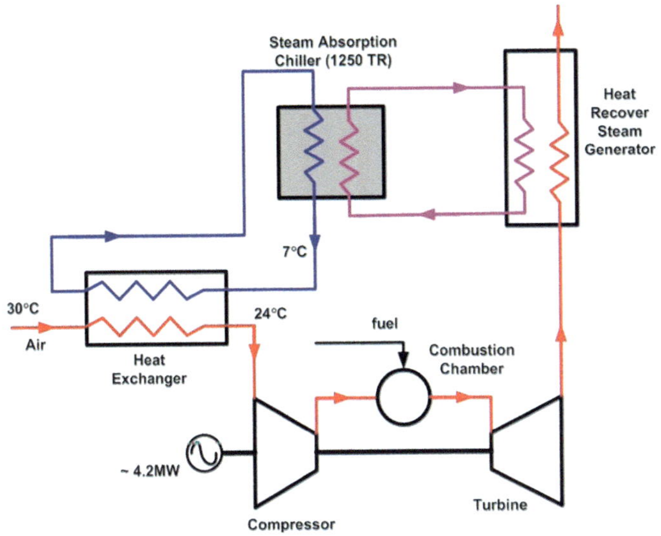

Fig. 2 SAC system to cool the air inlet of GT

the chilled water. The main objective of this paper is to determine the right size for a heat exchanger (plate-fin-tube type) for the inlet air cooling system which can be adapted into a GT in the UTP GDC. This includes a discussion on requirements of the overall heat transfer coefficient and face area of the HEX.

2 Methodology

The log mean temperature deficiency (LMTD) and the number of transfer units (NTU) methods were the basic approaches used to size and analyse the HEX. Based on the situations and information retrieved from the GDC and the Meteorology Department of Malaysia, the designs or determination of the HEX's suitability was conducted by using the effectiveness-NTU approach. The NTU methods require the calculation of the overall coefficient U, with some consideration for assumed conditions of the process.

Input data used in the research were categorized as climatic data, reference UTP GDC plant database and user defined input data, including standard parameters of components such as fins and tubing. Setiawan, Perak was selected as a reference of the ambient temperature as it is the nearest point of observation to the UTP GDC. Data of year the 2015 were gathered from the Meteorology Department of Malaysia to assess the influence of ambient temperatures and relative humidity on the IAC system. Throughout the year of 2015, 8760 h of ambient temperatures were recorded. The minimum ambient temperature was 25.5 °C in August 2015 and the highest was 32.8 °C in May, July and August 2015 as shown in Fig. 3. As the highest ambient

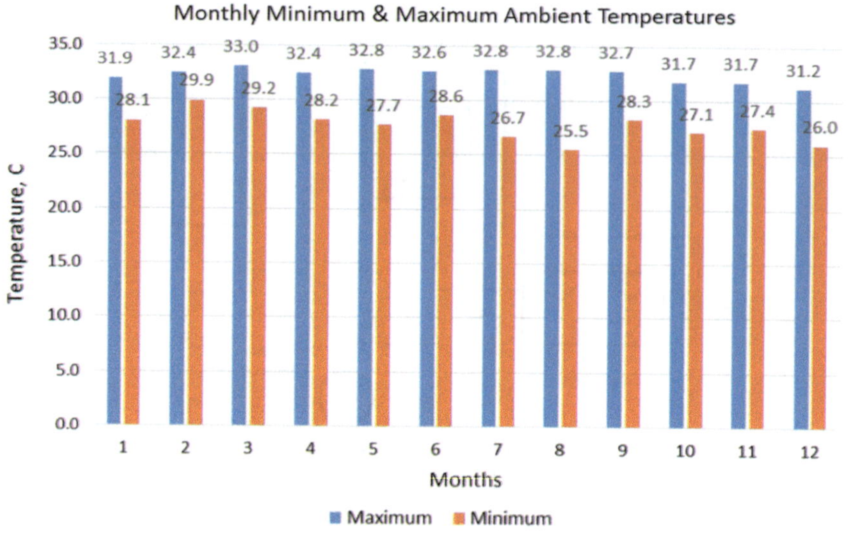

Fig. 3 Monthly minimum and maximum ambient temperatures

temperature was 32.8 °C, the air inlet was set at 34 °C, and for the determination of the HEX size and the desired temperature for the outlet of the HEX was set at 15 °C (ISO rated) as shown in Table 1. Entering chilled water was 6 °C and the temperature of water leaving was 12 °C with the velocity of the CHW at 4 ft/s. Specific Heat for both CHW and ambient air was set according to the temperature set. The circuit of water routing through tubes is Counter flow. The tubes were set according to the standard size given by ASHRAE Transactions [11], whereby a standard plate-fin-tube surface uses 1/2 in. tubes in a triangular layout as shown in Fig. 4, with the X_a of 1.25 in. and X_b of 1.083 in. The fin pitch dimensions follow ASHRAE standards [11], where the fin pitch is 8 fins/in and the fin thickness is 0.006 in.

The Heat Exchanger sizing determination which is performed can be classified into the 3 main parts, which are the air side, water side and fin efficiency. As the flow arrangement is crossflow, the NTU is given as in Eq. (1) and it is assumed to be of the same value of the effectiveness, ε [11]:

$$NTU = \frac{U_o A_o}{C_{min}} \tag{1}$$

Table 1 Parameter set for both water side and air side

Ambient temperature (°C)	34
GT inlet air temperature set (°C)	15
CHW inlet temperature (°C)	6
CHW return temperature (°C)	12

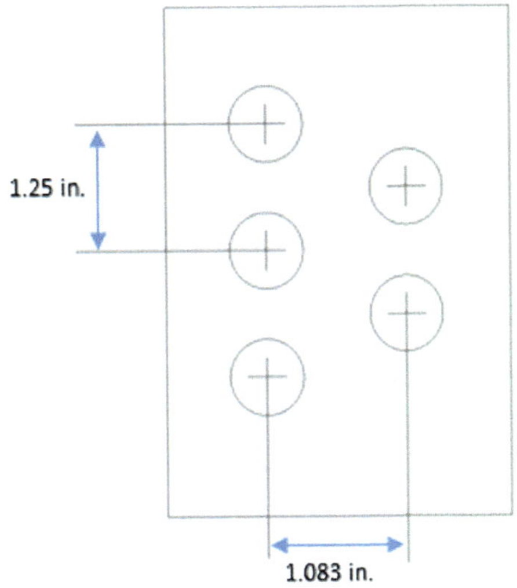

Fig. 4 Chilled Water Tubes arrangement

where U_o is overall heat transfer coefficients which are later introduced in Eq. (2), A_o is the area of air and C_{min} is the minimum value of unit thermal conductance.

The calculation of the overall heat transfer coefficient, U is based on the air-side area using Eq. (1) and is divided mainly into 3 main areas; water capacity, fluid capacity and fin efficiency.

$$\frac{1}{U_0} = \frac{1}{h_0 \eta_{s0}} + \frac{1}{h_i (A_i / A_0)} \tag{2}$$

where h_0 = Heat transfer coefficient on the outside, h_i = Heat transfer coefficient on the inside, η_{so} = Surface effectiveness, $\frac{A_i}{A_o}$ = Ratio of water side to air side areas.

Water Capacity

The Reynolds number then is calculated as shown in Eq. (3):

$$Re_D = \frac{\rho \overline{V} D}{\mu} \tag{3}$$

Water velocity is assume to 4 ft/s (to reduce high head loss), and ρ and μ is evaluate at 15 °C. The Prandtl number is given as Eq. (4):

$$Pr = \frac{\mu c_p}{k} \tag{4}$$

Equation (5) (Dittus-Boelter equation) can be used to calculate heat-transfer correlation if the $Re_D > 10{,}000$, and the Prandtl number is in the range of $0.7 < Pr < 100$:

$$\frac{\bar{h}D}{k} = 0.023\,(Re_D)^{0.8}\,(\text{Pr})^n \tag{5}$$

where n is 0.4, as the $t_{wall} > t_{bulk}$.

The NTU value was extracted from the effectiveness of cross-flow exchanger with fluids unmixed graph with the need of ratio of conductance value and the effectiveness, ε as below in Eq. (6):

$$\varepsilon = \frac{t_{co} - t_{ci}}{t_{hi} - t_{ci}} \tag{6}$$

Fluid Capacity

The Reynolds number is calculated as shown in Eq. (7):

$$Re_D = \frac{G_c\,D}{\mu} \tag{7}$$

J-factors were extract from heat transfer correlation graph from ASHRAE with the Correlating parameter, JP is given in Eq. (8):

$$JP = Re_D^{-0.4}\left[\frac{A}{A_t}\right]^{-0.15} \tag{8}$$

Hence, the heat transfer coefficient is determine as:

$$St\,Pr^{2/3} = \left(\frac{h_o}{G_c C_p}\right)\left(\frac{\mu C_p}{k}\right)^{2/3} \tag{9}$$

Fin Efficiency

For the Fin efficiency,

$$\eta = \frac{\tanh(mr\phi)}{mr\phi} \tag{10}$$

where m is given as,

$$m = \left(\frac{2h}{ky}\right)^{1/2} \tag{11}$$

ϕ is obtain as,

$$\phi = \left(\frac{R}{r} - 1\right)[1 + 0.35\ln(R/r)] \tag{12}$$

Table 2 HEX sizing

Front area (m^2)	40
Nr of rows of tubes	5
Face area (ft^2)	320
Head losses (Pa)	187.55

3 Results and Conclusions

As seen in Table 2, the determinations of HEX sizing have been carried out using the 34 °C as the ambient reference. The heat exchanger has been sized in order to guarantee the desired cooling demand during the peak hour of the GT process. The existing inlet size of the Gt which is directly attached to the compressor of the gas turbine is 30 ft^2 while the calculation of the new size of the heat exchanger is 117 ft^2. The wider size of the heat exchanger is due to the workload needed to cool down the ambient temperature to 15 °C as the climatic conditions were hot and humid. The hot and humid climate contributed to the higher temperature of the latent heat which the heat exchanger had to cope with. As a consequence, the cooling load for the system will be higher. For these surfaces of heat exchanger, a loss of 187.55 of pressure Pa will be experienced. The pressure loss took into consideration the entrance and the exit losses resulting from abrupt contraction and expansion.

Acknowledgements Authors would like to acknowledge the support of Universiti Teknologi PETRONAS (UTP) and Yayasan Universiti Teknologi PETRONAS (YUTP) for supporting this research study.

References

1. Shukla AK, Singh O (2016) Performance evaluation of steam injected gas turbine based power plant with inlet evaporative cooling. Appl Therm Eng 102:454–464
2. El-Shazly AA et al (2016) Gas turbine performance enhancement via utilizing different integrated turbine inlet cooling techniques. Alexandria Eng J 55(3):1903–1914
3. Breeze P (2016) Gas-Turbine power generation. Academic Press
4. Saghafifar M, Gadalla M (2015) Innovative inlet air cooling technology for gas turbine power plants using integrated solid desiccant and Maisotsenko cooler. Energy 87:663–677
5. Barigozzi G et al (2015) Techno-economic analysis of gas turbine inlet air cooling for combined cycle power plant for different climatic conditions. Appl Therm Eng 82:57–67
6. Al-Ibrahim AM, Varnham A (2010) A review of inlet air-cooling technologies for enhancing the performance of combustion turbines in Saudi Arabia. Appl Therm Eng 30(14–15):1879–1888
7. Farzaneh-Gord M, Deymi-Dashtebayaz M (2011) Effect of various inlet air cooling methods on gas turbine performance. Energy 36(2):1196–1205
8. Najjar YSH, Abubaker AM (2015) Indirect evaporative combined inlet air cooling with gas turbines for green power technology. Int J Refrig 59:235–250
9. Al-Ansary HA, Orfi JA, Ali ME (2013) Impact of the use of a hybrid turbine inlet air cooling system in arid climates. Energy Convers Manag 75:214–223

10. Yunus AC (2003) Heat transfer: a practical approach. MacGraw Hill, New York
11. McQuiston FC, Parker JD, Spilter J (2005) Heating, ventilating, and air conditioning: design and analysis. Wiley, NY

Techno-economic Assessment of Turbine Inlet Air Cooling for Small Scale Combined Cycle Power Plant in Malaysia Climate

Adzuieen Nordin, Didi Asmara Salim, Mohamad Asyraf Bin Othoman, Shahrul Nahar Omar Kamal, Danny Tam Hong Khai and Mohd Kamarul Yusri Yusof

Abstract A combined cycle power plant produces power and steam which the steam is used to generate additional power. The main source of the power and steam was the fuel entering the gas turbine. Hence the power and the steam being produced are related to the temperature inlet of the gas turbine. This study is to investigate the performance and economic effect of with and without using turbine inlet air cooling at the combined cycle power plant. The techno-economic formulations were used for evaluating the performance and the cost rates. Results indicate that with using of turbine inlet air cooling, the performance of gas turbine increased while the performance of steam turbine decreased. However, the power output of the power plant in Malaysia climate is increased due to contributed by performance of aero-derivative gas turbine. The results also demonstrate that the used of turbine inlet air cooling in combined cycle power plant is more profitable due to higher Net Present Value and Internal Rate of Return compared to without turbine inlet air cooling.

Keywords Techno-economic assessment · Combined cycle power plant · Turbine inlet air cooling

1 Introduction

The combined cycle power plant is one of the thermal energy systems which plays an important role in the production of energy. The combined cycle power plant is a cogeneration system that produces power and steam simultaneously from a single energy source [1]. A typical combined cycle power plant consists of a gas turbine (GT), a heat recovery steam generator (HRSG) and a steam turbine (ST). The combined cycle power plant is more thermodynamically efficient compared to

A. Nordin (✉) · D. A. Salim · M. A. B. Othoman · S. N. O. Kamal
Centre of Air Conditioning and Refrigeration, Politeknik Ungku Omar Ipoh, Perak, Malaysia
e-mail: adzuieen@puo.edu.my

D. T. H. Khai · M. K. Y. Yusof
Shinryo (Malaysia) Sdn. Bhd., Kuala Lumpur, Malaysia

C. L. Saw et al. (eds.), *Advancement in Emerging Technologies and Engineering Applications*, Lecture Notes in Mechanical Engineering,
https://doi.org/10.1007/978-981-15-0002-2_38

a typical power plant which uses more energy in the waste heat from the GT, which otherwise when released to the environment causes air pollution.

The power produced by the combined cycle power plant supplies power to its district area with the steam being used to generate additional power. The most important advantage derived from the combined cycle power plant is the saving in capital expenditure and operating costs. The management strategy and the size of the facility have a strong influence on the economic production of the combined cycle power plant [1].

In Malaysia, the warm and humid climate with the high ambient air temperature has an adverse effect on the performance of the gas turbine [2]. Therefore, it is important to reduce the inlet air temperature of the gas turbine in order to increase the performance of the power plant. One of the approaches to reduce the inlet air temperature is using turbine inlet air cooling (TIAC) technology. For the past few decades, studies of the TIAC have been undertaken by many authors [3–5]. De Pascale et al. [6] mentioned that evaporative cooling proved to be a reliable option for a power augmentation in integrated gasification combined cycle. Power augmentation from the introduction of TIAC to the gas turbine is quite significance to the annual power output of the power plant [7]. Yang et al. [8] reported that Gas Turbine Combined Cycle inlet fogging is superior in power efficiency at ambient temperature between 15 and 20 °C though it gains smaller profit margin than inlet chilling. Espanani et al. [9] found that by using fogging method, outlet power of turbine is increased about 6 MW however construction cost per KW reduced USD 4.G. Comodi [10] indicated that an almost linear trend can be obtained both in the electric power increase and in the electric efficiency increase as a function of the inlet air temperature when the chiller operates under nominal working conditions.

Although, this approach has been widely implemented in Middle East countries [5], in countries with a tropical climate, particularly Malaysia, very little focus has been given to the TIAC approach for combined cycle power plant [10]. Therefore, it is proposed that the techno-economic approach should be undertaken into account to determine the performance and economic evaluation by using TIAC in combined cycle power plant.

2 System Description

The combined cycle power plant diagram is shown in Fig. 1. The plant supplies power to the facilities. Fuel is used as the input for the GT to generate electricity. The exhaust gas is diverted into the HRSG to produce steam which is then used to generate additional power by the steam turbine. For power augmentation purposes, the TIAC is applied for the gas turbine. The cooling system used for the TIAC is chilled water generated by the electric chillers.

The summary of operating parameters of the combined cycle power plant and economic parameters is shown in Table 1.

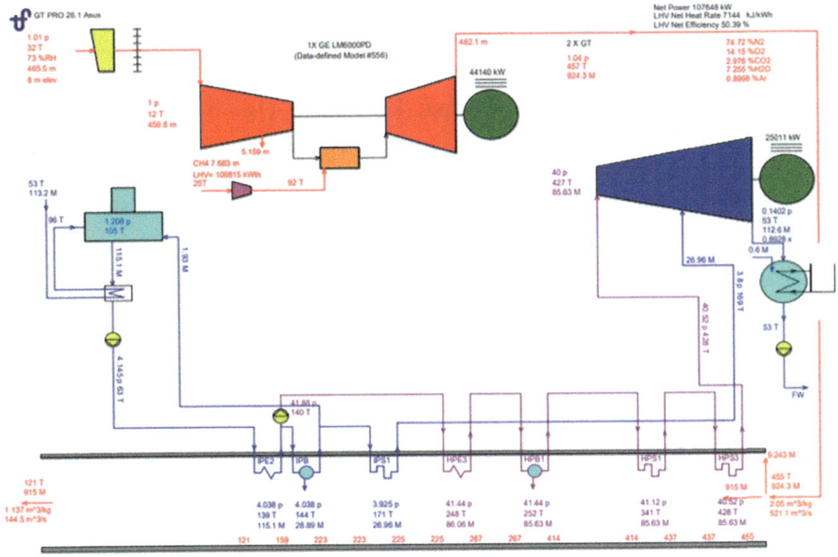

Fig. 1 Combined cycle power plant

Table 1 Operating parameters for combined cycle power plant [2]

	Non TIAC	TIAC with EC
Ambient temperature (°C)	24.3–28.3	24.3–28.3
Relative humidity (%)	78.2–89.1	78.2–89.1
GT inlet temperature set point (°C)	–	12
Inlet air cooling daily operational hours	–	24
Chiller COP	–	5.4
Inlet cooling system sizing	–	98% of operating hours in a year
Cooling capacity (tons)	–	1893.2

3 Methodology

Integrating of performance and economic analyses were adopted for this study. The techno-economic analyses is widely used to evaluate economic performance [11]. The following steps are used to evaluate economic performance at a combine cycle power plant:

(i) Determine the design point temperature from the Malaysia climatic data in order to identify the cooling capacity of the electric chillers.
(ii) Evaluate the performance of the combined cycle power plant with and without TIAC using GT Pro software.
Combined heat and Power Plant (CHP) thermal efficiency, $\eta_{thermal}$

$$\eta_{thermal} = \frac{W_E + \sum Q_{TH}}{Q_{Fuel}}. \quad (1)$$

where W_E is total net power output (kW) and $\sum Q_{TH}$ is total net process heat output (kW).

(iii) Economic analysis which Net Present Value (NPV), Internal Rate of Return (IRR) and payback time (PBT) approaches are adapted for this study.

$$NPV = \sum_{n=0}^{N} \frac{C_n}{(1+r)^n}. \quad (2)$$

where C_n is the net cash flow at year n, r is the discount rate (12%) and N is the time period for investment return. NPV is thus a function both discount rate r and investment time period N. IRR is then computed as the rate of return r for which NPV function is zero. PBT is computed as the year (n) at which the cumulative cash flow (NPV) equals the capital costs.

The analysis is based on the following assumptions:

i. The gas turbine models GE LM6000PD (aero-derivative) is used.
ii. Sitiawan's weather data is analyzed.
iii. Operating hours of the analysis are 8756 h.
iv. Public utility tariff for water consumption is RM2.28/m^3.
v. Average public utility tariff for electricity consumption is RM0.337/kWh.
vi. Maximum demand charged is RM29.70/kW per month.
vii. Natural gas consumption tariff is RM27.30 GJ.
viii. Steam consumption tariff is RM100/ton.
ix. The lifetime of the combined cycle power plant is 25 years.

4 Results and Discussions

Plant simulations have been carried out for the whole year using the GTPro software. Results for technical performance are presented in Table 2. The performance analyses of a combined cycle power plant were performed. In the performance evaluation, the results of the efficiency were calculated 38.94% and 41.32% for without TIAC and with TIAC respectively. This means that the used of TIAC increased the efficiency of the gas turbine.

While, for the steam turbine, the efficiency was decreased due to low temperature of exhaust gas resulting in low capacity of steam generated. However, the power output of the combined cycle power plant is increased due to the type of the gas turbine which is aero-derivative gas turbine is used in this study. The aero-derivative gas turbine responds more favourably to turbine inlet air cooling technology [12].

Table 2 Performance of combine cycle power plant comparison

	Non TIAC	TIAC
GTG quantity	2	2
Power output per GTG	33,565	44,140
GTG heat rate (LHV) (kJ/kWh)	9245	8712
GTG efficiency (LHV) (%)	38.94	41.32
Exhaust flow per GTG (t/h)	378.4	462.1
GTG exhaust temperature (°C)	480.6	456.6
HRSG quantity	2	2
HRSG efficiency (%)	78.63	77.87
Inlet energy per HRSG (kW)	103,531	59,095.5
STG qty	1	1
Power output per STG (kW)	22,939	25,011
STG efficiency (%)	28.18	27.18
Plant gross heat rate (LHV) (kJ/kWh)	6891	6789
Plant nett heat rate (LHV) (kJ/kWh)	7086	7144
Plant gross power output (kW)	90,070.00	113,290.00
Plant nett power output (kW)	87,588.00	107,648.00
Nett electricity export per annum (GWh)	803.90	948.70
Fuel consumption per annum (HHV) (TJ)	6300.00	7471.00
Water consumption per annum (m^3)	864,200.00	1127,000.00
Plant EPCC cost (RM)	469,533,000.00	511,059,000.00

Table 3 Economic analysis on combine cycle power plant

Descriptions	Unit	Non TIAC	TIAC with ECC
Cash inflow			
Maximum demand charges	RM/annum (Million)	31.11	38.24
Electricity charges	RM/annum (Million)	270.91	319.71
Cash outflow			
Fuel charges	RM/annum (Million)	171.99	203.96
Maintenance	RM/annum (Million)	18.78	20.44
Net revenue	RM/annum (Million)	70.25	82.19
Payback period			
NPV	RM (Million)	111.25	133.55
IRR	%	21.25	23.28
PBT	Years	–	1.86

Table 3 shows the results of economic analyses. The results of this study showed that highest NPV is RM133.55 Million for TIAC with electrical chillers compared to RM111.25 for combined cycle power plant without TIAC. The results also indicate that the IRR for combined cycle power plant with TIAC is higher around 23% and 21% for without used the TIAC. The results demonstrate that by using TIAC gave 1 year 10 months relative to non TIAC.

5 Conclusions

The techno-economic approach was used in order to evaluate the performance and economic feasibility of combined cycle power plant in Malaysia climate. Performance analysis shows that by using TIAC with electric chillers proved to be efficient although it decreased the efficiency of the steam turbine. However, the power generated from the combined cycle power plant is increased due to the used of aero-derivative gas turbine which is more responsive to the gas turbine inlet air temperature. For economic analyses, the used of TIAC in combined cycle power plant is more profitable due to higher NPV and IRR.

Acknowledgements The authors gratefully acknowledge Shinryo (Malaysia) Sdn. Bhd. for providing the financial support to complete the work in this project.

References

1. Klaassen RE, Patel MK (2013) District heating in the Netherlands today: a techno-economic assessment for NGCC-CHP (Natural Gas Combined Cycle combined heat and power). Energy 54:63–73
2. Omar Kamal SN (2017) Feasibility study of turbine inlet air cooling using mechanical chillers in Malaysia climate. Energy Proc 138:558–563
3. Al-Ansary HA, Orfi JA, Ali ME (2013) Impact of the use of a hybrid turbine inlet air cooling system in arid climates. Energy Convers Manag 75:214–223
4. De Sa A, Al Zubaidy S (2011) Gas turbine performance at varying ambient temperature. Appl Therm Eng 31(14–15):2735–2739
5. Al-Ibrahim AM, Varnham A (2010) A review of inlet air-cooling technologies for enhancing the performance of combustion turbines in Saudi Arabia. Appl Therm Eng 30(14–15):1879–1888
6. De Pascale A, Melino F, Morini M (2014) Analysis of inlet air cooling for IGCC power augmentation. Energy Proc 45:1265–1274
7. Jaber Q, Jaber J, Khawaldah M (2007) Assessment of power augmentation from gas turbine power plants using different inlet air cooling systems. JJMIE 1(1):7–15
8. Yang C, Yang Z, Cai R (2009) Analytical method for evaluation of gas turbine inlet air cooling in combined cycle power plant. Appl Energy 86(6):848–856
9. Espanani R, Ebrahimi S, Ziaeimoghadam H (2013) Efficiency improvement methods of gas turbine. Energy Environ Eng 1(2):36–54
10. Comodi G (2015) Enhancing micro gas turbine performance in hot climates through inlet air cooling vapour compression technique. Appl Energy 147:40–48

11. Akbulut A (2012) Techno-economic analysis of electricity and heat generation from farm-scale biogas plant: Çiçekdağı case study. Energy 44(1):381–390
12. Nordin A, Salim DA, Othoman MA, Kamal SO, Tam D, Yusof MK (2017) Turbine inlet air cooling for industrial and aeroderivative gas turbine in Malaysia climate. In: IOP conference series: Earth and environmental science, vol 104, no 1, p 012003. IOP Publishing

Feasibility of a District Cooling System in Technical and Vocational Institutions

Adzuieen Nordin, Zulhikmi Zali, Mohamad Asyraf Bin Othoman, Shahrul Nahar Omar Kamal and Didi Asmara Salim

Abstract The paper presents an economic comparative study on electricity consumption between using a district cooling system and a conventional system for air-conditioning purposes. This study adopted the approach of comparative evaluation. The results of the study show that the district cooling system is more economically feasible when compared to the conventional air-conditioning system. The cost of energy consumption using the district cooling system is 83% more profitable compared to that of the conventional system..

Keywords District cooling · Cooling load · Thermal energy storage

1 Introduction

District cooling (DC) is the centralized production and distribution of cooling energy. Chilled water is delivered through an underground insulated pipeline to the office, industrial and residential buildings in order to cool the indoor air of the buildings within a district. Specially designed units in each building then use the chilled water to obtain the lower temperature of air that passes through the building's air conditioning system.

DC can be operated by using electricity or natural gas, and regular water or seawater can be used. Along with electricity and water, district cooling constitutes a new form of energy service. Gas-fueled DC systems in Malaysia have been installed since 1997. Kuala Lumpur City Centre (KLCC) is among one of the earliest developers

A. Nordin (✉) · M. A. B. Othoman · S. N. O. Kamal · D. A. Salim
Department of Mechanical Engineering, Ungku Omar Politechnic, Jalan Raja Musa Mahadi, 31400 Ipoh, Perak, Malaysia
e-mail: adzuieen@puo.edu.my

Z. Zali
Department of Mathematics, Science and Computer, Ungku Omar Politechnic, Jalan Raja Musa Mahadi, 31400 Ipoh, Perak, Malaysia

C. L. Saw et al. (eds.), *Advancement in Emerging Technologies and Engineering Applications*, Lecture Notes in Mechanical Engineering,
https://doi.org/10.1007/978-981-15-0002-2_39

of the gas-fueled district cooling system in Malaysia. Other developers that followed were Universiti Teknologi Petronas (UTP), Kuala Lumpur International Airport (KLIA), Putrajaya International Convention Centre (PICC) and Wisma Putrajaya in Kuala Lumpur. The advantages in the use of gas-fueled DC system are its economic feasibility, space conservation, noise reduction and flexibility. In the economic advantage, the gas-fueled DC system has overall lower total capital cost compared to the split cooling system that requires its own cooling equipment [1]. The space used for the cooling equipment can be made vacant for the other purposes of a gas-fueled DC system. Other than that, the noise that is produced by the cooling machines can be prevented in the consumer buildings; moreover, the system is flexible enough to employ a wide range of inter-related thermal storage technologies [2].

The DC system has been studied by many researchers. Among them Ito et al. [3] found in his study that the operation costs are relatively stable by day and season when the district heating and cooling system was used. The results of a study on the potential and limitations for DC of buildings by Ruesch and Haller [4] show a correlation between the chilled water temperature generated by chillers and the weather data. This indicates that the temperature of the chilled water is one of the main factors in a DC system. Other studies have also addressed the benefits of the DC system [5–9].

With respect to energy usage, the DC system plays an important role in potentially reducing the overall cost incurred by the region when meeting its growing electricity demand. The current approach used in Technical and Vocational (TVET) institutions in Malaysia is the application of the split air-conditioning unit system in office buildings and classrooms. The major factor of the higher cost in electricity consumption in educational institutions is the use of air-conditioning. It has be found that electricity consumption increases especially in the peak period which between 7.00 a.m. and 7.00 p.m. whereas consumption is lower in the off-peak period which is from 7.00 p.m. and 7.00 a.m. Therefore, it is important to study the feasibility of applying DC in TVET institutions. This research proposes the use of Thermal Energy Storage (TES) for the purpose of storing the chilled water produced by the electric chiller during an off-peak period so that the chilled water will be distributed to the offices and classrooms during the peak period for air-conditioning purposes. This study will cover one of the TVET institutions located in Ipoh, Perak.

2 Methodology

In order to study the feasibility of applying DC in TVET institutions, the approach used covers the following:

(i) Study the building layout of the TVET institution.
(ii) Calculate the cooling load required for the institution.
(iii) Calculate the amount of refrigeration tonnage required.

(iv) Determine the size and type of electric chiller based on what is provided in the catalogue.
(v) Determine the size of TES.
(vi) Estimate the current electricity consumption.
(vii) Calculate the consumption of DC electricity.
(viii) Analyze the results and findings.

(i) *Study the building layout of the TVET institution.*

One of the TVET institutions in Ipoh, Perak has been taken as a case study. The complex comprised 23 buildings which approximately with an approximate area of 2112,633 ft^2.

(ii) *Calculate the cooling load required for the institution.*

The amount of cooling load required is determined by using the Daikin Method equations (1)–(6), where the Coefficient of B and F refer to Daikin's table.

$$Cooling\,load_{window/wall/roof/ceiling/floor}\,(\mathrm{m}^2) = (Area) \times (Coefficient\ B) \times (Coefficient\ F)\,[\mathrm{kW}] \quad (1)$$

$$Cooling\,load_{outdoor\,air}\,(\mathrm{m}^3) = (Volume) \times (Coefficient\ B) \times (Coefficient\ F)\,[\mathrm{kW}] \quad (2)$$

$$Cooling\,load_{human} = (Number\ of\ human) \times (Coefficient\ B) \times (Coefficient\ F)\,[\mathrm{kW}] \quad (3)$$

$$Cooling\,load_{occupants} = (Number\ of\ human) \times (Coefficient\ B) \times (Coefficient\ F)\,[\mathrm{kW}] \quad (4)$$

$$Cooling\,load_{lights} = (Number\ of\ light) \times (power) \times (Coefficient\ B) \times (Coefficient\ F)\,[\mathrm{kW}] \quad (5)$$

$$Cooling\,load_{electrical\,appliances} = (Number\ of\ electric\,appliance\,used) \times (power) \times (Coefficient\ B) \times (Coefficient\ F)\,[\mathrm{kW}] \quad (6)$$

(iii) *Calculate the amount of refrigeration tonnage required.*

The amount of refrigeration tonnage required was calculated based on the Eq. (7), where BTU is British Thermal Unit and hr is hour.

$$1\,refrigeration\,ton = 12{,}000\,\mathrm{BTU/hr} \quad (7)$$

(iv) *Determine the size and type of electric chiller based on what is provided in the catalogue.*

For this phase, the size and type of electric chiller was determined by using the electric chiller catalogue according to the brand chosen.

(v) *Determine the size of TES.*

The TES size was determined based on Eq. (8).

$$Capacity\,of\,TES = Cooling\,load(tonnage) \times Safety\,Factor \tag{8}$$

(vi) *Estimate the current electricity consumption.*

As now, the conventional approach in the TVET institution is the use of the split unit for the air-conditioning system. By knowing the ampere rate for the system, the power was calculated based on Eq. (9), where P is power (watt), I is and V is voltage.

$$P_{without\,DC} = IV\,[\text{kW}] \tag{9}$$

$$Electricity\,Tariff(without\,DC) = P_{without\,DC} \times Electric\,Tariff_{(peak\,hours)} \tag{10}$$

(vii) *Calculate the consumption of DC electricity.*

From method 4, the electricity consumption using the DC system was calculated as shown in Eq. (11), where $\boldsymbol{P}_{\boldsymbol{with\,DC}}$ refers to what is in the catalogue;

$$Electricity\,Tariff(with\,DC) = P_{with\,DC} \times Electric\,Tariff_{(off-peak\,hour)} \tag{11}$$

(viii) *Analyze the results and findings.*

The analysis on the feasibility is based on the total difference in the electricity consumptions between using the conventional system and the DC system.

3 Case Study

Politeknik Ungku Omar (PUO) is taken as a case study. PUO which is located in Ipoh, Perak consist of:

(i) 1 block of 4 floors of administration buildings.
(ii) 1 block of 2 floors of lodge.
(iii) 1 block of 2 floors of research center.
(iv) 7 blocks of 4 floors of classrooms.

(v) 2 lecture halls.
(vi) 2 multi-purpose halls.
(vii) 1 block of library.
(viii) 1 block of prayer hall.
(ix) 1 block of store.
(x) 1 block of workshop.
(xi) 5 blocks of lecturers' office.

The analysis is based on the following assumptions:

(i) The peak period of electricity rate is RM 0.355 [10].
(ii) The off-peak period is RM 0.219 [10].
(iii) The split air-conditioning units are operated from 7.00 a.m. to 5.00 p.m.
(iv) The efficiency of the equipment is negligible.
(v) Only one campus was considered in the calculation.
(vi) Only air-conditioning was taken into account in the calculations.

4 Results and Discussions

Using Eqs. (1)–(7), the estimated cooling load for the PUO campus buildings is shown in Table 1.

From the results in Table 1, the total cooling load is estimated at about 1233 RTh. It is denoted by percentage 37.7% for lecturer's offices, 15.8% for multi-purpose hall and 1.6% for lodge. Therefore, the capacity of TES required is around 2000 RTh and the capacity of electric chillers is 6 units of 325 RT. Using Eqs. (9)–(10), the results of electricity consumption without DC is shown in Table 2.

The results have shown the estimated power consumption and electricity tariff with using DC on a daily basis. It is estimated that the cost of using electricity in

Table 1 Estimated cooling load for PUO

Block	Cooling load Daikin's Method (RTh/day)
Administrative	85
Lodge	20
Classrooms	152
Lecture hall	47
Multi-purpose hall	195
Library	52
Prayer hall	19
Store	30
Lecturers' offices	465
Workshop	168
Total	1233

Table 2 Electricity consumption without DC

Block	$P_{without\ DC}$ [kWh/day]	Electricity tariff (RM/day)
Administrative	2500	888.00
Lodge	240	86.00
Classrooms	1260	447.00
Lecture hall	4280	1520
Multi-purpose hall	1110	395.00
Library	1120	398.00
Prayer hall	100	40.00
Store	60	25.00
Lecturers' offices	11,140	3960.00
Workshop	2800	1000.00
Total	24,610	8759

the campus is around RM8 759 per day, resulting in a monthly charge of RM262 770/month for air-conditioning purposes.

On the other hand, by using DC system, the results show the electricity consumption to generate chilled water by electric chillers during the off-peak period is 6858 kWh/day, resulting in a cost of RM1501/day and RM45 030/month.

From the comparative study, the results show the electricity consumption without applying DC is higher that than using DC system as shown on Fig. 1. The results, RM 262 770 and RM45 057 respectively shows that by using the DC system for air-conditioning purposes, it is more profitable by 83%.

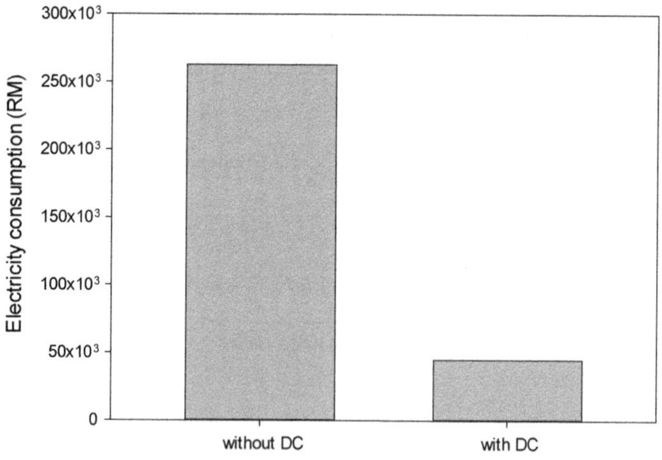

Fig. 1 Electricity consumption with and without DC

5 Conclusion

With the increasing demands for electricity, the DC system can be one of the options for the sustainable use of energy. The comparative study of the economic feasibility was carried out in order to identify a profitable method to reduce electricity consumption. The study used two approaches one using DC system and another without using the DC system. The results showed the electricity consumption of using electric chillers to generate chilled water during the off-peak period and store in TES 83% more profitable than that using the split unit air-conditioning system. The chilled water is distributed during the day time for air-conditioning purposes. However, for future study, the energy calculations from pump, cooling tower and AHU unit will take into account. In conclusion, it is preferable use the DC option.

References

1. Sulaiman SA, Majid A, Amin M (2011) Gas district cooling in Malaysia. Lambert Academic Publishing
2. Abdullah MO et al (2013) Electricity cost saving comparison due to tariff change and ice thermal storage (ITS) usage based on a hybrid centrifugal-ITS system for buildings: a university district cooling perspective. Energy Build 67:70–78
3. Ito M et al (2017) Electricity adjustment for capacity market auction by a district heating and cooling system. Appl Energy 206:623–633
4. Ruesch F, Haller M (2017) Potential and limitations of using low-temperature district heating and cooling networks for direct cooling of buildings. Energ Proc 122:1099–1104
5. Marugán-Cruz C et al (2014) District cooling using central tower power plant. Energy Proc 49:1800–1809
6. Rezaie B, Rosen MA (2012) District heating and cooling: review of technology and potential enhancements. Appl Energy 93:2–10
7. Werner S (2017) International review of district heating and cooling. Energy 137:617–631
8. Wu X, Chen Z (2017) Performance analysis of a district cooling system based on operation data. Proc Eng 205:3117–3122
9. Dominković DF et al (2017) Potential of district cooling in hot and humid climates. Appl Energy 208:49–61
10. Tenaga S (2013) Review on electricity tariff in Peninsular Malaysia under the incentive-based regulation mechanism (FY2014–FY2017)

Evaporation of River Water Using Pyramid Shape Solar Still

Abdul Aziz Abdullah and Chun Lin Saw

Abstract Solar desalination using solar still is the simplest method to produce clean distilled water but it yield low potable water. Hence, a pyramid shape glazing is modified on the solar distiller is used in this study. River water in front of Politeknik Ungku Omar Ipoh, Perak Malaysia was collected for distillation process for duration of 1 month. The experiment maximum yield of the distilled water output is about 3 l of distilled water from 20 l waste water capacity that comply with WHO clean water requirements. The maximum yield of solar still desalination performance is approximately 15%.

Keywords Solar still · Distilled water · Pyramid shape · Solar energy

1 Introduction

Water is essential for the maintenance of life and also the key to human's prosperity. Nowadays, water is a natural sources that being polluted by human activities, urbanization and industrialization. There is a critical need to improve the safe drinking water and adequate water supply globally. The ground water is often over exploited to meet the increasing demand of the people. Less than 1% of earth's water is available for human consumption and more than 1.2 billion people still have no access to safe drinking water. Estimated more than 50% of the world population is residing in urban areas, and almost 50% of mega cities that having population over 10 million are heavily dependent on ground water [1]. Most of the rural people still live in absolute poverty and often lack access to clean drinking water. When all viable sources of water have been utilized, extracting drinkable water from other water resources can help to solve this problem.

Desalination is one of the method to produce clean water by using sunlight. Thus, it is a method or process of obtaining pure and drinkable water from waste/brackish or saline water. Various desalination techniques are used to purify the water. Solar

A. A. Abdullah (✉) · C. L. Saw
Politeknik Ungku Omar, Jalan Raja Musa Mahadi, 31400 Ipoh, Perak, Malaysia
e-mail: azizulah@puo.edu.my

C. L. Saw et al. (eds.), *Advancement in Emerging Technologies and Engineering Applications*, Lecture Notes in Mechanical Engineering,
https://doi.org/10.1007/978-981-15-0002-2_40

distillation is an easy and cost effective method to provide pure drinking water in rural areas without affecting the nature. Solar distillation process is carried out both in passive and active mode. Solar collector with different inclination angle is seems to contribute to different performance of solar collector [2]. Hence, solar still with pyramid shape is proposed instead of flat glazing that yield 25% more from the conventional solar still [3, 4]. In this paper, pyramid shape solar still experimental work is presented.

The purpose is to analyse the performance of solar still and to analyse the treated water or distilled water contents produced whether it comply with WHO clean water standard requirements.

2 Methodology

Flow chart depicted in Fig. 1 shows the method on designing and performance evaluation of the solar still. Experimental measurement will be carried out for 1 month to collect water temperature data and amount of distilled water produced data based various climate condition. The experimental measurement will be carried out from 7AM to 7PM.

3 Conceptual Design

Figure 2 shows the finalized conceptual design for solar still with pyramid shape glazing. The purpose of glazing designed using pyramid shape was to improve the evaporation and condensation of distilled water production. The 45° inclination angle chosen due to better flow of condensed water into the ducting drainage solar still. Table 1 listed the materials used to build the solar still prototype.

3.1 Working Prototype

Figure 3a shows the solar still prototype that is used for experimental data collection. A portable data logger is used to measure the hourly data of distilled water and waste water inside the solar still. Distilled water produced was collected in the plastic bottle placed at the outlet pipe of the solar still. A weighting device is used to measure the weight of distilled water hourly to convert to litres. While, Fig. 3b shows the inner compartment that is painted with black paint to absorbed more heat from solar radiation. Polystyrene is of inch thick wrap with aluminium foil is attached to the collector to reduce the heat loss. Waste water will be poured into the inner compartment at 7 AM before sunrise then data collection was performed hourly until sunset at 7 PM.

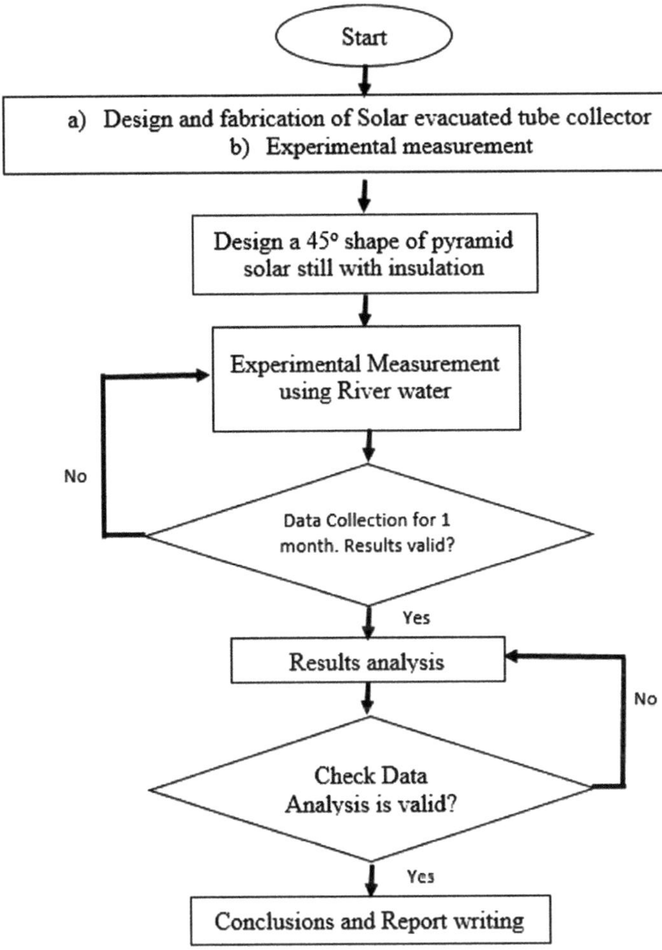

Fig. 1 Solar still design and data collection process

4 Results and Discussions

Distilled water collected by desalinating the river water as shown in Fig. 4. It shows that a clear and clean distilled water was produced by the solar still. The sample of river water and distilled water was analysed at Department of Environmental, Ipoh Perak. Parameters that were analysed was pH, turbidity, salinity and conductivity. The focus on those 4 parameters are to comply Ministry of Health Malaysia clean raw water level standard. Ministry of Health Malaysia raw water for pH is range between 5.5 and 9.0 that neutral to alkaline value, turbidity is 1000 nephelometric turbidity units (NTU) and below that gauge the clearness of water while, salinity or total dissolve solid shows salt contents in the water is less than 1500 mg/L and

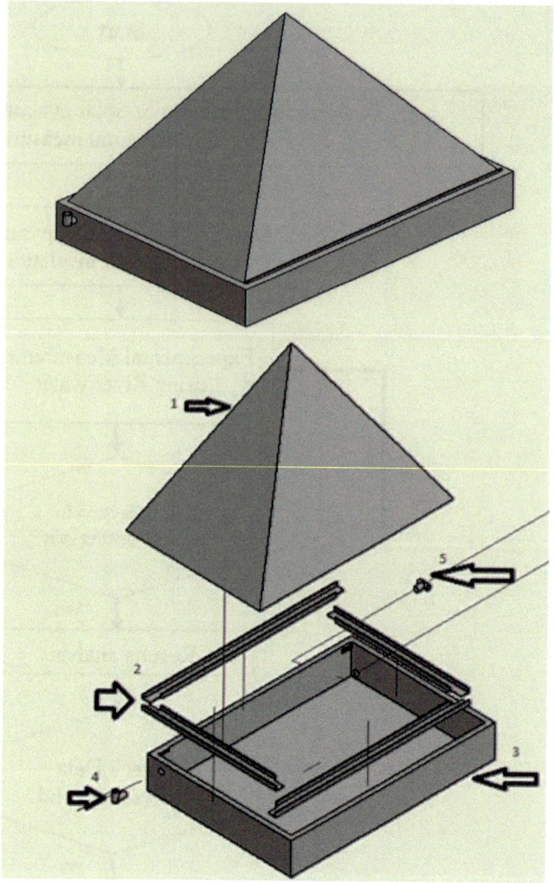

Fig. 2 Exploded view of solar still design

Table 1 Material selection for the solar still

No.	Part description	Material
1	Pyramid	Fiber glass
2	Drainage flow pipe	Plastic PV link
3	Water container	Plastic basin
4	Inlet pipe	PVC elbow pipe
5	Outlet pipe	PVC tee pipe

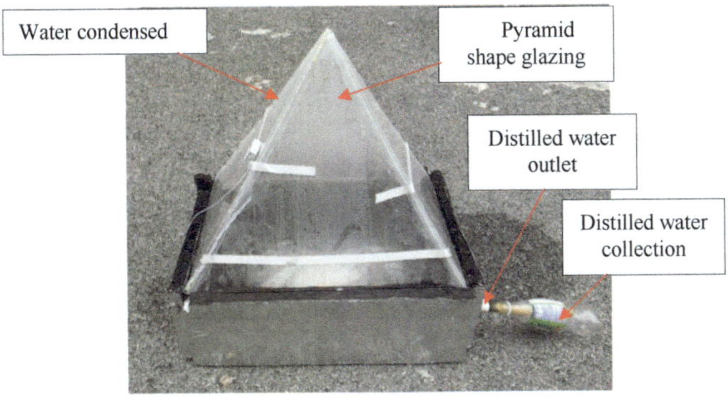

(a) Solar still prototype.

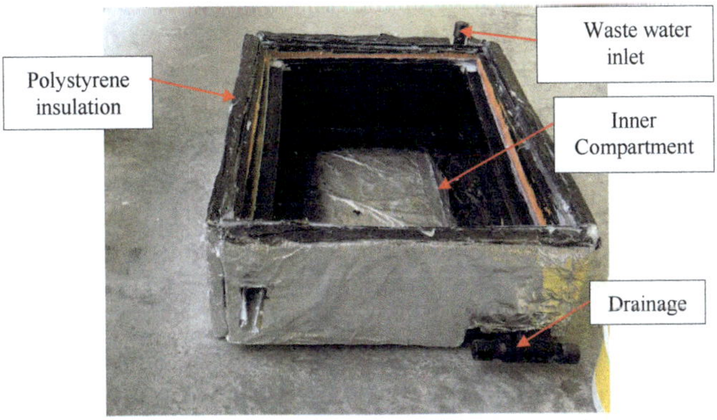

(b) Inner compartment of solar still.

Fig. 3 **a** Solar still prototype. **b** Inner compartment of solar still

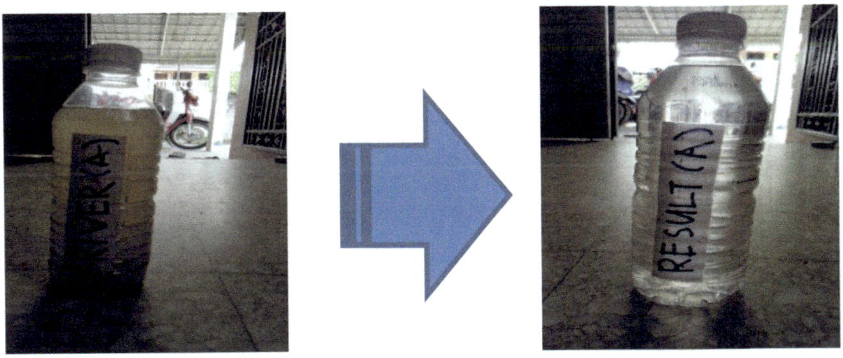

Fig. 4 Distilled water

Table 2 Distilled water analysis

Parameters		
Type of water	Water from river	River water after distillation
Temperature (°C)	29.6	28.7
Dissolved oxygen (DO %)	95.0	93.5
Conductivity (μmhos/cm)	0.073	0.001
Salinity (mg/L)	0.03	0.00
pH value (pH)	8.25	7.26
Turbidity (NTU)	8.0	6.2

conductivity is to measure the dissolve ions such as manganese, chrome, copper etc. inside the water that the value should be below 250 μmhos/cm [5].

Table 2 shows the analysis result from Lembaga Air Perak (LAP), the authority in charge of water utility in Perak state. The distilled water produced after desalination process shows it comply with Ministry of Health Malaysia standard where conductivity is 0.001 μS/cm, salinity is 0.00 mg/L, pH value is 7.26 pH and turbidity is a bit higher from 5.0 NTU mentioned by Ministry of Health since the water has not being treated for drinking yet with value of 6.2 NTU but it is still a clean raw water to be released back to river. Although the results show there is quality improvement of distilled water, however the water still need to be boiled before consumable to kill all the bacteria.

It was observed that the distilled water production yield range between 1.2 and 3 l daily from 20 l of river water used for the experiment. Hence, in a month the average of distilled water produced is 78 l. Figure 5 shows the production of distilled water on 5th of June 2018.

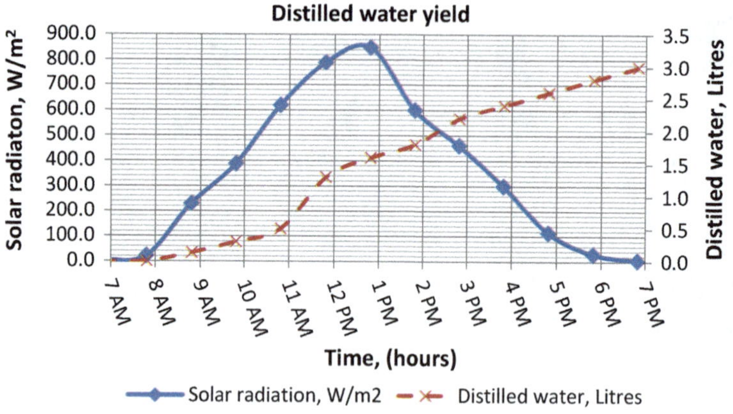

Fig. 5 Experimental data of desalination on 5th of June 2018

5 Conclusions

These results show that the distilled water production yield range between 1.2 and 3 l daily and shows the maximum performance of solar still is about 15%. The reason for the low productivity is the water mass (water depth), higher water depth has an intense effect on the distilled output of the solar still system. Metal absorber too has the effect of absorbed more heat from solar radiation since the current model use plastic absorber.

References

1. El-Sebaii AA, Al-Ghamdi AA, Al-Hazmi FS, Adel SF (2009) Thermal performance of a single basin solar still with PCM as a storage medium. Appl Energy 86:1187–1195
2. Al-Kayiem HH, Saw CL (2014) Performance evaluation of solar water heater integrated with PCM nanocomposite TES at various inclinations. Sol Energy 109(1):82–92
3. Winfred RDD, Iniyan S, Suganthi L, Davies PA (2016) Solar stills: a comprehensive review of designs, performance and material advances. Renew Sustain Energy Rev 63:464–496
4. Taamneh Y, Taamneh M (2012) Performance of pyramid-shaped solar still: experimental study. Desalination 291:65–68
5. Ministry of Health Malaysia (2018) Drinking water quality standard, http://kmam.moh.gov.my/public-user/drinking-water-quality-standard.html, last accessed 26 July 2018

Simulation of Flow Distribution Inside Small Cavity at Two Way Radio by CFD (Ansys Fluent)

Wan Masrurah Hairudin, Norilmi Amilia Ismail and Zaidi Mohd Ripin

Abstract This research is to investigate flow distribution involved in the small cavity at two-way radio that contribute to the occurrence of noise during a windy situation. A Computational Fluid Dynamic (CFD)—ANSYS Fluent approach is utilized to understand the relationship of vortex flow and noise inside the small cavity at a different position. The results showed that the recirculation was managed to occur inside the cavity P3 and decreased at a further cavity position (P4 and P5). The vortices were occurred at the leading edge, trailing edge and downstream of the cavity of P3 and P4. It concluded that position 5 is a suitable position to avoid the high noise produced.

Keywords Flow analysis · Wind-induced noise · Computational fluid dynamic (CFD)

1 Introduction

A two-way radio is a mobile telecommunication device that can both receive and transmit radio signals. It has been used at large sites business such as hospitals, security, military, oilrig plant and others to eliminate the obscure communications [1]. Besides, it also has broad application including for personal used and hobbies. However, they faced a problem of unclear audio with the background noise when exposed to the windy condition affecting the quality of the signal receiver. The received signal is unclear due to wind noise level is higher than the speech signal. The situation is hazardous in a critical situation especially when it used for a rescue team in operation to save human life. Wind is one of the sources that contributed to the occurrence of noise. The wind noise mainly occurs at low frequencies of 0–500 Hz [2]. Wind-induced noise is generated by wind flowing over an element or

W. M. Hairudin · N. A. Ismail (✉)
School of Aerospace Engineering, Universiti Sains Malaysia, 14300 Nibong Tebal, Malaysia
e-mail: aenorilmi@usm.my

Z. M. Ripin
School of Mechanical Engineering, Universiti Sains Malaysia, 14300 Nibong Tebal, Malaysia

C. L. Saw et al. (eds.), *Advancement in Emerging Technologies and Engineering Applications*, Lecture Notes in Mechanical Engineering,
https://doi.org/10.1007/978-981-15-0002-2_41

through gaps and cavities between elements causing a formation of vortex shedding, stationary recirculation, and other flow instabilities [2–4]. The phenomenon of wind-induced noise in cavities has been studied by numerous investigations in the past especially in aerospace and automotive area [2]. The noise sources of two-way radio commonly come between the electrical part and ambient [5, 6]. The electrical part is referred to the mechanism of microphones meanwhile the ambient is the atmospheric condition. In the atmospheric condition, the wind speed and atmospheric stability are the factors that contribute to the wind noise where the sound pressure level (SPL) can be affected by the noise induced by the wind at screened microphone [7, 8].

Study on the two-way radio is recently conducted in [9–13]. The relationship between the flow speed and noise level in two way radio has been studied by [10] and showed that the higher the flow speed had increased the noise level. The study also shows a significant increase in noise level when the radio was oriented to the critical angle of attack at 15°. Another work in [11] performed an investigation of sound quality of two-way radio by measuring the effect of angles orientation, wind speed and threshold hearing level intelligibility of the received speech signal. Motorola group [12] also studied on noise in a small cavity and found the source of noise result from the CFD simulation was correlated with flow visualization experiment. Based on their model they strongly stated that this model could be further strengthened according to Rossiter's Theory.

In a windy condition, the two-way radio has a cavity that receives flow, hits the diaphragm and the microphone, and then converts the flow to a signal. The mechanism of noise cavity flow is shown in Fig. 1. The air flowing over the open cavity and creates the vortices. These vortices flow until they reach and impact the other end point B thus forming the strong acoustics propagates in the environment [9]. The noise reaches the point A where it stimulates the formation of new vortices that will repeat the process and creating even more noise. The noise is generated when wind flow interacts with disturbance, then creates the vortex and the phenomenon is called wind-induced noise. These vortex and cavity flows are the results of complex interactions between the wind flow field and geometry of the two-way radio. This condition can cause significant disruption and may cause severe problems in the structures and performances [2].

In this study, we have carried out a simulation using ANSYS Fluent to analyse and understand the wind flow pattern generated inside the cavity model of two-way radio under different cavity position with the wind speed of 3.75 m/s.

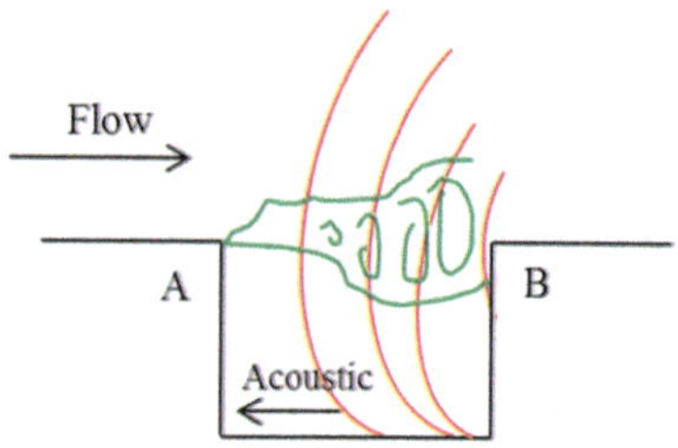

Fig. 1 The mechanism of noise cavity flow

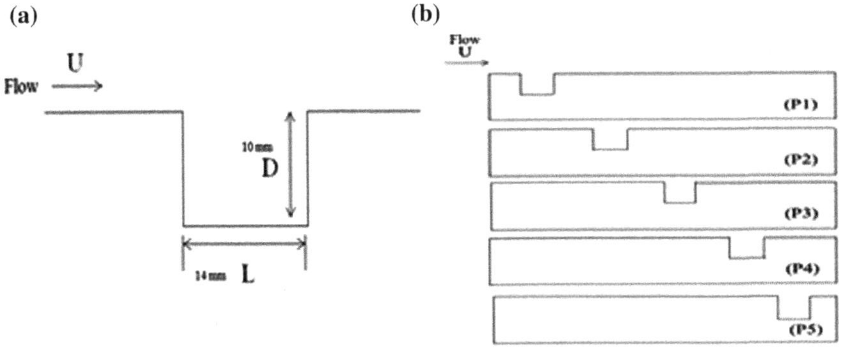

Fig. 2 **a** Schematic view of the cavity. **b** The cross sectional at upper cavities position (i) P1:10 mm (ii) P2:44 mm (iii) P3:78 mm (iv) P4:112 mm (v) P5: 146 mm

2 Computational Fluid Dynamic (CFD) Model

2.1 Cavity Model

A schematic view of a rectangular cavity model shows in Fig. 2. Cavities can be classified based on the length to depth ratio (L/D), where L is the length and D is the depth of the cavity. There are three types of cavities; 'open' if $L/D < 10$, 'closed' if $L/D > 13$ and transitional for $10 \leq L/D \leq 13$ [3]. The dimensions of the cavity with L/D ratio is 1.4 is shown in Fig. 2a. The five cavity position are namely: (a) P1 (10 mm), (b) P2 (44 mm), (c) P3 (78 mm), and (d) P4 (112 mm) (e) P5 (146 mm) are created as shown in Fig. 2b.

2.2 Mesh and Boundary Condition

The simulation was utilised in ANSYS Fluent 16 using Reynold Averaged Navier-Stokes (RANS) equation approach. A laminar state regime was chosen to simulate the flow characteristics when passing over the small cavity in two-way radio in a single direction. Pressure based approach has been selected for low-speed incompressible flows where the air density is assumed constant throughout the simulation. A semi-implicit method for pressure-linked equations (SIMPLE) algorithm of the pressure-velocity coupling was selected to solve the governing equations. The flow conditions were specified as Re = 43,471 corresponding to $U = 3.75$ m/s. Figure 3 shows the fine mesh of the position 5. The simulations are calculated with 0.01 s time step size and performed for 350-time steps. The fluid is set as air. The other simulation parameter is shown in Table 1.

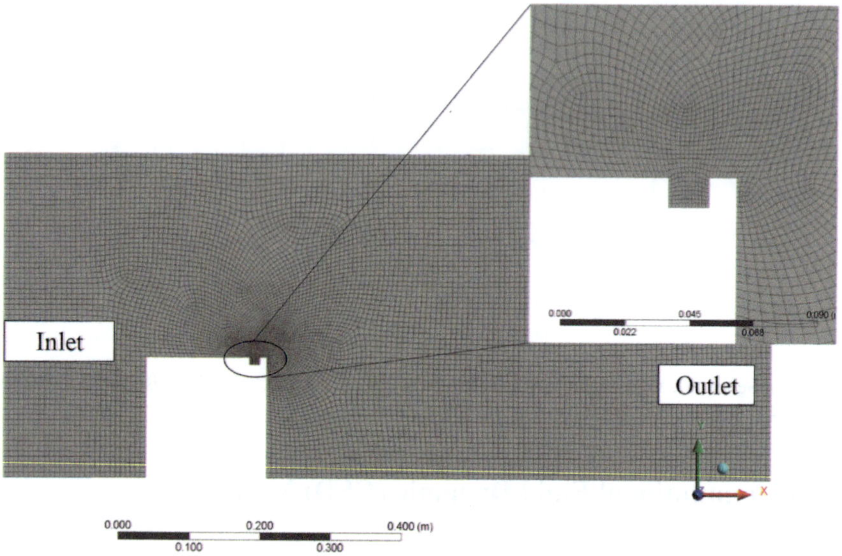

Fig. 3 The mesh and boundary condition of the cavity model

Table 1 Simulation parameters of the cavity model

Simulation parameters	Value
Air density (ρ)	1.225 kg/m^3
Kinematic viscosity (μ)	1.7×10^{-5} kg/ms
Pressure (Pa)	101,325 Pa
Time step (t)	0.01 s

3 Result and Discussion

3.1 Effect of Cavity Position

Figures 4 and 5 show the velocity and vorticity contour inside the different cavity (P3, P4 and P5) under Reynold number of 43,471. High recirculation flow region is observed at cavity P3, P4 and compared with cavity P5 under wind speed of 3.75 m/s. The phenomenon happened due to the higher velocity has produced flow thick separation boundary layer that occurred earlier at upstream of the cavity. Therefore, those positions received more airflow compared to P5. The velocity contour showed high vortex start produced at position P3. It can be seen the recirculation has evolved inside the cavity of P3. With the increase of the wind speed and farther position, a weak swirl begins to form inside the cavity. The airflow fully enters inside the cavity position (P3–P5) at wind speed 3.75 m/s. The velocity flow at P4 and P5 are found to be blue color region since both positions are located quite far from the wind source.

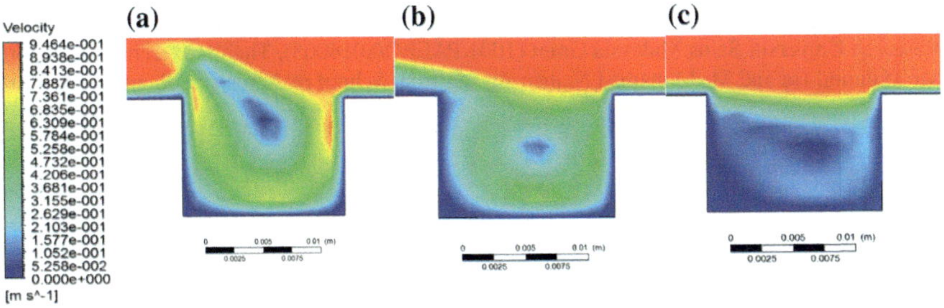

Fig. 4 Velocity distribution at wind speed of 3.75 m/s for **a** P3 **b** P4 **c** P5

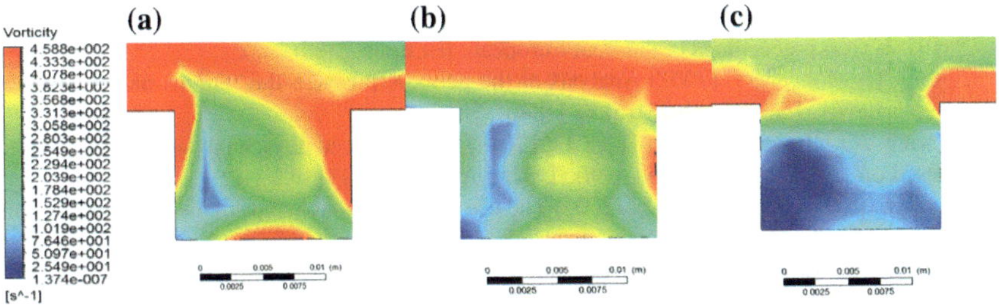

Fig. 5 Vorticity contour at wind velocity of 3.75 m/s for **a** P3 **b** P4 **c** P5

It is shown that the vortex formation flow increases when wind flow moves through the farther distance. Thus, this phenomenon influences the wind noise inside the cavity. Figure 5 shows the high (red color region) vortices are formed at leading edge, trailing edge, and beneath of the cavity, P3 compared P4 and P5. The longer cavity distance from the wind source, the flow separated, and vortices exist. The existence of vortices probably showed those were a source of noise generated.

4 Conclusion

In this paper, CFD simulation was used to predict airflow distribution inside a small cavity in two-way radio. The simulation of cavity model at speed of 3.75 m/s with the varied position (P3, P4 and P5) using ANSYS Fluent was presented. It shows that the vortex formation increases more at P3 compared with the other position of P4 and P5. Thus, a weak vortices flow produced inside the cavity. It can be concluded that the wind flow gave a significant effect on the cavity position prediction due to the pressure decreased at the farther position. A wind tunnel experiments will be carried out in the future to study the comparative of those varying parameter effect using flow visualisation approach and validate the finding of a simulation study. The outcome of this study will help to understand the flow fields produced inside the small cavity to reduce the wind noise when using a two-way radio at the windy situation.

Acknowledgements The authors acknowledge the research work is financially supported by the Research Universiti Sains Malaysia Grant (1001/PAERO/8014005). Thanks to technical supports, Mr Mahmud Isa and Mr Mohamad Najhan Awang from School of Aerospace Engineering for the help.

References

1. Rogers AL, Manwell JF, Wright S (2006) Wind turbine acoustic noise. Renewable Energy Research Laboratory. University of Massachusetts at Amherst
2. Nelke CM, Nawroth N, Jeub M, Beaugeant C, Vary P (2002) Single microphone wind noise reduction using techniques of artificial bandwidth extension. In: Signal processing conference (EUSIPCO) proceedings of the 20th European. IEEE, pp 2328–2332
3. Kook H, Mongeau L (2002) Analysis of the periodic pressure fluctuations induced by flow over a cavity. J Sound Vib 251(5):823–846
4. Zhang YO, Zhang T, Ouyang H, Li TY (2014) Flow-induced noise analysis for 3D trash rack based on LES/Lighthill hybrid method. Appl Acoust 79:141–152
5. Leclercq D, Cooper J, Stead M (2008) The use of microphone windshields for outdoors noise measurements. In: Proceedings of acoustics. Geelong
6. Morgan S, Raspet R (1992) Investigation of the mechanisms of low-frequency wind noise generation outdoors. J Acoust Soc Am 92(2):1180–1183
7. Xu Y, Zheng ZC, Wilson DK (2011) A computational study of the effect of windscreen shape and flow resistivity on turbulent wind noise reduction. J Acoust Soc Am 129(4):1740–1747
8. Nakasako N, Sakamoto K, Itoh H, Takakuwa Y (2005) A practical countermeasure to wind-induced noise in environmental noise measurement based on linear and/or non-linear model using wind speed information. School of Biology-Oriented Science and Technology
9. Ecotiere D (2012) Estimation of uncertainties due to the wind-induced noise in a screened microphone. In: Acoustics, 23 Apr 2012
10. Hashim MKR, Saad AA, Ripin ZM (2016) Psychoacoustic annoyance analysis for two way radio under wind noise. ARPN J Eng App Sci 11(3):2120
11. Fisol UMM, Ripin ZM Ismail et al (2013) Wind noise analysis of a two-way radio. In: 2013 IEEE international conference on smart instrumentation, measurement and applications (ICSIMA). IEEE, Nov 2013, pp 1–5
12. Hashim MKR, Ismail NA, Saad AA Sound quality analysis for two-way radio under wind noise. In: 2013 IEEE international conference on smart instrumentation, measurement and applications (ICSIMA). IEEE, Nov 2013, pp 1–5
13. Yow HT, Hii D, Saw A, Tan CH, Fisol UM, Ripin ZM, Ismail NA, Saad AA, Hashim MK, Yi CP (2013) Prediction of wind-induced noise over bodies and small cavity. J Acoust Soc Am 134(5):4222

Influence of Partially Standing Waves on Offloading Operations for Shuttle Tanker

Mohammed Shihab Patel, Mohd. Shahir Liew, Zahiraniza Mustaffa, Abdurrasheed Said Abdurrasheed and Andrew Whyte

Abstract The side-by-side configuration of a shuttle tanker undergoes more complex hydrodynamic behavior. This is more critical in extreme weather conditions. The gap between the FPSO and shuttle tanker adds to complexity of hydrodynamic behavior. Past studies have been performed to understand the gap resonance and other behavior of water column between the FPSO and shuttle tanker. This paper aims to study the effect of partially standing waves on the offloading operation, which occurs between the gap of FPSO and shuttle tanker pertaining to different encountering time period of waves for shuttle tanker. The velocity potential and wave kinematics of partially standing waves are derived and subsequently the different percentage of reflected wave height on the wave kinematics are studied. The objective of studying the influence of partially standing waves offloading operation is achieved by studying two types of waves as moderate and extreme waves. The higher wave height affects the kinematics of waves between the vessels. Also, the higher percentage of reflected wave height contributes to greater kinematics of waves. The conclusions of study and scope of future work is presented at the end.

Keywords Side-by-side · Partially standing waves · FPSO · Offloading · Shuttle tanker

1 Introduction

With continuous and monotonous efforts in searching oil fields, most of the Floating Production Storage and Offloading systems (FPSO) are now converting to side-by-side configuration [1]. The offloading is one of the key operations during production.

M. S. Patel (✉) · Mohd. S. Liew · Z. Mustaffa · A. S. Abdurrasheed
Universiti Teknologi PETRONAS, 32610 Seri Iskandar, Perak, Malaysia
e-mail: shihabpatel91@gmail.com

Mohd. S. Liew
e-mail: mohammed_17001672@utp.edu.my

A. Whyte
Curtin University, Bentley Campus, Perth, Australia

C. L. Saw et al. (eds.), *Advancement in Emerging Technologies and Engineering Applications*, Lecture Notes in Mechanical Engineering,
https://doi.org/10.1007/978-981-15-0002-2_42

The shuttle tanker is usually stationed adjacent to the FPSO for 8–12 h. The safe offloading operation is a concern as the shuttle tanker is usually connected through hawsers and is comparatively unrestrained compared to a moored FPSO [2]. In an event of extreme weather conditions and longitudinal waves attacking the vessel, there can be possibility of extreme vessel response and hydrodynamic behavior leading to non-feasibility of safe offloading operations [3, 4]. The gap between the vessels play a very crucial role in hydrodynamic interactions [5]. Due to close proximity of side vessel, strong hydrodynamic interactions are believed to take place which affects the wave forces on the bodies [6]. Moreover, sharp peaks are observed between the gaps of side-by-side vessels which affect the hydrodynamic parameters [7].

The dynamic characteristics and wave-body interactions differ with respect to configuration. The roll response of a vessel is usually attributed to resonance and tends to shoot higher in presence of closed vessels in head seas [8, 9]. Furthermore, the gap between the vessels is sensitive to Hemholtz type resonance as well [10]. The present study tends to evaluate the effect of partially standing waves on the offloading operation by studying the wave kinematics between the gap. The effect of percentage of wave reflection and different water depth on wave kinematics between the vessels gap is studied.

2 Mathematical Modeling and Analysis

A partially standing wave is made of incident wave height Hi and reflected wave height *Hr*. Due to closing spacing of the FPSO and adjacent shuttle tanker in close proximity, there would be waves generated between the gap of the vessels. When an incident wave of certain height from one vessel propagates and reaches the other vessel, part of the incident wave is reflected and the reflected wave travels in the opposite direction. Such repetitive occurrences of this event causes next incoming wave height and already reflected wave height superimposes to generate partially standing waves [11–13]. However, the wave kinematics of partially standing wave depends on the percentage of reflected wave height. The wave elevation for partially standing waves is given as follows:

$$\eta_t = \frac{H_i}{2}\cos(kx - \omega t) + \frac{H_r}{2}\cos(kx + \omega t + \epsilon) \tag{1}$$

where,

- η_t is the wave elevation.
- H_i is the incident wave height.
- H_r is the reflected wave height.
- k is the wave number and ω is the wave frequency.

The velocity potential for the wave surface ($z = 0$) [14],

$$\phi = g \int \eta_t \tag{2}$$

Further solving of Eq. (2) with the assumption that wave elevation has zero spatial and temporal mean yields the final velocity potential as

$$\phi_{z=0} = \frac{g}{2\omega}[H_r \sin(kx + \omega t + \epsilon) - H_i \sin(kx - \omega t)] \tag{3}$$

$$\phi_{z>0} = \frac{g}{2\omega}\frac{\cosh(k(d+z))}{\cosh(kd)}[H_r \sin(kx + \omega t + \epsilon) - H_i \sin(kx - \omega t)] \tag{4}$$

where,

- d is the wave depth.

The equations of wave kinematics for partially standing waves are as follows:

$$u = -\frac{gk}{2\omega}\frac{\cosh(k(d+z))}{\cosh(kd)}[H_r \cos(kx + \omega t + \epsilon) - H_i \cos(kx - \omega t)] \tag{5}$$

$$w = -\frac{gk}{2\omega}\frac{\cosh(k(d+z))}{\cosh(kd)}[H_r \sin(kx + \omega t + \epsilon) - H_i \sin(kx - \omega t) \tag{6}$$

The horizontal velocity is given by Eq. (5) while the vertical velocity is given by Eq. (6). Further derivative of the velocities with respect to time and space will yield the local and convective accelerations respectively as represented from Eq. (7) to Eq. (12)

$$\dot{u}_x = -\frac{gk^2}{2\omega}\frac{\cosh(k(d+z))}{\cosh(kd)}[H_i \sin(kx - \omega t) - H_r \sin(kx + \omega t + \epsilon) \tag{7}$$

$$\dot{u}_z = -\frac{gk^2}{2\omega}\frac{\sinh(k(d+z))}{\cosh(kd)}[H_r \cos(kx + \omega t + \epsilon) - H_i \cos(kx - \omega t) \tag{8}$$

$$\dot{w}_x = -\frac{gk^2}{2\omega}\frac{\cosh(k(d+z))}{\cosh(kd)}[H_r \cos(kx + \omega t + \epsilon) - H_i \sin(kx - \omega t) \tag{9}$$

$$\dot{w}_z = -\frac{gk^2}{2\omega}\frac{\cosh(k(d+z))}{\cosh(kd)}[H_r \sin(kx + \omega t + \epsilon) - H_i \sin(kx - \omega t) \tag{10}$$

$$\dot{u}_t = \frac{gk}{2}\frac{\cosh(k(d+z))}{\cosh(kd)}[H_r \sin(kx + \omega t + \epsilon) + H_i \sin(kx - \omega t) \tag{11}$$

$$\dot{w}_t = -\frac{gk}{2}\frac{\sinh(k(d+z))}{\cosh(kd)}[H_r \cos(kx + \omega t + \epsilon) + H_i \sin(kx - \omega t) \tag{12}$$

In this study, the spacing of the vessel is considered for percentage of reflected wave height. Furthermore, it is based on the assumption that greater spacing of the vessel would relate to lower percentage of reflected wave height. An incident wave propagating from a distantly spaced vessel will gradually reduce in the momentum before getting reflected and the percentage of reflected wave would be less as opposed to a closely spaced vessel which would correspond to a higher reflected wave. The above wave kinematics equations from Eqs. (5)–(12), were solved in a multi-paradigm numerical computing software, MATLAB, by evaluating the kinematics for different conditions of wave through direct substitution as explained in the succeeding section.

3 Results and Discussion

Two cases of reflected wave height for the initial phase of our present study. A wave reflected percentage of 10% relates to widely spaced vessel while 40% relates to closer spaced vessels respectively. The water depth for our study is 50 m and as the derived wave equations are sinusoidal, observations are made for initial 1000 s only. The discussion is further subdivided into two cases as following.

3.1 Partially Standing Waves: Moderate Waves

An incident wave of 3 m is considered for initial phase of analysis. The result is plotted for two different time periods of wave as given in Table 1.

From the plots of kinematics in Fig. 1, it is clearly observed that the wave elevation for 3.5 s time period wave, is higher for greater percentage of reflected wave height while the horizontal velocity and horizontal acceleration is comparatively higher for lesser percentage of reflected waves respectively. This nature is repeated for 16 s wave as well. Moreover, the vertical velocity and vertical acceleration remains to be higher for greater percentage of reflected wave heights.

The wave elevation for the 16 s wave stretches longer over time and the time taken to reach peak values of velocity is more. It is further observed that for the same time of observation of 1000 s, the wave with lower time period undergoes higher number of transitions in peak velocities. Hence, a wave with a lower time period is more critical and prone to cause higher wave turbulence. It can also be related to greater motion response of vessel.

Table 1 Encountering wave time period on shuttle tanker

Parameter	Case 1	Case 2
Time period (s)	3.5	16

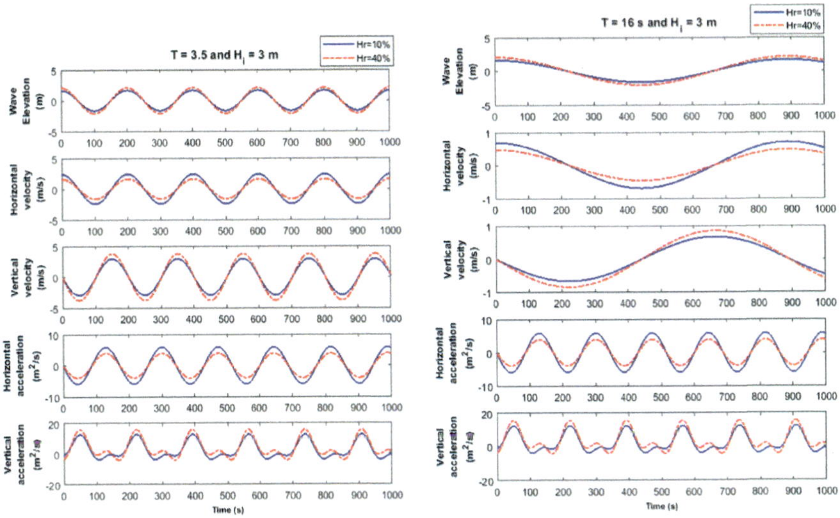

Fig. 1 Wave kinematics of partially standing waves for $H_i = 3\,\mathrm{m}$

3.2 Partially Standing Waves: Extreme Waves

The study on wave kinematics of partially standing waves was extended to higher incident wave heights of 6 m and 10 m respectively. Two major case was studied as tabulated in Table 1. An exact similar nature is observed for the wave kinematic profile as observed for moderate wave case. The only difference lies in the peak values. From Figs. 2 and 3, it can be concluded again that higher percentage of

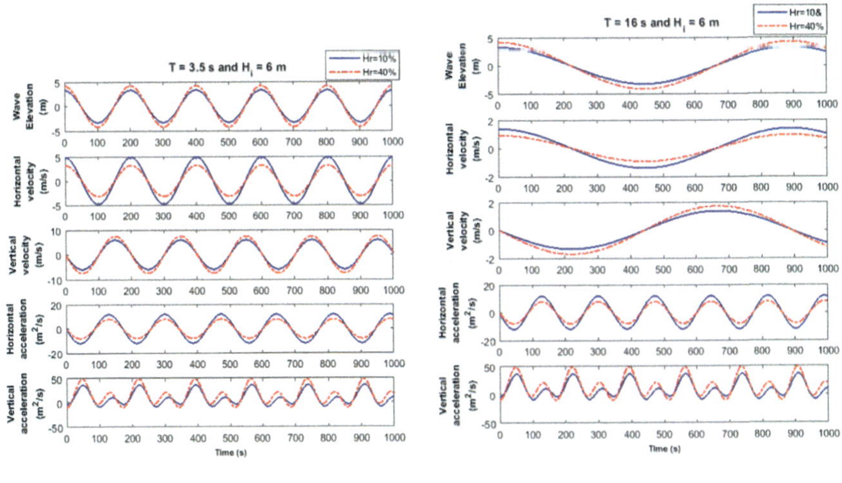

Fig. 2 Wave kinematics of partially standing waves $(H_i = 6\,\mathrm{m})$

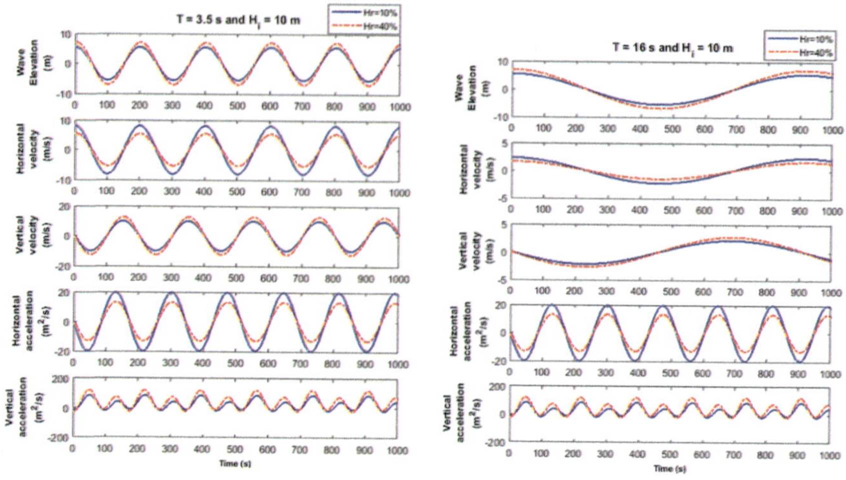

Fig. 3 Wave kinematics of partially standing waves ($H_i = 10\,\text{m}$)

reflected wave heights corresponds to comparatively increase in wave elevation, vertical velocity and vertical acceleration. Furthermore, the time period of the wave affects the number of transitions of peak values respectively. Hence there is more wave turbulence involved for a wave of lower time period. But, it is also clearly seen that increased wave height of 6 m and 10 m has more peak values of wave kinematics.

It is further seen that greater percentage of reflected wave height is more effective on wave kinematics parameters. Thus, it can be said that closer the spacing of vessels, greater is the percentage of reflected waves which further influences the wave kinematics. Since the effect of partially standing waves are higher with 40% reflected wave height and a smaller time period wave is more critical, Fig. 4 represents the 40% reflected wave height for two cases of higher incident wave height but, for a smaller time period wave. It is observed clearly that effect of wave kinematics is dominant for greater height of incident wave. There is a drastic increase in the vertical acceleration. Finally, it can be concluded that higher percentage of reflected wave height for a smaller time period wave is more critical for offloading operations. These waves kinematics can be related to higher vessel behavior and stronger hydrodynamic interaction.

4 Conclusion

The study primarily focused on the contribution of wave height as an influential parameter on the wave kinematics between the floating vessels. The study of partially standing waves between the gap of FPSO and shuttle tanker can be concluded as:

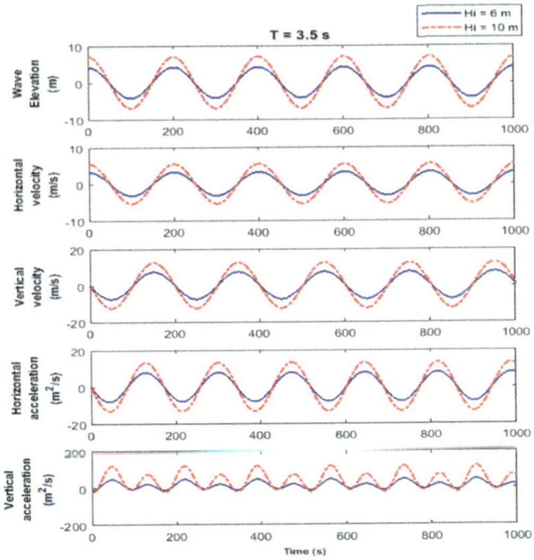

Fig. 4 Wave kinematics of partially standing waves ($H_i = 6\,\text{m}$ and $10\,\text{m}$)

- The influence of wave kinematics in partially standing wave is higher for higher incident wave height.
- The wave kinematics interactions are more susceptible for a wave of lower time period with greater percentage of reflected wave, 40% in the present study as compared to 10%, respectively.

Future Works The future scope of work involves calculating downtime cost of offloading operation under the influence of partially standing waves. A graphic user interface (GUI) would be developed for linking the wave kinematics to cost of production.

References

1. Zhao WH, Yang JM, Hu ZQ (2012) Hydrodynamic interaction between FLNG vessel and LNG carrier in side by side configuration. J Hydrodyn 24(5):648–657
2. Zhao W, Yang J, Hu Z, Tao L (2014) Prediction of hydrodynamic performance of an FLNG system in side-by-side offloading operation. J Fluids Struct 46:89–110
3. Buchner B, Van Dijk A, De Wilde J (2001) Numerical multiple-body simulations of side-by-side mooring to an FPSO. In: The eleventh international offshore and polar engineering conference. International Society of Offshore and Polar Engineers
4. Pessoa J, Fonseca N, Soares CG (2016) Side-by-side FLNG and shuttle tanker linear and second order low frequency wave induced dynamics. Ocean Eng 111:234–253
5. Pauw WH, Huijsmans RH, Voogt A (2007) Advances in the hydrodynamics of side-by-side moored vessels. In: 26th international conference on offshore mechanics and arctic engineering. American Society of Mechanical Engineers, pp 597–603
6. Jeong H, Kim M, Lee J, Kim B, Ha M (2010) Offloading operability analysis of side-by-side Moored LNG FPSO. In: The ninth ISOPE Pacific/Asia offshore mechanics symposium. International Society of Offshore and Polar Engineers

7. Xu X, Yang JM, Li X, Xu L (2014) Hydrodynamic performance study of two side-by-side barges. Ships Offshore Struct 9(5):475–488
8. Hong SY, Kim JH, Kim HJ, Choi YR (2002) Experimental study on behavior of tandem and side-by-side moored vessels. In: The twelfth international offshore and polar engineering conference. International Society of Offshore and Polar Engineers
9. Perwitasari RN (2010) Hydrodynamic interaction and mooring analysis for offloading between FPSO and LNG shuttle tanker. Master's thesis: Norges teknisk-naturvitenskapelige universitet, Fakultet for ingeniørvitenskap og teknologi, Institutt for Marin Teknikk
10. Ha MK, Kim MS, Park JJ, Lee JH (2004) First-and second-order hydrodynamic forces and moments on two offshore floating structures in waves. In: The fourteenth international offshore and polar engineering conference. International Society of Offshore and Polar Engineers
11. Dean RG, Dalrymple RA (1991) Water wave mechanics for engineers and scientists, vol 2. World Scientific Publishing Company
12. Journée JMJ, Massie WW (2001) Introduction in offshore hydromechanics (OT3600). Delft University of Technology
13. Chen XB, Orozco JM, Malenic S (2005) Evaluation of wave and current loads on offloading FPSOS. In: Offshore technology conference
14. Chakrabarti SK (1987) Hydrodynamics of offshore structures. WIT Press, London

Effects of Motion Responses and Drift Forces on Side-by-Side Offloading Operations of FPSO

Mohammed Shihab Patel, Mohd. Shahir Liew, Zahiraniza Mustaffa, Abdurrasheed Said Abdurrasheed and Andrew Whyte

Abstract Offloading operations involves transfer of oil or gas from the Floating Production Storage and Offloading system (FPSO's) to a nearby stationed shuttle tanker. The hydrodynamic interaction between the FPSO and shuttle tanker is an important subject of study. The actual estimate of the vessels behavior is necessary for offloading operation. This paper aims to study the hydrodynamic behavior of side-by-side stationed vessels for offloading operations. The investigation of hydrodynamic behavior is achieved through a diffraction module of ANSYS. The present study includes studying the motion responses and drift forces of side-by-side configured FPSO and shuttle tankers in three different water depth. The obtained results are found consistent with the published work.

Keywords Side-by-side · Offloading · Hydrodynamics · FPSO

1 Introduction

The recent survey [1] of Floating Production Storage and Offloading (FPSO) revealed that, there are total of 178 FPSOs working around the globe. The worldwide distribution of FPSO [1] shows that 14 FPSOs are currently operating in Australian waters while 51 FPSOs are stationed in Southeast Asia. Malaysia alone has 6 FPSOs. The side-by-side configuration is of recent origin and is a topic of interest for many reasons like stability and lower cost of production [2]. A typical FPSO consists of turret area, process area, storage and offloading systems. The stability is one of the key concerns of design. A side-by-side vessel is more complicated and tends to undergo complex behavior [2]. The common system of transporting oil from FPSO is by using a shuttle tanker. The shuttle tanker is usually placed adjacent to the moored FPSO and the stability of shuttle tanker is achieved by tugs or hawser connected

M. S. Patel (✉) · Mohd. S. Liew · Z. Mustaffa · A. S. Abdurrasheed
Universiti Teknologi PETRONAS, 32610 Seri Iskandar, Perak, Malaysia
e-mail: shihabpatel91@gmail.com

A. Whyte
Curtin University, Bentley Campus, Perth, Australia

C. L. Saw et al. (eds.), *Advancement in Emerging Technologies and Engineering Applications*, Lecture Notes in Mechanical Engineering,
https://doi.org/10.1007/978-981-15-0002-2_43

between the two floating bodies. Arrangement of all risers on one end and shuttle tanker interface on the other end reduces the need of complicated riser systems.

The offloading may be done by a direct transfer using a hose or by using a separate offloading system like pipeline or riser. The FPSO and the shuttle tanker in the side-by-side configuration is subjected to continuous wave forces, wind forces, current forces as well as instability due to low lying resonant or second order forces [3], a shuttle tanker is stationed to transport the oil from the FPSO. In case of occurrence of extreme wave conditions, there could be cease of offloading operations due to increase in mooring forces than permissible limits, increase in responses of vessels, greater hawser loads than the limiting value [4]. Such an event can be attributed to downtime. The offloading operation is then re-continued when feasible conditions prevail [5].

The offloading operation is affected by large hydrodynamic forces and higher vessel response [6]. It is necessary to determine the actual vessel behavior and response for a safe offloading operation. The authors in Ref. [7] has experimentally investigated the basic interaction characteristics of both tandem and side-by- side moored vessels. The experimental results have been numerically compared using higher order boundary element method. The authors in Ref. [8] has predicted the motion and hydrodynamic force between two floating structures using three dimensional potential theory and source distribution technique. The obtained results are validated with experiments. The probability for collisions and risk are found and explained. The authors in Ref. [9] have studied offloading operation in specific location of Brazilian waters. The study revolves around the tug force for safe operation and hawsers forces respectively. Offloading assistance was then defined. The FPSO and shuttle tanker in a side-by-side configuration are analyzed in ANSYS AQWA by subjecting them to diffraction analysis in three different water depths of 50 m, 100 m and 250 m respectively. The response motion and drift forces are studied for offloading operations. The findings of the result are found to be consistent.

2 Hydrodynamic Analysis and Geometric Modeling

Integration of pressure over wetted surface area yields forces due to interaction of fluids on the body [10]. These forces due to fluid can be further bifurcated into reactive and active components. The active force which is also known as wave exciting force is sum of Froude-Krylov force and the diffraction force [10]. The body motion induces radiation waves which can be attributed to radiation force, also known as reactive force. The real and imaginary parts of radiation wave potential ϕ_{rk} is substituted to produce the added mass and wave damping coefficients as given in Ref. [11].

$$F_{rjk} = \omega^2 A_{jk} + i\omega B_{jk}$$

where, A_{jk} and B_{jk} are added mass and damping coefficients respectively. A set of linear algebraic equations are solved in the AQWA solver to obtain the harmonic

Table 1 Geometric details of FPSO and shuttle tanker

Category	Detail	FPSO	Shuttle tanker
Geometric details	Total structural mass	58.2572e6 kg	23.466e6 kg
	Kxx	10.948 m	8.806 m
	Kyy	45.71 m	37.9 m
	Kzz	47.53 m	39.416 m
	Ixx	69.8263e8 kg-m^2	18.1968 kg-m^2
	Iyy	12.1722e10 kg-m^2	3.37e10 kg-m^2
	Izz	13.1655e10 kg-m^2	3.6452kg-m^2
Details of mesh	Tolerance	1 m	1 m
	Longest period	30 s	30 s
	Shortest period	3 s	3 s
	Interval period	0.5 s	0.5 s

response of the body in regular wave. These response characteristics are known as Response Amplitude Operators (RAO) and are proportional to wave amplitude. The set of linear equations of motion with frequency dependent coefficients are given in Refs. [12, 13]. The FPSO and the shuttle tanker were first subjected to hydrodynamic analysis alone without any side-by-side configuration. Subsequently, a side-by-side hydrodynamic analysis was performed under the same wave conditions. The time period of analysis was from 3 to 30 s. The wave directions considered were 180° to +180°. The details of the FPSO and shuttle tanker are displayed in Table 1. Figure 3 displays the side-by-side configuration of the models in AQWA.

3 Results and Discussions

The FPSO and shuttle tanker were subjected to hydrodynamic diffraction analysis. The results obtained then were compared to the finding of the results in Refs. [3, 4] for establishing consistency of present work with established past work respectively. It can be said that the present work was in good agreement with the findings of the past work. The prime focus of the present work is studying the behavior of vessels with and without side-by-side configuration respectively. The drift forces and responses in the six degree of freedom are primarily studied.

It is observed from Fig. 1 that the trend is almost similar for smaller time period. There is lesser shielding effect on the FPSO by the shuttle tanker for the small time period. However, shielding effect is higher for higher time period. The response in surge for quartering seas does not exhibit any shielding effect. Moreover, the presence of adjacent shuttle tanker causes the response to be more than the FPSO alone. Hence, there is effect of side vessel on sway response as shown in Fig. 1 respectively. A significant shielding effect is seen in the heave response for the

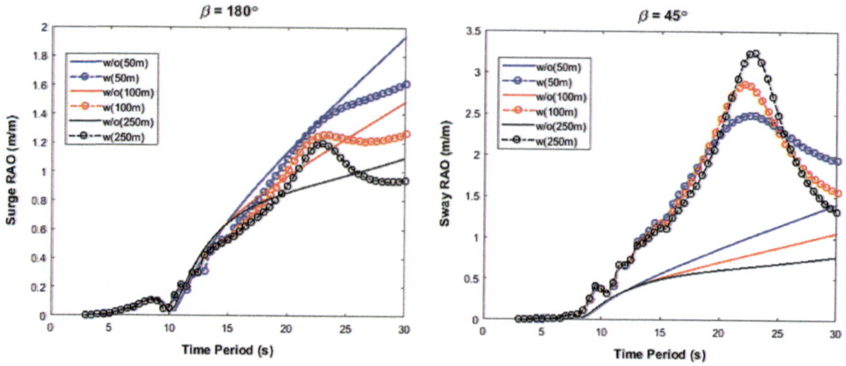

Fig. 1 Surge (180°) and sway (45°) responses of FPSO

FPSO in beam sea. There is considerate decrease in the heave response for lee side vessel but disappears eventually for higher time period as shown in Fig. 2. For the roll response, it is observed that there is no roll without shuttle tanker in head seas. The roll response is governed by resonance. Furthermore, presence of shuttle tanker causes roll significantly which clearly explains that presence of adjacent shuttle tanker causes resonance and thereby causes roll response to shoot up as displayed in Fig. 2. The vessels are very closely spaced and the water column present between the gap resonates which can be explained by Helmholtz type resonance. However, in beam seas there is significant shielding effect for the roll response due to presence of shuttle tanker which effectively reduces the roll response of FPSO as shown in Fig. 3.

Drift forces are induced due to interaction of the vessels. Large fluctuations are seen for surge drift forces due to presence of shuttle tanker as shown in Fig. 3. An interesting phenomenon is observed that there is no lateral and yaw drift forces in the absence of shuttle tanker. However, from Figs. 4 and 5 it can be seen that strong hydrodynamic interactions cause lateral and yaw drift forces. The presence of shuttle

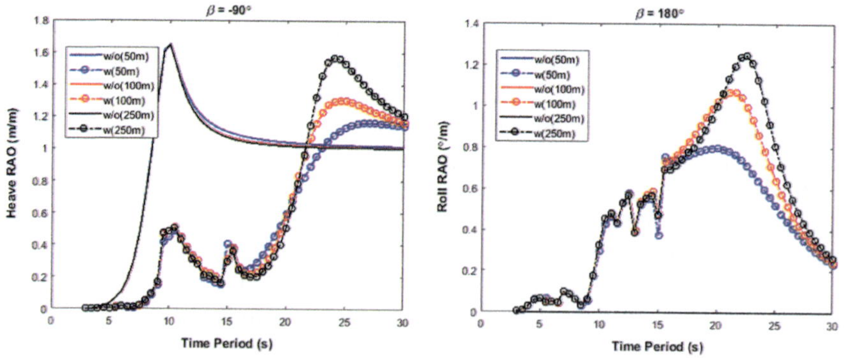

Fig. 2 Heave (−90°) and roll (180°) responses of FPSO

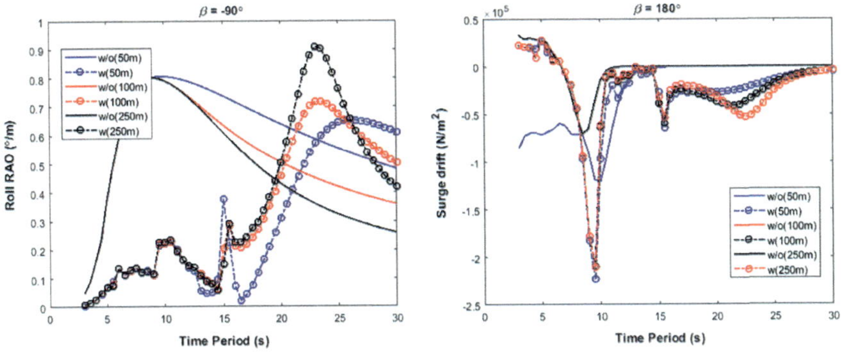

Fig. 3 Roll response (−90°) and longitudinal drift force (180°) for FPSO

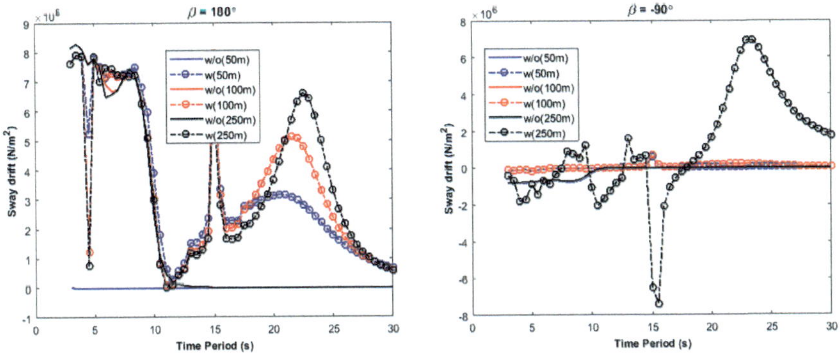

Fig. 4 Lateral drift force (180° and −90°) of FPSO

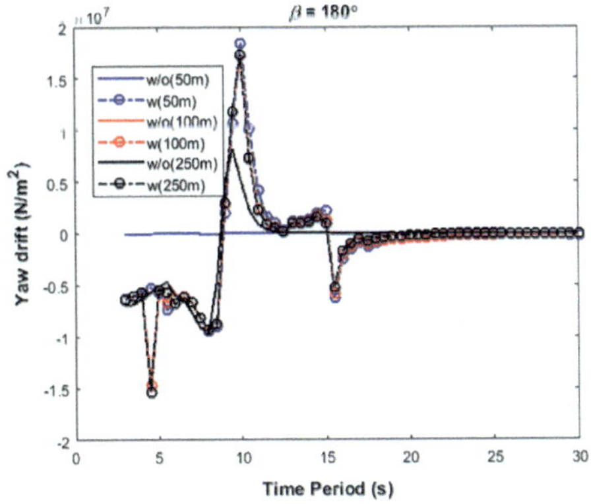

Fig. 5 Yaw drift force (180°) on FPSO

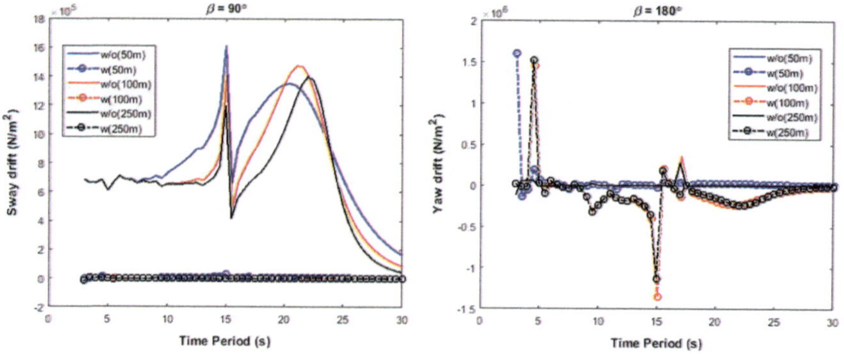

Fig. 6 Lateral drift force (90°) and yaw drift force (180°) on shuttle tanker

tanker causes the FPSO to drift in positive ‘y’ direction. The shuttle tanker was also observed for drift forces under the influence of FPSO. There is similar display of nature as that of FPSO. The sway and yaw drift forces at beam and head seas are shown in Fig. 6.

4 Conclusion

From the present study on the hydrodynamics of side-by-side vessels, following conclusions can be deduced as follows

1. The surge response on FPSO is affected by shielding effect for higher time period for head seas while quartering seas has no shielding effect predominantly.
2. There is significant increase in the sway response of FPSO is due to presence of shuttle tanker for quartering seas.
3. The heave response is shielded by the shuttle tanker in beam seas.
4. There is remarkable effect on the roll response of FPSO in side-by-side configuration for head seas while beam seas tend to exhibit shielding effect.
5. The presence of adjacent vessel influences the drift forces to a great extent.

Future Work The future work includes downtime analysis of offloading operations in Malaysian water conditions and developing GUI for linking motion responses to cost.

References

1. Kurt AE, Chris B, Heather H (2017) Worldwide survey of floating production, storage and offloading (FPSO) units: organized by FPSO vessel owners and operating vessels, Aug 2017

2. Fujiwara T, Yukawa K, Sato H, Otsubo K, Taniguchi T (2012) Wind effect estimation in side by side offloading operation for FLNG and LNG carrier ships. In: 31st international conference on ocean, offshore and arctic engineering. American Society of Mechanical Engineers (ASME), pp 71–78
3. Buchner B, Van Dijk A, De Wilde J (2001) Numerical multiple-body simulations of side-by-side mooring to an FPSO. In: The eleventh international offshore and polar engineering conference. International Society of Offshore and Polar Engineers
4. Vinnem JE, Utne IB, Schjølberg I (2015) On the need for online decision support in FPSO–shuttle tanker collision risk reduction. Ocean Eng 101:109–117
5. Djatmiko EB, Murdjito Ika P (2006) Operability analysis of an FPSO on the basis of its slamming and green water performances. In: 5th biennale conference on marine technology. MARTEC, Indonesia
6. Zhao W, Yang J, Hu Z, Tao L (2014) Prediction of hydrodynamic performance of an FLNG system in side-by-side offloading operation. J Fluids Struct 46:89–110
7. Hong SY, Kim JH, Kim HJ, Choi YR (2002) Experimental study on behavior of tandem and side-by-side moored vessels. In: The twelfth international offshore and polar engineering conference. International Society of Offshore and Polar Engineers
8. Ha MK, Kim MS, Park JJ, Lee JH (2004) First-and second-order hydrodynamic forces and moments on two offshore floating structures in waves. In: The fourteenth international offshore and polar engineering conference. International Society of Offshore and Polar Engineers
9. Tannuri EA, Torres FG, Igreja H, Masetti IQ (2009) FPSO and mono-buoy offloading operation with a conventional shuttle tanker: dimensioning of tugboat based on numerical simulation. In: IFAC proceedings volumes, vol 42(18), pp 134–139
10. Journée JMJ, Massie WW (2001) Introduction in offshore hydromechanics (OT3600). Delft University of Technology
11. Kim M, Jeong H, Kwak H, Kim B, Eom J (2012) Improvement method on offloading operability of side-by-side moored FLNG. In: The twenty-second international offshore and polar engineering conference. International Society of Offshore and Polar Engineers
12. Perwitasari RN (2010) Hydrodynamic interaction and mooring analysis for offloading between FPSO and LNG shuttle tanker. Master's thesis: Norges teknisk-naturvitenskapelige universitet, Fakultet for ingeniørvitenskap og teknologi, Institutt for Marin Teknikk
13. Chen XB, Orozco JM, Malenic S (2005) Evaluation of wave and current loads on offloading FPSOS. In: Offshore technology conference

Wave-in-Deck Force on Fixed Jacket Platforms by Silhoutte Method and Detailed Component Method

L. A. Pangestu, C. Y. Ng, A. E. Kajuputra, M. K. Muzammil and S. Sabtu

Abstract Wave-in-deck creates extra forces to the platform by increasing its moment arm which cause instability in terms of overturning. Fixed offshore platforms are generally not intended to withstand the substantial forces produced by wave-in-deck loads. Nonetheless, the requirement to accurately predict the magnitude of wave-in-deck forces and the platform global response of a jacket platforms in regards to current engineering knowledge and variance jacket platforms configuration remains limited. The research aims to compare API procedure, i.e. the silhouette method for predicting wave-in-deck force with detailed component method. Simulation study has been carried out and evaluate the wave-in-deck forces on two fixed jacket platforms installed in South China Sea using SESAM Software (GeniE and USFOS) by DNV GL and to quantify wave forces and the effect of wave-in-deck in terms of RSR on the jacket platforms. The results indicate that detailed component method has smaller reduction of RSR for platform A by 2.65% and platform B by 9.25% whereas silhouette method's reduction by 18.58% and 32.91% respectively.

Keywords Wave-in-deck · Fixed offshore platform · Ocean wave-structure interactions · Silhouette method · Detailed component method · RSR · Reserve strength ratio · Airgap · Pushover analysis

1 Introduction

There are more than 9000 fixed offshore platform for hydrocarbon production all around the world [1], about 200 jacket platform currently operating in South China Sea [2]. More than half of these platform in the area have exceeded their service life of 25 years. Therefore, reassessment of platform has to be performed to comply or ensure that the platform safety. The factor that affecting most jacket platforms

L. A. Pangestu (✉) · C. Y. Ng · M. K. Muzammil · S. Sabtu
Universiti Teknologi PETRONAS, Seri Iskandar, Perak, Malaysia
e-mail: ligar.a_17780@utp.edu.my

A. E. Kajuputra
PETRONAS Global Technical Solutions, Kuala Lumpur, Malaysia

C. L. Saw et al. (eds.), *Advancement in Emerging Technologies and Engineering Applications*, Lecture Notes in Mechanical Engineering,
https://doi.org/10.1007/978-981-15-0002-2_44

are seabed subsidence caused by reservoir compaction, increased topside weight or operational loads, revised environmental criteria, reduced capacity due to damage, corrosion and deterioration. Seabed subsidence occurrence causing the platform to settle down and causing the wave-in-deck phenomenon. Wave-in-deck occur when the wave crest hit the top deck, this scenario is likely to occur when the platform has negative airgap, the space between the lower deck of the platform and the water surface. Therefore, this phenomena is becoming more vulnerable to deck wave inundation.

The process of reassessment of the offshore platforms in an area often starts with a simplified evaluation of a larger number of platforms, proceeding to more detailed analyses for those platforms that do not fulfill relevant code requirements when being subject to simplified evaluation methods. If a platform fails to fulfill the requirements during the reassessment process, there are several alternatives for mitigation, such as increase topside elevation, removal of weight from topside or removal of conductors or risers, marine growth etc. to reduce environmental loads.

2 Wave-in-Deck

Wave-in-deck creates extra forces to the platform by increasing its moment arm which cause instability in terms of overturning. Fixed offshore platforms are generally not intended to withstand the substantial forces produced by wave-in-deck loads. On the off chance that a wave yet hits the deck, the deck legs, which are not estimated to exchange shear forces of this magnitude from the deck into the jacket platform, might be excessively loaded. Likewise, substantial upwards and downwards acting vertical loads might be presented in the platform, additionally decreasing the deck legs' ability to convey the transverse load. At last may likewise apply to the jacket legs. Subsequently, other failure modes than those considered during design phase can governing for platforms exposed to wave-in-deck loads.

So far there is no common engineering practice on how to model impact loading from waves on topside platforms. Several methods are previously used for this task, some verified against experimental data and some not. Reassessment of offshore platforms, particularly where wave-in-deck load is expected to simplified methods for wave-in-deck loads estimation recommended [3], are widely accepted in industry practice. The American Petroleum Institute (API) standard has the most commonly utilized methods and numerical formulations for wave-in-deck forces. The methods are divided into two fundamental groups, to be specific component models and global or silhouette models. The silhouette models are again subdivided into two groups, those based on loss-of-momentum formulation and on drag formulation.

There have been researches to find the most accurate method on how to determine wave-in-deck forces. Murray [4] conducted both experimental and analytical studies of wave impact forces to the Ekofisk platform, operated by Phillips Petroleum Company Norway. The fitness-for-purpose analysis of a platform using nonlinear methods or equivalent linear methods to determine the platforms global ultimate capacity, often referred to as a pushover analysis [3] can also be utilized to determine wave-in-deck forces. Raaij and Gudmestad [5] studied both global and local approach to find the realistic time-load histories for jacket platforms in the North Seas.

In 2011, Incremental Wave Analysis (IWA) was introduced to produce an estimation of various limit states and precise performance of jacket platforms against wave loadings. Golafshani [6] claimed that this approach is more accurate to represent the actual behavior of a jacket platform against environmental wave loadings as alternative to the conventional pushover practice. The pushover analysis using 100-year storm return period and RSR parameter is unable to determine the real failure mechanism against wave loads, as well as the platform's response towards various wave hazard level.

Suyuthi and Haver [7] reported that, these simplified method could significantly underestimated the impact load in extreme conditions. Experimental model tests are arguably the best approach to access the wave-in-deck load [8]. However, model test also has its own limitations, such as scale effects and significantly high cost [9].

Variety of models exist, some of which have been more extensively examined than others [5]. The methods can be divided into two main groups; firstly the global or silhouette method and secondly the detailed component methods where the load on each single members is calculated separately. The 'detailed' methods require a detailed deck model and allow for calculation of wave-in-deck loads on component level [10, 11]. Where the silhouette method implies that no detailed deck model is needed formulated by API [3], ISO formulation [12], DNV slamming formulation [13].

2.1 Detailed Component Method

The Reserve Strength Ratio (RSR) from pushover analysis is the ratio between collapse load and the design load of 100 years storm. The design load is scaled up on the platform until it collapses, which realistically should be related to increment of wave height that may reach the deck and generates wav-in-deck forces. However, USFOS is not capable modeling of such phenomena of wave height increment during load increment in pushover analysis.

Wave-in-deck generates additional load from the deck with equipment being exposed and also increases moment arm to the platform. These consequently increase the overturning moment that influence the collapse load or capacity, which decreases the RSR value at the end.

The limiting RSR in extreme airgap analysis is the lower bound of the true ultimate RSR. The actual collapse RSR with wave-in-deck is obviously higher than the lower bound value but lower than the upper bound value of RSR collapse taken from pushover analysis without considering wave-in-deck.

This method which examine loading on each individual component on the topside platform. Higher drag coefficient (Cd) value applied on deck members inundated by the waves following recommendation given by both API RP 2SIM [3] and ISO 19902 [12] in order to capture the additional loads from the equipment on the deck. The flow process of the numerical simulation is shown in the Fig. 1.

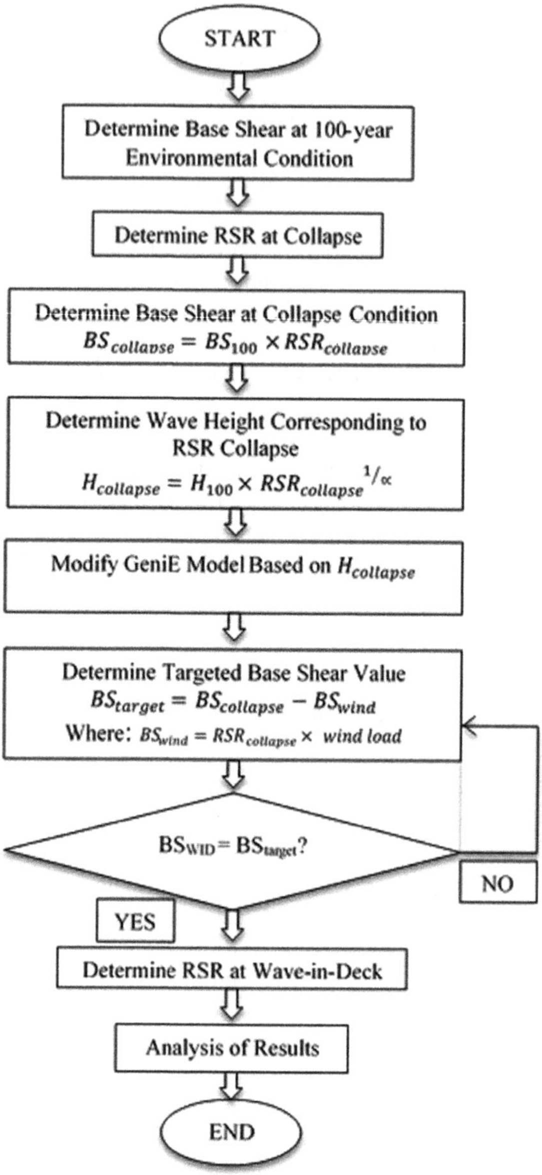

Fig. 1 Detailed component method

2.2 Silhouette Method

The drag based silhouette wave-in-deck models have in common the typical drag formulation known from Morison equation. The absolute value of the load as follows:

$$F_x = \frac{1}{2}\rho A C_d u_w^2 \tag{1}$$

where ρ is sea water density u_w is the water particle velocity, $A = s_d\ b$ is the exposed area, s_d is the inundation (height) and b the width of the inundated area. The drag factor Cd is chosen to account for different loading scenarios. These models are typically intended for calculation of maximum load, but the simple formulation makes them also easy to apply for time domain analyses, noting that the particle velocity u_w and the inundated area A are time dependent variables. In API RP 2SIM [3], the Cd values is varied between 1.2 and 2.5 according to the wave direction and the equipment density on the deck. The water particle velocity u_w contains a sum of current velocity and wave induced particle velocity, as well as a current blockage factor and a wave kinematics factor.

The deck force procedure relies on a calculated crest height. The crest height should be calculated using the wave theory using Two-Dimensional Wave Kinematics and the ultimate strength analysis wave height, associated wave period, and storm tide.

Given the crest height, compute the wetted "silhouette" deck area, (A) projected in the wave direction, (θw). The full silhouette area for a deck is defined as the shaded area in Fig. 2, the area between the bottom of the scaffold deck and the top of the "solid" equipment on the main deck. The area, A, is computed as follows:

$$A = A_x \cos\theta_w + A_y \sin\theta_w \tag{2}$$

where, θ_w, A_x, and A_y are defined in Fig. 2.

The wave-in-deck force, can be computed by the following:

$$F_{dk} = \frac{1}{2}\rho A C_d \left(a_{wkf} \cdot V + a_{cbf} \cdot U\right)^2 \tag{3}$$

where,

U	Current speed in-line with the wave
a_{wkf}	Wave kinematics factor (0.88 for hurricanes and 1.0 for winter storms)
a_{cbf}	Current blockage factor for the jacket
ρ	Mass density of seawater
C_d	Drag coefficient, given in the Table 1.

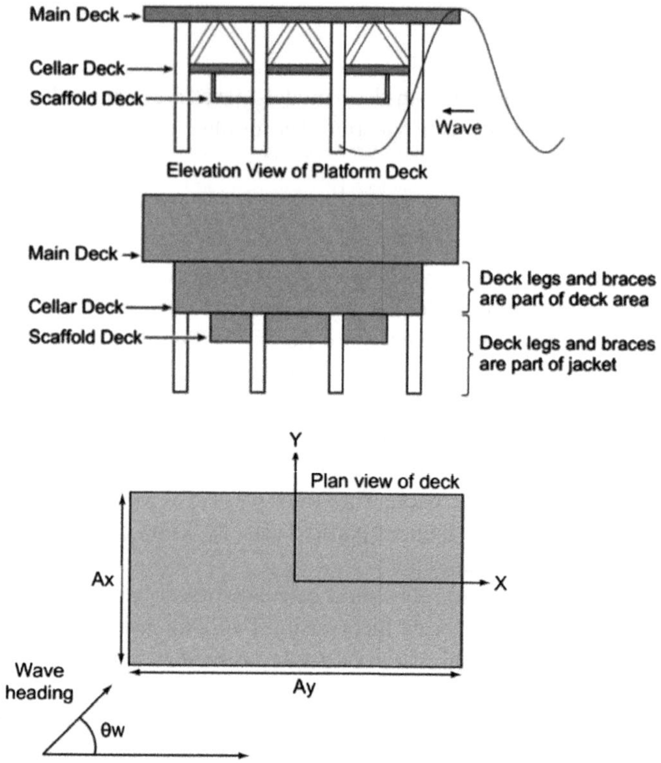

Fig. 2 Silhouette area definition & wave heading and direction convention [3]

Table 1 Drag coefficient (Cd) for wave-in-deck forces [3]

Deck type	Cd end-on and broadside	Cd diagonal (45°)
Heavily equipped (solid)	2.5	1.9
Moderately equipped	2.0	1.5
Bare (no equipment)	1.6	1.2

3 Data Collection

Data collection of the jacket platforms for the parameters identification and data selection are presented in Table 2. Table 3 shows the data collection of extreme airgap analysis spreadsheet of Platform A and B. The direction of 225° was chosen as critical direction based on the lowest RSR value of 1.83 with −3.11 m of airgap

Table 2 Platforms description

Model	A	B
Water depth (m)	29.87	75.4
Design service category	Production	Wellhead
Design safety category	Unmanned	Unmanned
Installed	1968	2016
No. of legs	4	4
No. of piles (mm Ø)	4 main (610) and 5 shear (1067)	4 (1372)

for Platform A and 315° as the critical direction of Platform B with −2.93 m of airgap.

4 Pushover Analysis

Figure 3 shows the isometric view of platform model Platform A and B. Simulations were done using SESAM GeniE and USFOS by modifying wave height and wave period based on the process shown in Fig. 1.

4.1 Detailed Component Method

Platform A has extreme airgap analysis results as shown in Table 3, that negative airgap appears at 0°, 90°, 135°, 225°, 270° and 315° metocean direction at RSR collapse condition. Negative airgap has already occurred prior to the approach of 100 years storm wave height for the critical direction (225°) with −3.11 m airgap and for 270° and 315° direction with −3.31 m and −3.01 m airgap, respectively. The storm 100 years wave height for 225° is 7.20 and 6.50 m for 270° and 315°. The limiting RSR after considering control to avoid wave-in-deck is 0.56 (reduced from 1.83) for 225° direction, which does not satisfy the minimum criteria 1.32 for unmanned platform.

Similarly for Platform B, the extreme airgap analysis results show that negative airgap appears at 0°, 180°, 270° and 315° metocean directions at collapse condition or at RSR value. The RSR value is then limited to the value when the zero airgap is reached to avoid wave-in-deck forces. The lowest RSR at 315° direction is then reduced from 3.89 to 2.40.

Wave-in-deck analysis using detailed component method were done to follow up the unsatisfactory results from extreme airgap analysis. Higher Cd value 1.9 is applied to the topside members for both platform considering heavy equipped topside as recommended for diagonal direction by ISO 19902 and API RP 2SIM.

Table 3 Extreme airgap analysis

Platform	A								B							
Direction	N (45°)	NE (0°)	E (315°)	SE (270°)	S (225°)	SW (180°)	W (135°)	NW (90°)	SE (0°)	E (45°)	NE (90°)	N (135°)	NW (180°)	W (225°)	SW (270°)	S (*315°*)
100-year wave height (m)	1.20	5.30	6.50	6.50	7.20	1.20	1.20	1.20	8.50	5.80	5.60	6.20	6.60	6.00	9.00	12.60
RSR	5.97	2.23	2.12	2.3	*1.83*	9.3	13.34	19.56	7.91	6.38	10.19	7.55	10.00	7.30	7.74	*3.89*
Cellar deck elevation (m)	6.17	6.17	6.17	6.17	6.17	6.17	6.17	6.17	12.50	12.50	12.50	12.50	12.50	12.50	12.50	12.50
1/2 of deck girder height (m)	0.47	0.47	0.47	0.47	0.47	0.47	0.47	0.47	0.17	0.17	0.17	0.17	0.17	0.17	0.17	0.17
Water depth from MSL (m)	29.87	29.87	29.87	29.87	29.87	29.87	29.87	29.87	75.40	75.40	75.40	75.40	75.40	75.40	75.40	75.40
HAT (m)	1.14	1.14	1.14	1.14	1.14	1.14	1.14	1.14	1.00	1.00	1.00	1.00	1.00	1.00	1.00	1.00
Storm surge	0.60	0.60	0.60	0.60	0.60	0.60	0.60	0.60	0.60	0.60	0.60	0.60	0.60	0.60	0.60	0.60
Wave crest height factor	0.60	0.60	0.60	0.60	0.60	0.60	0.60	0.60	0.55	0.55	0.55	0.55	0.55	0.55	0.55	0.55
Wave height exponent	1.70	1.70	1.70	1.70	1.70	1.70	1.70	1.70	2.00	2.00	2.00	2.00	2.00	2.00	2.00	2.00

(continued)

Table 3 (continued)

Platform	A								B							
Direction	N (45°)	NE (0°)	E (315°)	SE (270°)	S (225°)	SW (180°)	W (135°)	NW (90°)	SE (0°)	E (45°)	NE (90°)	N (135°)	NW (180°)	W (225°)	SW (270°)	S (315°)
Initial air-gap(m)	3.06	3.06	3.06	3.06	3.06	3.05	3.06	3.06	10.74	10.74	10.74	10.74	10.74	10.74	10.74	10.74
MSL + HAT + Storm surge (m)	31.61	31.61	31.61	31.61	31.61	31.61	31.61	31.61	77.00	77.00	77.00	77.00	77.00	77.00	77.00	77.00
Bottom of cellar deck elevation (m)	34.67	34.67	34.67	34.67	34.67	34.67	34.67	34.67	87.74	87.74	87.74	87.74	87.74	87.74	87.74	87.74
Wave height corresponding to RSR (m)	3.43	8.50	10.11	10.61	10.27	4.46	5.51	6.90	23.91	14.65	17.88	17.04	20.87	16.21	25.04	24.85
Wave crest height (m)	2.06	5.10	6.07	6.37	6.16	2.67	3.31	4.14	13.15	8.06	9.83	9.37	11.48	8.92	13.77	13.67
Crest elevation (m)	4.71	7.75	8.72	9.02	8.81	5.32	5.96	6.79	14.75	9.66	11.43	10.97	13.08	10.52	15.37	15.27

(continued)

Table 3 (continued)

Platform	A								B							
Direction	N (45°)	NE (0°)	E (315°)	SE (270°)	*S (225°)*	SW (180°)	W (135°)	NW (90°)	SE (0°)	E (45°)	NE (90°)	N (135°)	NW (180°)	W (225°)	SW (270°)	*S (315°)*
Airgap (m)	1.00	−2.04	−3.01	−3.31	*−3.11*	0.38	−0.25	−1.08	−2.41	2.68	0.90	1.37	−0.74	1.82	−3.04	*−2.93*
Limiting RSR	11.68	0.93	0.66	0.66	*0.56*	11.68	11.68	11.68	5.27	11.32	12.15	9.91	8.75	10.58	4.70	*2.40*

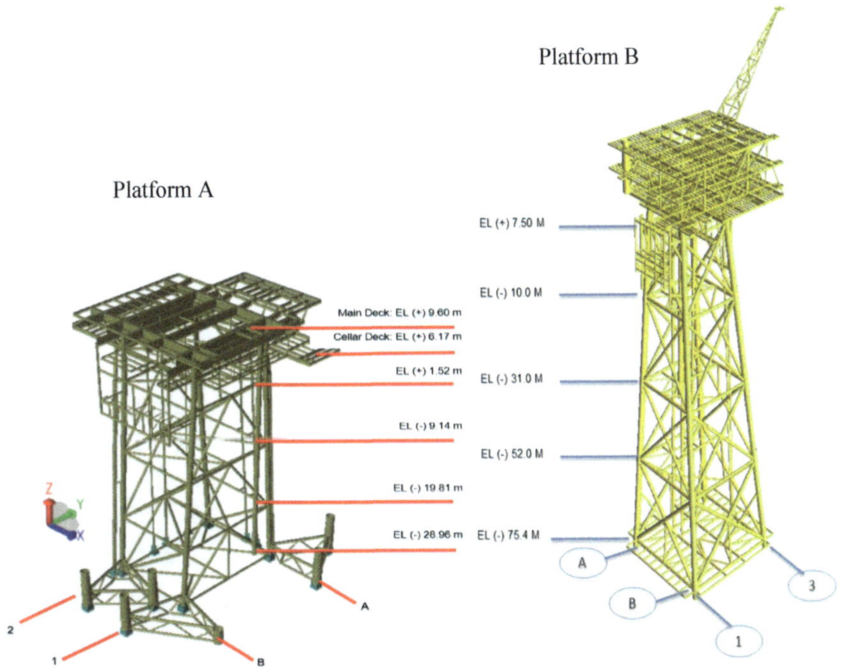

Fig. 3 Isometric view platform A and platform B

Joint probability metocean data (wave govern) is also applied in this wave-in-deck analysis considering the probability of occurrence of wave and current 100 years return period accordingly. The analysis results shows that RSR ultimate for Platform A at wave-in-deck condition is 1.78 with reduction ratio of 0.97 to the RSR collapse 1.83, and 3.54 with reduction ration of 0.91 from RSR collapse of 3.89 for Platform B.

4.2 Silhouette Method

Figures 4 and 5 show the comparison between the collapse pushover analysis and wave-in-deck pushover analysis using Sihlouette Method for Platform A at 225° direction.

Based on Eq. (2), Platform A has silhouette area of 68.65 m^2 and wave force generated from 10.27 m of wave height taken from Table 3 is 357.07 kN, which has applied equally to each topside legs. The analysis result shows a quite significant reduction of RSR to 1.49 as shown in Fig. 5, compare to RSR collapse of 1.83 and RSR ultimate using detailed component method which is 1.78.

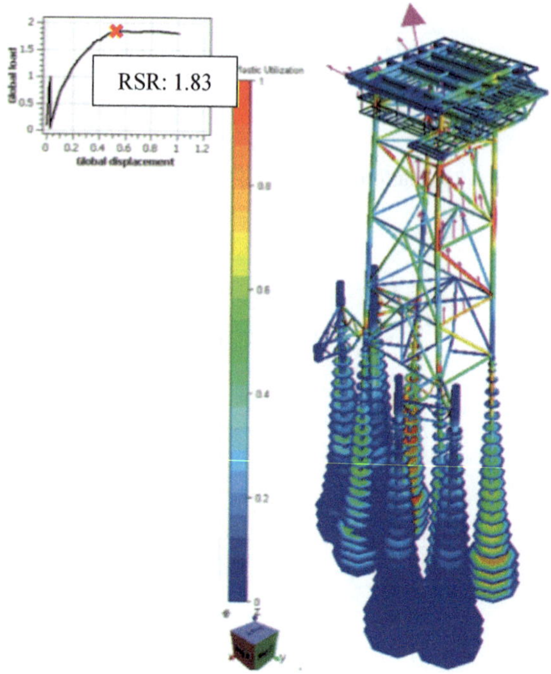

Fig. 4 Platform A 225° pushover analysis plot

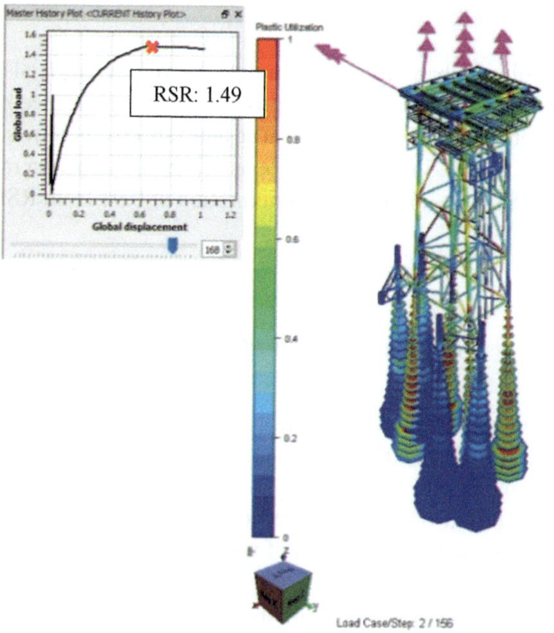

Fig. 5 Silhouette method platform A 225° wave-in-deck pushover analysis plot

On the other hand, Platform B received 981.34 kN to each legs from silhouette area of 79.34 m^2 with 24.85 m corresponding wave height, reducing RSR from 3.89 to 2.61 as shown in Figs. 6 and 7 respectively.

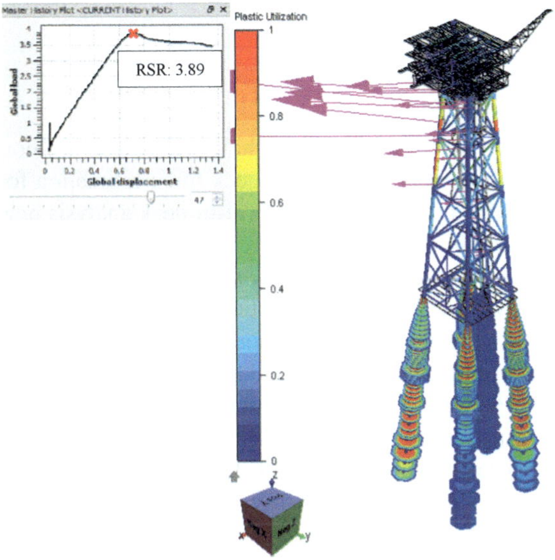

Fig. 6 Platform B 315° pushover analysis plot

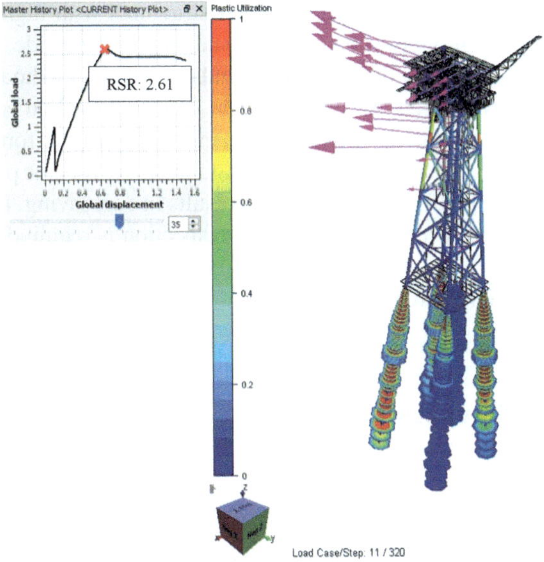

Fig. 7 Silhouette method platform B 315° wave-in-deck pushover

Table 4 RSR wave-in-deck results

Platform	A	B
$RSR_{Collapse}$	1.83	3.89
$RSR_{Ultimate}$ DCM	1.78	3.54
$RSR_{Ultimate}$ SM	1.49	2.61

5 Conclusions

This research studies the significance between silhouette method and detailed component method on wave-in-deck force estimation for fixed jacket platforms. It was identified that the current wave-in-deck analysis practice has no mutual concession approach.

As described on Table 4, the detailed component method result of Platform A shows there was RSR reduction by 3% from 1.83 to 1.78 and Platform B by 9% to RSR 3.54. Compared to silhouette method which has more substantial reduction for both platforms, RSR reduced from 1.83 to 1.49 for Platform A and RSR 3.89 to 2.61 for Platform B. Although, it is still satisfied the minimum acceptance criteria of 1.32 for unmanned platform.

The difference between silhouette method being lower than detailed component method is due to the overestimated wave force acting on topside as wave-in-deck force. The wave-in-deck forces computed in silhouette methods assumed that the deck exposed to the wave as a whole of block rectangular of silhouette.

As compared to detailed component method which simulated the wave exerted to the topside similar to jacket structures with higher drag coefficient (Cd) value representing the component installed on the topside.

Therefore, this results can indicate which method is more efficient and reliable. Industries can have more insight on which method is more reliable because there is no general consensus on analyzing wave-in-deck force towards offshore structure.

Other than that, by knowing the actual RSR of the platform without over estimation of the platform strength it can results in cost saving. It is because the RSR value will indicate whether underground inspection is required or otherwise. Hence, knowing real RSR value can save millions of ringgit with regards to maintenance works of offshore platform.

References

1. Raaij KV (2005) Dynamic behaviour of jackets exposed to wave-in-deck forces. PhD Thesis in Mechanical and Structure Engineering and Material Science. Stavanger, Norway, University of Stavanger
2. Kurian VJ, Voon MC, Wahab MMA, Liew MS (2014) System reliability assessment of existing jacket platforms in Malaysian waters. Res J Appl Sci Eng Tech 8(23):2305–2314
3. API (2007) Recommended practice for planning, designing and constructing fixed offshore platforms–working stress design, 2A-WSD, 2007. American Petroleum Institute

4. Murray JJ, Kaplan P (1995) Experimental and analytical studies of wave impact forces on ekofisk platform structures. In: Proceedings of Offshore Technology Conference, Paper No. 7782
5. Van Raaij K, Gudmestad OT (2007) Wave-in-deck loading on fixed steel jacket decks. Mar Struct 20(3):164–184
6. Golafshani AA, Bagheri V, Ebrahimian H, Holmas T (2011) Incremental wave analysis and its application to performance-based assessment of jacket platforms. J Constr Steel Res 67(10):1649–1657
7. Suyuthi A, Haver SK (2009) Extreme loads due to wave breaking against platform column. In: The nineteenth international offshore and polar engineering conference. International society of offshore and polar engineers
8. Scharnke J, Vestbøstad T, de Wilde J, Haver, S (2014) Wave-in-deck impact load measurements on a fixed platform deck. In: ASME 2014 33rd international conference on ocean, offshore and arctic engineering. American society of mechanical engineers digital collection
9. Kim J, Jang H, Izarra R, Martin D, Dalane O (2014, May) CFD-FE simulation of wave slamming on an offshore platform in extreme sea states. In: Offshore technology conference. Offshore technology conference
10. Kaplan P, Murray JJ, Yu WC (1995) Theoretical analysis of wave impact forces on platform deck structures. In: Proceedings of the 14th international conference on offshore mechanics and arctic engineering (OMAE). Copenhagen, Denmark
11. Pawsey S, Driver D, Gebara J, Bole J, Westlake H (1998) Characterization of environmental loads on subsiding offshore platforms. In: Proceedings of the 17th international conference on offshore mechanics and arctic engineering (OMAE). Lisbon, Portugal
12. ISO (2007) ISO 19902:2007. Petroleum and natural gas industries–Fixed steel offshore structures. British Standards
13. DNV (2000) Environmental conditions and environmental loads. Oslo, Norway

Crude Oil Fouling Mitigation by Products Thermal Management in Heat Exchangers

Ogboo Chikere Aja, Reuben Brandon Huan Chung Lee and Saw Chun Lin

Abstract The management and maintenance of process plants involves huge amount of money. In the maintaining the heat exchangers, fouling effect has been identified as one of the major cost which also contribute to loss of production when the exchanger is out of service. This is applicable in the oil and gas industries which involves fouling of crude preheat train as a result of asphaltene precipitation and other fouling factors. The rate of fouling in the preheat train has been identified to be a function of temperature difference in the heat exchanger between the thermal source and the crude flow. This study focus on reducing the rate of fouling by varying the flow rate of the high temperature products through the heat exchangers in order to mitigate the fouling rate of occurrence. The study was conducted using MATLAB SIMULINK. The crude preheat train of four heat exchanges network was modeled in SIMULINK. The model was simulated under various flow rates and temperatures. The results showed that at fixed products mass flow rate of 2 kg/s with flow velocity of 0.06 m/s while the crude inlet temperature were varied from 303 to 312 K, the rate of fouling drops with increase in inlet temperature. A consideration of constant crude inlet temperature of 303 K with varied mass flow rate of the products from 1.6 kg/s in the increment of 0.2 to 2.4 kg/s showed that the rate of fouling dropped most at lower product mass flow rate. The simulation is done over a period of 10,000 min. The fouling factor that is obtained from the simulation period is lower than the actual result. The average fouling factor for the crude oil through the heat exchanger network is approximately 0.003.

Keywords Crude oil fouling · Crude preheat train · Fouling factor

---

O. C. Aja (✉) · R. B. H. C. Lee
Department of Mechanical Engineering, Curtin University, CDT 250, 98009 Miri, Sarawak, Malaysia
e-mail: aja.ogboo@curtin.edu.my

S. C. Lin
Mechanical and Marine Engineering Department, Politeknik Ungku Omar, Jalan Raja Musa Mahadi, 31400 Ipoh, Negeri Perak, Malaysia

C. L. Saw et al. (eds.), *Advancement in Emerging Technologies and Engineering Applications*, Lecture Notes in Mechanical Engineering,
https://doi.org/10.1007/978-981-15-0002-2_45

1 Introduction

Fouling in the crude preheat train is caused by the inverse precipitation of asphaltene at varying temperatures [1]. Fouling as a result of precipitation of asphaltene affects the energy required to heat the fluid as well as create pressure build-up along the crude flow line as a result of the deposits in the internal surface of the tube [2]. This is a major challenge in the refining industry as it reduces the production rate thus, increase cost and time for maintenance.

In a shell and tube heat exchanger where two fluids of varying thermo-physical properties and temperatures are transported with the aim of heat recovery, the variation in their temperature results in fouling as in the case of crude oil at low temperature and refined products at high temperature leading to asphaltene precipitation in the crude oil [3]. The greater the temperature difference the lower the solvating power of crude, which increases the precipitation of asphaltene. With the precipitation of fouling in a heat exchanger tube, the pressure increases, thus requiring high pumping power which then lead to increase in cost of production as well as loss of production. On cost consideration, fouling effect lead to loss of USD 15 Billion in the USA and USD 2.5 Billion in the UK on mitigation and maintenance of preheat trains [4].

A significant amount on research have been done on fouling mitigation in crude preheat train that is mainly focused on the use of chemicals to reduce the precipitation effect of asphaltene in crude transport tube. At high temperature difference between the channel carrying the hot fluid and the channel carrying the low temperature crude, the chemical for asphaltene precipitation mitigation tends to be ineffective. Temperature difference between the crude transport tube section and the product transport shell section has been identified as a major factor causing fouling [5, 6] but the management has not been proposed. Thus, this study investigated the management of the temperature difference and the rate of fouling by mass flow rate control and temperature variations at the different heat exchangers. The mass flow rate will also play an important part for the efficiency of the heat exchanger. So, with the suitable temperature range, the flow process and the thermal field can be analyzed to obtain suitable mass flow rate range to increase the efficiency of the heat exchanger thus develop an optimized model for the heat exchanger network (HEN).

The main objective of the research is to stimulate the flow integration between the cold fluid, crude oil and hot fluid, product to reduce the temperature difference (ΔT) with an aim to mitigate fouling in the heat exchanger. Besides, the research is to investigate the impact of mass flow rate ($\dot{m}$) of the product (shell side) of the heat exchanger to mitigate fouling. Thus a model of the preheat train heat exchanger is modelled to control the temperature difference and enhance the production efficiency with the identification of the fouling factor at different inlet temperature of the crude oil and the different mass flow rate of the product that is flowing through the shell side of the heat exchanger.

2 Methodology

Figure 1 indicates the overall heat exchanger network. The heat exchanger network consists of 4 different heat exchanger block. Each heat exchanger will have different product with different properties and temperature flowing through it. The mass flow rate of the crude oil is set to be 1 kg/s throughout the entire simulation. It is also important to bear in mind that the output temperature of the first heat exchanger will be the input temperature of the second heat exchanger and so forth. With the energy equation, the heat flux for the crude oil and the product is equal as shown in Eq. 1.

$$\dot{Q}_{Crude} = \dot{Q}_{product}$$
$$\dot{m}_{Crude} * C_{P\,crude} * \Delta T_{Crude} = \dot{m}_{product} * C_{P\,Product} * \Delta T_{product} \quad (1)$$

The overall heat transfer coefficient multiply the total fluid area (UA) value for each individual heat exchanger can be calculated by using its total thermal resistance (R_{total}). The total resistance equation for the heat exchanger is shown in Eq. 2.

$$R_{total} = \frac{1}{h_{crude} 2\pi r_{in} L} + \frac{\ln\left(\frac{r_{out}}{r_{in}}\right)}{2\pi k_{tube} L} + \frac{1}{h_{product} 2\pi r_{out} L} \quad (2)$$

After the total resistance is found, the total number of tubes is divided with the total resistance to obtain the UA value as shown in Eq. 3.

$$UA = \frac{n}{R_{total}} \quad (3)$$

From Eq. 2, there is an unknown value for the convective heat transfer, h. We assume that the flow of the fluid is fully developed and is in laminar flow, we can easily determine the convective heat transfer value with the Eq. 4.

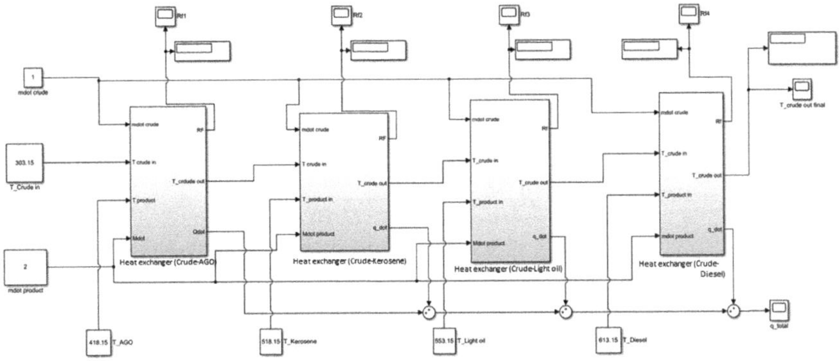

Fig. 1 Heat exchanger network

$$h_c = \frac{Nu.K}{D} \quad (4)$$

where, Nu = Nusselt Number, K = thermal conductivity of different fluid and D is the characteristic length (in our case the diameter).

In addition, to obtain the heat flux of the system, the E-NTU method is applied, Eq. 5.

$$\dot{Q}_{product} = \in C_{min}\left(T_{product} - T_{crudeoil}\right) \quad (5)$$

where, $(\dot{m}_{product}) * (C_{P\ Product})$ and $(\dot{m}_{crude}) * (C_{P\ crude})$ whichever product with the lesser amount will be the $C = C_{min}$ and the maximum will be the C_{max}.

The effectiveness of the heat exchanger, ϵ can be calculated with Eq. 6 and is being insert back to Eq. 5 to obtain the heat flow.

$$\epsilon = \frac{1 - \exp(-NTU(1 - C_r))}{1 - C_r * \exp(-NTU(1 - C_r))} \quad (6)$$

where, $C_r = \frac{C_{min}}{C_{max}}$ while $NTU = \frac{UA}{C_{min}}$.

The SIMULINK diagram of the heat flow governed by Eq. 5 is shown in Fig. 2.

Once the heat flow is found, it is now possible to determine the output temperature of the crude oil through a heat exchanger with the energy equation. The standard heat energy equation from Eq. 1 is modified to calculate the output temperature with respect to time as shown in Eq. 7.

$$\int_0^{t_f} \left[(T_i - T_o)_{crude} * \dot{m}_{crude} * \left(C_{p\,crude}\right) + \dot{Q}_{product}\right] * [W_{Crude} * UA]dt = T_o(t) \quad (7)$$

where, $t_f = final\,time\,(minutes)$, W_{Crude} = mass of crude = $\rho_{crude} * Volume\,of\,crude$.

The SIMULINK model of the energy balance which is defined by Eq. 7 is shown in Fig. 3.

However, to determine the fouling factor (Rf), it would require the temperature difference of the crude and the heat flow ($\dot{Q}$). Equation 8 is used to determine the fouling factor.

$$R_f = \frac{1}{U(t)A} - \frac{1}{U(0)A} \quad (8)$$

where, $(t)A = \frac{\Delta T_{Crude}}{\dot{Q}}$.

$U(t)A$ is the overall heat transfer at a particular time multiply the Area of heat transfer. Thus when $U(0)A$ is at the initial time of $t = 0$. In other term, the term $\frac{1}{U(t)A}$ simply means when the heat exchanger is dirty and when $\frac{1}{U(0)A}$ signifies the heat exchanger is clean. The SIMULINK model that prescribes Eq. 8 is shown in Fig. 4.

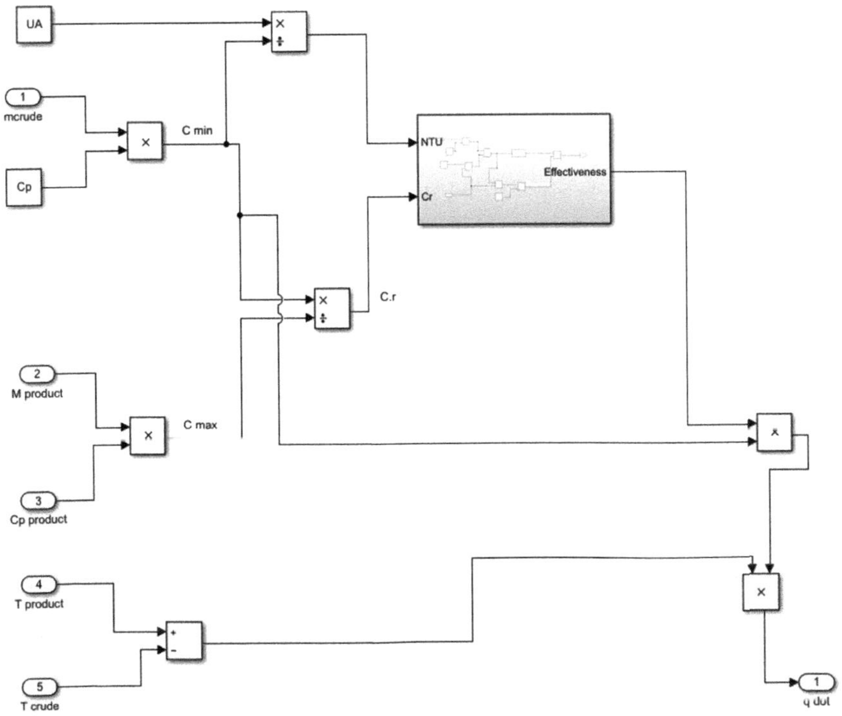

Fig. 2 Thermal energy flow field

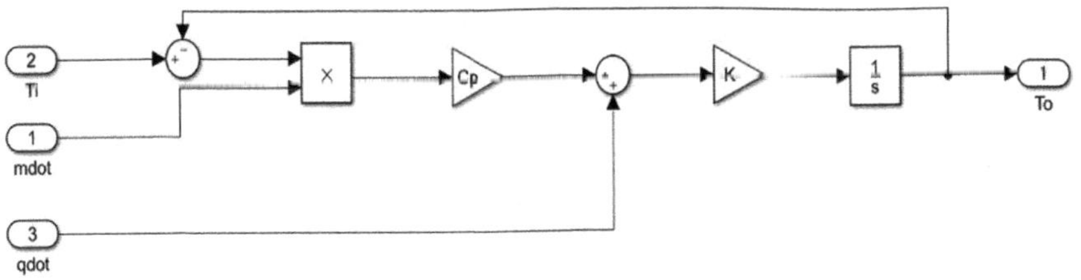

Fig. 3 Energy balance

3 Results and Discussions

The results are presented with respect to the temperature effects and the effect of mass flow rates on the rate of fouling in the preheat train.

(a) Exit Crude Temperature Gradients Considering Inlet Crude Temperature of 303.15 K

The first simulation was done with the inlet temperature of 303.15 K and the mass flow rate of 2 kg/s. In Fig. 5, it indicates the final temperature of the crude when it

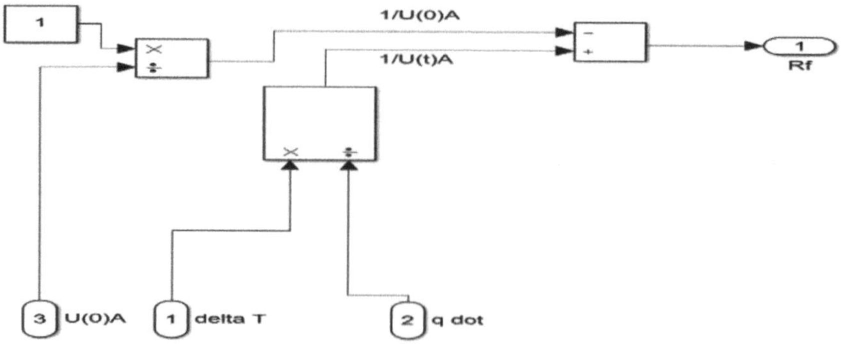

Fig. 4 Fouling factor

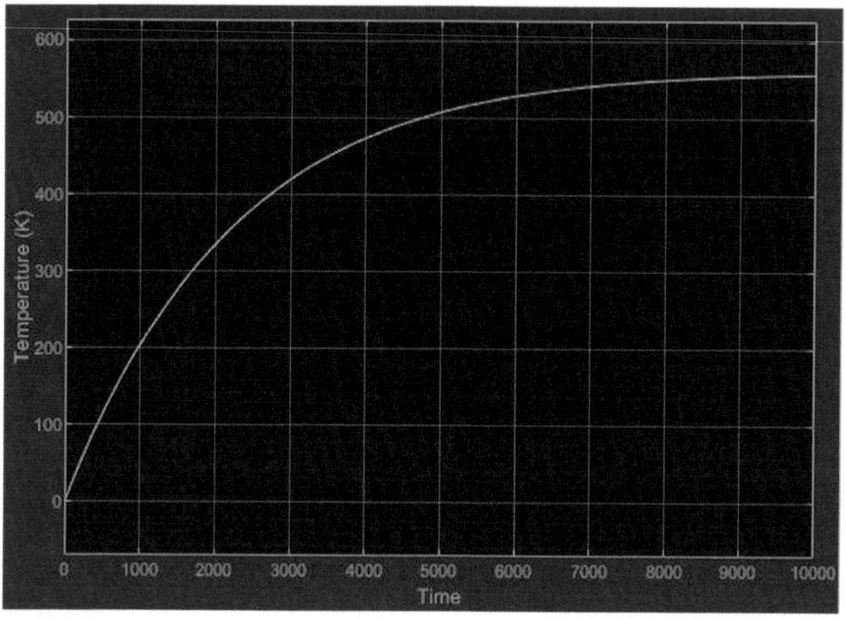

Fig. 5 Crude oil final outlet temperature

passes through all four heat exchanger in the heat exchanger network with the output temperature of 557.78 K. It increases drastically with time until it reaches 5000 min (83.33 h). The gradient of the curve before 5000 min is approximately 0.1. After 5000 min, the gradient of the curve is approximately 0.01. This indicates that the crude oil temperature is increasing lesser as compared to the first 5000 min as the crude tends to be thermally saturated.

In addition, Fig. 6, shows the total heat flow through the four heat exchangers. It can be seen that the heat flow is decreasing over time. For the first 5000 min, the

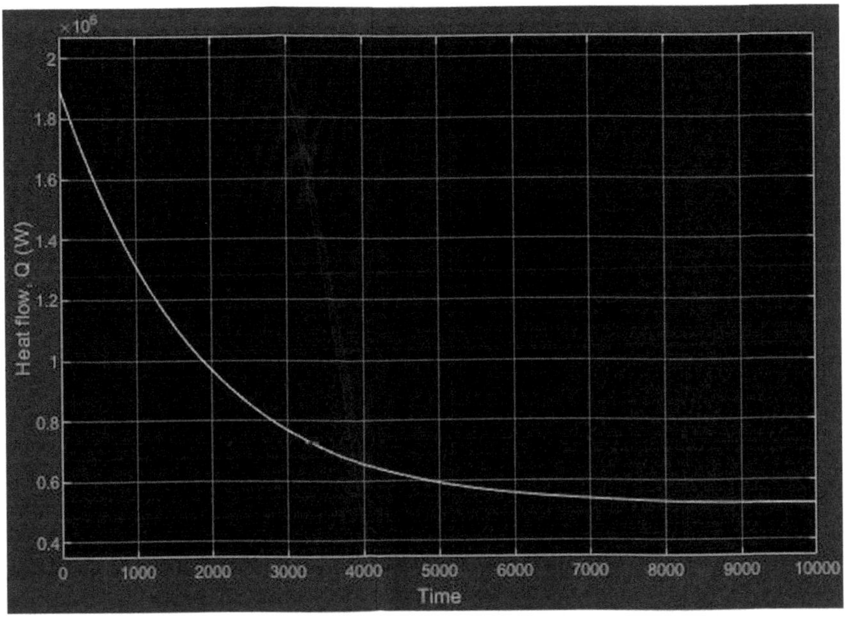

Fig. 6 Total heat flow

heat flow decreases drastically with a gradient of −261.14. This means that a large amount of heat exchange is happening for the first 5000 min. However, for the last 5000 min the gradient of the graph is −14.61. As compared to the first 5000 min, the final 5000 min have lesser heat transfer.

Figures 5 and 6 are correlated. As the temperature gradient is smaller, this indicates that the heat flow is slower and thus the heat energy in both liquids is almost equilibrium. After 5000 min, the temperature gradient is approximately 0.01 and the heat transfer gradient is −14.61 which both parameter's gradient is lesser than the first 5000 min. However, the fouling factor of each individual heat exchanger is shown in Fig. 7. It can be seen that the heat exchanger with the product Automotive Gas Oil (AGO) will undergo fouling first at approximately 2933 min. It would take until 6933 min until the Rf value reaches 0.000292. The Rf value will continue to rise till it reaches 0.000306 at approximately 9682 min. For the Rf of kerosene, the starting point would be close to the Rf of AGO. The final Rf after a series of time would also be close to the Rf of the AGO. The only difference between the two graph is the initial steepness of the graph. This is because the temperature difference for kerosene is smaller than the temperature difference in AGO. This also applies to the Rf of the light oil and the Rf of diesel. Both Rf begins at almost the same time but portraying different graph. This is also caused by the light oil heat exchanger have higher temperature difference of the crude inlet and the outlet as compared to the diesel heat exchanger.

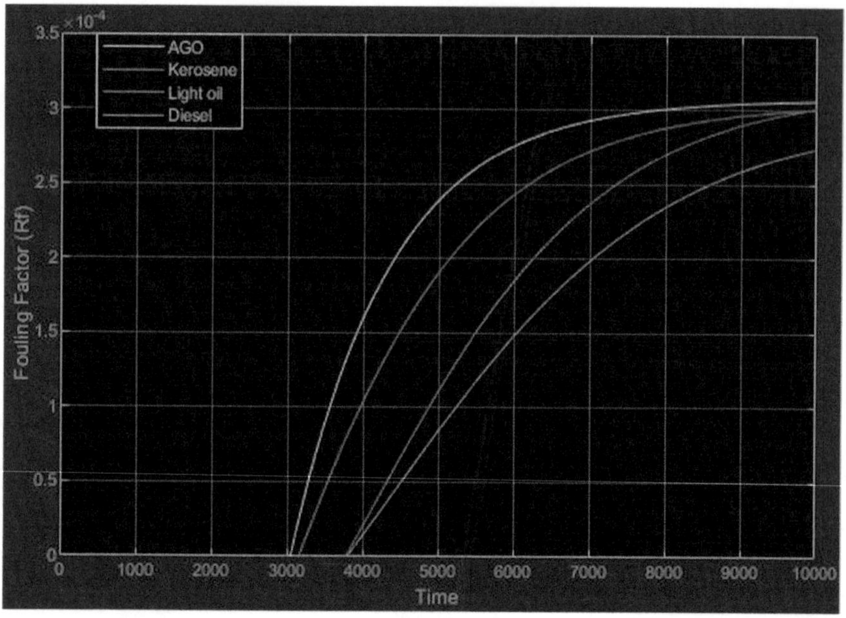

Fig. 7 Graph of fouling factor (Rf) against time

(b) Varying Mass Flow Rate of Products to 1.6 kg/s

The simulation was conducted also to see the effect of mass flow rate variation of the products into the individual heat exchanger at 1.6 kg/s. The graph of the fouling rate against time were computed and reported as shown in Fig. 8. The fouling factor for the crude passing through the heat exchanger with AGO starts from 3212 min (53.5 h). It increases drastically with time until it reaches 8000 min. After 8000 min, the Rf value tends to be a constant value of 0.000306. The final Rf value for the heat exchanger with AGO, kerosene and light oil are very close to each other. The final Rf value for these three are 0.000306, 0.00029985 and 0.0002993. The only difference that is visible from these 3 Rf curve is the time it starts and the time for it to reach its maximum value. The heat exchanger with kerosene will start fouling at approximately 3362 min and for the heat exchanger with light oil, the fouling will start at 3993 min. For the heat exchanger with diesel, the graph is still increasing but with the simulation run time of 10,000 min, the final Rf value of the heat exchanger is 0.00027. The heat exchanger with diesel will start fouling at 4032 min.

(c) Varying Mass Flow Rate of Products to 1.8 kg/s

The simulation was run with the mass flow rate of the product into the heat exchanger of 1.8 kg/s and other variables are kept constant. The graph of the fouling factor against time is in Fig. 9. The fouling factor for the crude passing through the heat exchanger with AGO starts from 3212 min. It increases with time until it reaches 8000 min. After 8000 min, the Rf value tends to be a constant value of 0.000306. The

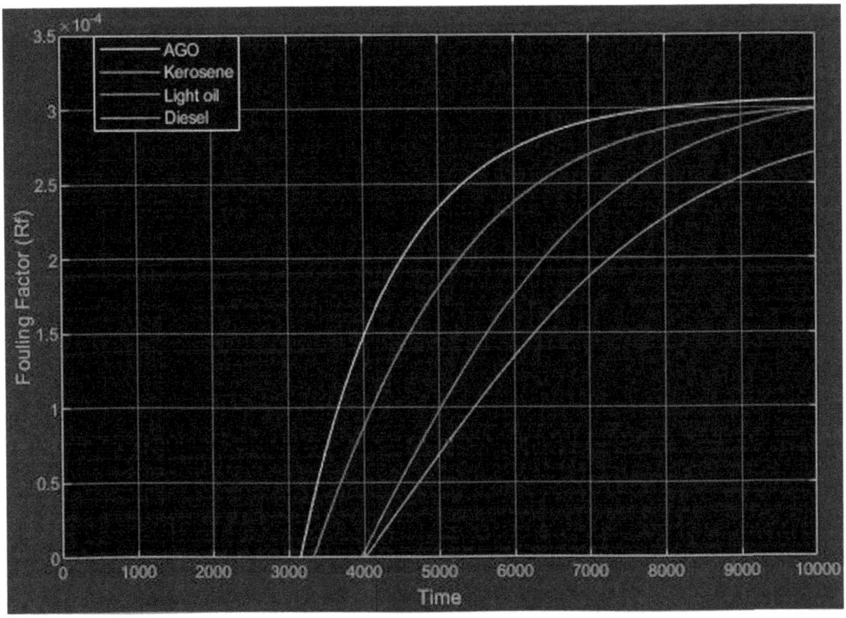

Fig. 8 Graph of fouling factor against time for $\dot{m}_{product} = 1.6\,\text{kg/s}$

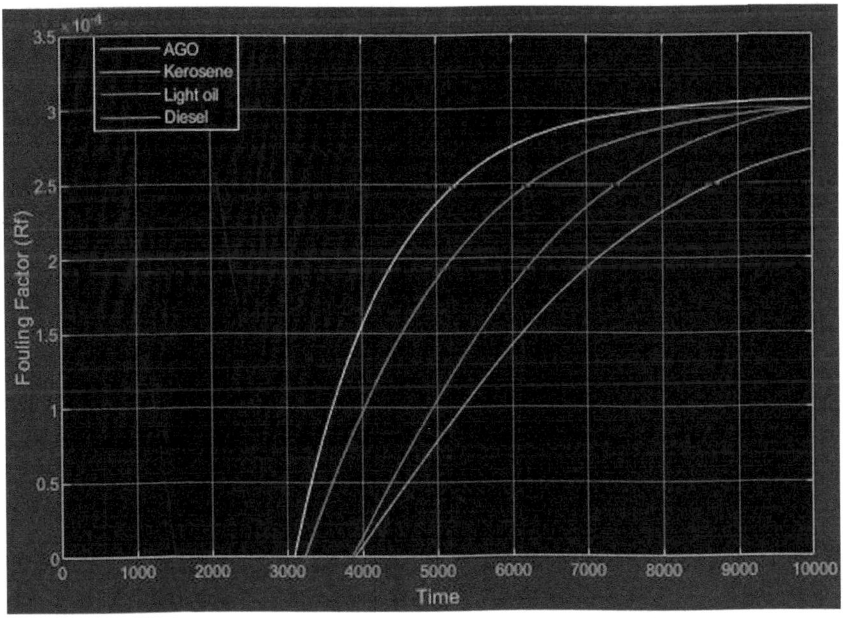

Fig. 9 Graph of fouling factor against time for $\dot{m}_{product} = 1.8$ kg/s

final Rf value for the heat exchanger with AGO, kerosene and light oil are close to each other. The final Rf value for these three are 0.000306, 0.0003 and 0.0003. The only difference that is visible from these 3 Rf curve is the start time of the fouling that is 3212 min for AGO, 3272 min for kerosene and 3892 for Light oil. For the heat exchanger with diesel, the graph is still increasing but with the simulation run time of 10,000 min, the final Rf value of the heat exchanger is 0.000272. The heat exchanger with diesel will start fouling at 3912 min.

(d) Varying Mass Flow Rate of Products to 2.2 kg/s

The mass flow rate of the product is changed to 2.2 kg/s with all other variables remains unchanged. The graph of the fouling factor against time is shown in Fig. 10. For the heat exchanger with AGO flowing through, the fouling will start at 3062 min. It then increases till 7082 min and the Rf value slowly increases with time till it reaches its maximum of 0.000306. The maximum Rf value for heat exchanger with kerosene and heat exchanger with light oil is slightly lower than AGO but both the Rf values are the same. The maximum Rf value of kerosene and light oil is 0.000301. The time for fouling to occur for kerosene is 3122 min and for light oil is 3742 min. However, for the heat exchanger with diesel as the product, the fouling would start at 3732 min and would have a maximum of 0.000275.

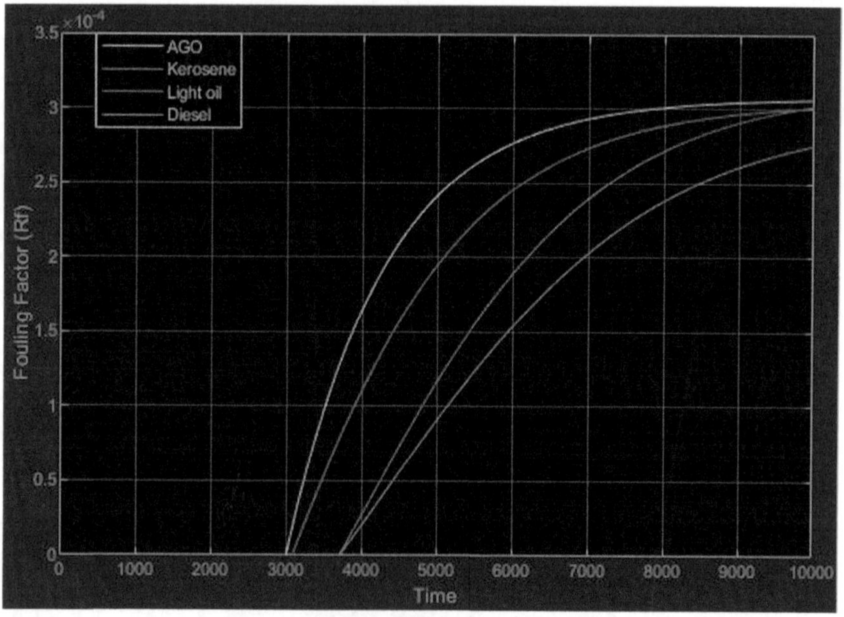

Fig. 10 Graph of fouling factor against time for $\dot{m}_{product} = 2.2$ kg/s

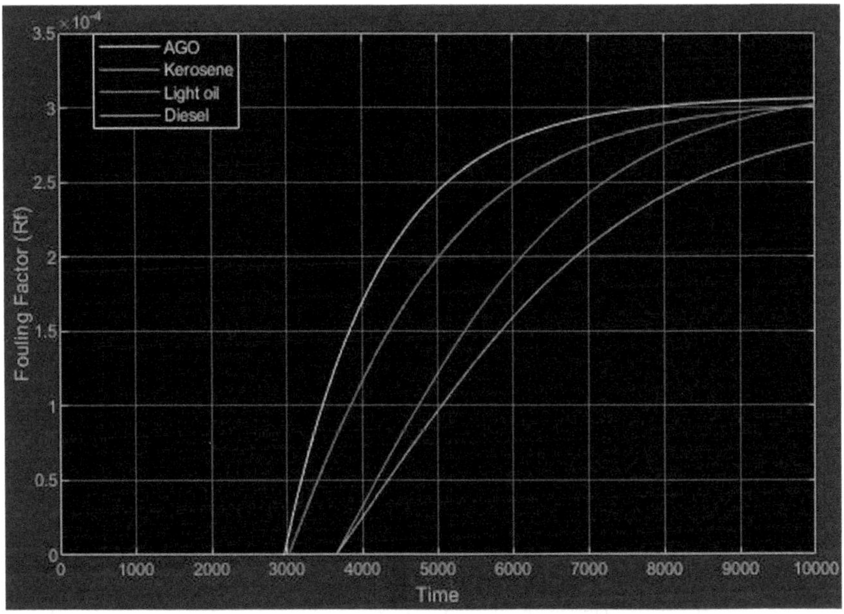

Fig. 11 Graph of fouling factor against time for $\dot{m}_{product} = 2.4$ kg/s

(e) Varying Mass Flow Rate of Products to 2.4 kg/s

The mass flow rate of the product is set to 2.4 kg/s and all other variables remain a constant. The graph of the fouling factor against time is shown in Fig. 11. For the heat exchanger with AGO flowing through, the fouling will start at 3032 min. It then increases until approximately 7500 min and the Rf value slowly increases with time till it reaches its maximum of 0.000306. The maximum Rf value for kerosene and light oil is slightly lower than AGO but both the Rf values are the close to each other. The maximum Rf value of the heat exchanger with kerosene 0.000301 and for the heat exchanger with light oil 0.000302. The time for fouling to occur for kerosene is 3062 min and for light oil is 3682 min. However, for the heat exchanger with diesel as the product, the fouling would start at 3662 min and would have a maximum of 0.000276.

4 Conclusions

This research indicates that fouling crude oil in the heat exchanger at the inlet temperature of 303.15 K will start at 3052 min. It will then increase with time until the maximum fouling factor for the entire heat exchanger network to be 0.000306. This will also give an output temperature of 577.78 K while maintaining at a very low fouling factor. The increase in the inlet temperature of the crude will cause the fouling factor to reduce and also the time for fouling to start will be slower. This is because the temperature between the crude and the product in which flows through the heat exchanger will be smaller. The mass flow rate of the product supplying heat to the crude also plays a very important role. As the mass flow rate of the product increases, the fouling factor increases. This is because the temperature of the product does not change much as the mass flow rate increases. The products are run through the heat exchanger faster causing the temperature of the product in the heat exchanger to be a constant. This will intern increase the temperature difference in the heat exchanger to be higher and thus increase the fouling factor. It is also noticeable that the time for fouling to occur at a lower mass flow rate of the product is slower as compared to the higher mass flow rate. Further research on crude oil fouling with respect to temperature and mass flow rate must be done by taking into consideration other factors and other different type of heat exchanger that is use in the industry.

References

1. Awad MM, Gad HE, Yousef A (2009) Effect of surface temperature on particulate and crystallization fouling. In: Thirteenth international water technology conference, IWTC 13 2009, Hurghada, Egypt
2. Bennett Christopher A (2012) A theory describing asphaltene adhesion fouling inside heat exchanger tubes. Heat Transfer Eng 33(15):1246–1250
3. Zhang J (2013) Optimize crude preheat train to balance efficiency and operability. Hydrocarbon processing June special report 2013
4. Tay FH, Sergei GK (2009) Study of petroleum heat-exchanger deposits with ATR-FTIR spectroscopic imaging. Energy Fuels 23(8):4059–4067
5. Gounder RM, Sampath E (2017) CFD simulations of crude oil fouling on heat transfer surfaces. In: Computational fluid dynamics-basic instruments and applications in science. InTech open. London, UK
6. Ibrahim Al-Haj H (2012) Fouling in heat exchangers. In MATLAB a fundamental tool for scientific computing and engineering applications, vol 3. InTech Open. London, UK

9789811500015